Harnessing AutoCAD® Civil 3D® 2011

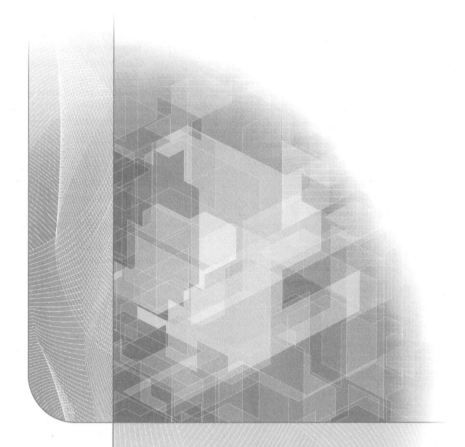

Harnessing AutoCAD® Civil 3D® 2011

autodesk Press

PHILLIP J. ZIMMERMAN

DELMAR
CENGAGE Learning™

Australia • Brazil • Japan • Korea • Mexico • Singapore • Spain • United Kingdom • United States

DELMAR
CENGAGE Learning

Harnessing AutoCAD® Civil 3D® 2011
Phillip J. Zimmerman

Vice President, Career and
Professional Editorial: Dave Garza

Director of Learning Solutions: Sandy Clark

Acquisitions Editor: Stacy Masucci

Managing Editor: Larry Main

Senior Product Manager: John Fisher

Editorial Assistant: Andrea Timpano

Vice President, Career and
Professional Marketing: Jennifer Baker

Marketing Director: Deborah Yarnell

Marketing Manager: Katie Hall

Associate Marketing Manager: Mark Pierro

Production Director: Wendy Troeger

Senior Content Project Manager: Glenn Castle

Senior Art Director: David Arsenault

Technology Project Manager: Joe Pliss

Cover image: View of a modern bridge at sunset.
Image copyright Pinosub 2011. Used under
license from Shutterstock.com

Library of Congress Control Number: 2010935107

ISBN-13: 978-1-111-13791-5

ISBN-10: 1-111-13791-9

Delmar
5 Maxwell Drive
Clifton Park, NY 12065-2919
USA

Cengage Learning is a leading provider of customized learning solutions with
office locations around the globe, including Singapore, the United Kingdom,
Australia, Mexico, Brazil, and Japan. Locate your local office at: **international.
cengage.com/region**

Cengage Learning products are represented in Canada by Nelson Education, Ltd.

To learn more about Delmar, visit **www.cengage.com/delmar**

Purchase any of our products at your local college store or at our preferred
online store **www.cengagebrain.com**

Printed in the United States of America
1 2 3 4 5 6 7 13 12 11 10

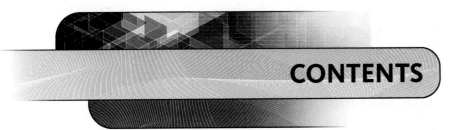

CONTENTS

CHAPTER 3 SITE AND PARCELS 55

CHAPTER 4 SURFACES 87

CHAPTER 5 ALIGNMENTS 146

CHAPTER 6 PROFILE VIEWS AND PROFILES 178

CHAPTER 7 ASSEMBLIES AND CORRIDORS 204

CHAPTER 8 CROSS-SECTIONS AND VOLUMES 223

CHAPTER 10 GRADING AND VOLUMES 288

CHAPTER 11 PIPE NETWORKS 322

CHAPTER 12 CIVIL 3D SHORTCUTS 343

CHAPTER 13 HYDRAULICS AND PIPE DESIGN 347

CHAPTER 14 SURVEY BASICS 382

CHAPTER 15 SURVEY AND TRAVERSE ADJUSTMENTS 400

INTRODUCTION

Civil 3D is the successor to the Autodesk Land Desktop product line (Land Desktop, Civil Design, and Survey). Like Autodesk Land Desktop (LDT), the program is a powerful drafting tool and unlike Autodesk (LDT), Civil 3D is a dynamic engineering environment. Civil 3D does not create data based on drawing entity snapshots, but drawing objects. The dynamic engineering environment is the single fundamental change that radically alters the CADD design environment. Changes in data produce changes in documentation.

Autodesk (LDT) with its hidden data and definitions disappears in Civil 3D. Prospector and Data Shortcuts provide open access to object data and its properties. Prospector manages an object and provides the interface to the object's data. In Civil 3D, data is only a few clicks away. Data Shortcuts allow data sharing.

The Settings Toolspace presents a unique way of managing a company's "look" that is consistent irrespective of which user develops a design. This is where the journey starts, with Civil 3D styles that accomplish its design drafting and documentation. Where Land Desktop was a take-it or draft-it environment, Civil 3D is a la carte. With Civil 3D you either live with its shipping content or you spend time developing your content. It is a new beginning for the Civil Industry using Civil 3D tools and styles.

Civil 3D's tool sets include points, surfaces, parcels, alignments, profiles, sections, pipes, and survey. To use Civil 3D's tool sets efficiently requires fundamental Civil Design and Survey process knowledge. This book addresses the use of and provides a basic understanding of the tool sets. This book, however, is NOT an engineering or surveying textbook nor is it a tips and tricks book. You can peruse the newsgroups for those needs. This book provides explanations of and exercises helping in the development of a basic understanding of Civil 3D's tools. Many examples are from people using the software while other examples contain difficulties to demonstrate capabilities.

Civil 3D 2010 introduces the ribbon and it is a departure from traditional AutoCAD menus. The ribbon focuses on tasks, whereas the menus focus on object types. I chose to use the ribbon for this book and find its focus on tasks rather than object types a more flowing work process.

When working in Civil 3D you have four potential tasks: creating objects (Home tab), analyzing their values (Analyze tab), editing them if wrong (Modify tab), and labeling (Annotate tab) their values. If using the ribbon and not panicking for menus, the ribbon will become an ingrained habit. The menus remain, as does selecting with the left mouse button, pressing the right mouse button, and selecting a command from a shortcut menu.

Working through the exercises, you will encounter issues with the software. I have not had a single version of Civil 3D complete all of the tasks in this book. With this release, Civil 3D accomplishes most of its goals. I cannot imagine going back to LDT after becoming familiar with Civil 3D.

Drawing files are included on the book's Student Companion Premium Website for use in exercises.

Text Element	Example
Step-by-Step Tutorials	1.1 Perform these steps
Text Element	Example
AutoCAD Command	PAN
Civil 3D Command	*CREATE ALIGNMENT FROM POLYLINE*
Tab, Icon, or Button	**Save**
User Input	**Bold**
Files Names	*Italic*

STUDENT COMPANION PREMIUM WEBSITE

The Student Companion Premium Website for this text includes the instructional portion of this text as chapter PDF files. These instructions are provided to augment the exercises in this printed text. Drawing files for the exercises are also located on this site. Follow the directions below to access the companion website.

Redeeming an Access Code

1. Go to http://www.cengagebrain.com
2. Enter the Access code in the Prepaid Code or Access Key field, **Redeem**
3. Register as a new user or log in as an existing user if you already have an account with Cengage Learning or CengageBrain.com, **Redeem**
4. Select **Go to My Account**
5. Open the product from the My Account page

Accessing a Premium Website from CengageBrain → My Account

1. Sign in to your account at http://www.cengagebrain.com
2. Go to My Account to view purchases
3. Locate the desired product
   - You will find an eTextbook (CLeBook), and
   - An Online Study Tool (Premium Website)
4. Click on the **Open** button next to the Premium Website entry

ACKNOWLEDGMENTS

Chapter 2's exercise files are from surveys processed by Jerry Bartels. The manual survey of Chapter 13 is from Gail and Gil Evans of Chicago. Chapter 14's network traverse loop is from Bill Laster. Chapter 3's lot exercise subdivision is from a plat developed by Intech Consultants, Inc. (Tom Fahrenbok, Scott, and Joe).

Thanks to the staff at Delmar Cengage Learning for patiently waiting for the manuscript's completion, especially John Fisher and Glenn Castle; and Preetha Sreekanth for PreMediaGlobal's edits and compositing. I want to thank my dear mom (Lorraine), sisters (Sharon and Dawn), and patient friends for their support and help. Also, I want to thank Imaginit, Carl, Len, Angela, Andrea, V, Brent, Jay, Mike Choquette, and many other friends for giving me advice and opportunities. I want to give Mark Martinez special thanks for his joyful and enthusiastic rereading of the manuscript for technical errors. Sadly, no matter how many times he and I reread the manuscript, there will be errors. Let me know where they are so I can fix them.

Phillip J. Zimmerman

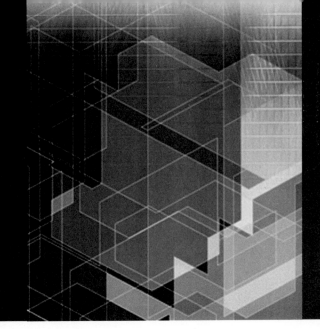

The Beginning

EXERCISE 1-1

After completing this exercise, you will:

- Become familiar with Toolspace and Prospector.
- Observe how Prospector dynamically manages objects.
- Change an alignment's path and view the changes to the road model.
- Dissolve and add parcels to a site design and view the changes to the Parcels data tree.

Toolspace Basics

1. If not in CIVIL 3D, double-click its desktop icon to start the application.
2. If the Toolspace does not show, at the Ribbon's top left (Home tab, Palettes panel) click the **Toolspace** icon.
3. Close all open drawings and do not save them.
4. In Toolspace, Prospector tab top right, under the Help icon, select the drop-list arrow and from the list MASTER VIEW.
5. In Prospector, expand Drawing Templates and AutoCAD branches, from the list select **_AutoCAD Civil 3D (Imperial) NCS**, press the right mouse button, and, from the shortcut menu, select CREATE NEW DRAWING.
6. If the Toolspace is not floating, click and hold the left mouse button down near Toolspace's heading, and drag it to the screen's center.
7. On Toolspace's mast, right mouse click and, from the shortcut menu, click Allow Docking to toggle it **OFF**.
8. Again on Toolspace's mast, right mouse click and, in the shortcut menu if it is toggled on, click Auto-hide to toggle it **OFF**.
9. On the Toolspace's mast, press and hold the left mouse button and move the Toolspace to the screen's right and left sides.

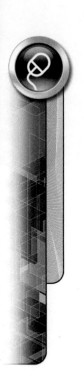

Toolspace's mast switches from side to side, but preview remains below Prospector.

10. On the Toolspace's mast, press the right mouse button and, from the shortcut menu, click Auto-hide, toggling it **ON**. Move the cursor away from Toolspace.

The Toolspace hides Prospector under its mast.

11. On the Toolspace's mast, press the right mouse button and, from the shortcut menu, click Allow Docking, toggling it **ON**. At the screen's left side, dock the Toolspace.

12. If necessary, click the *Prospector* tab and scroll to the top until viewing the Open Drawings branch.

13. If necessary, to the left of the drawing's name, click the expand tree icon (plus sign), expanding the drawing's object type hierarchy.

14. Adjust preview's size by placing the cursor at the Prospector and preview boundary, pressing and holding down the left mouse button, and sliding it up or down.

15. At the Toolspace's top, under the Help icon, select the drop-list arrow and select ACTIVE DRAWING VIEW.

16. In the display's upper right, select the *Close* icon and do not save the drawing.

Prospector Object Management

The remainder of the exercise uses drawings from this textbook's CD, Chapter 01 folder.

Prospector is Civil 3D's control center and it dynamically updates its object list as changes occur in the drawing. Prospector also manages an object's data and makes its data available.

1. At Civil 3D's top left, Quick Access Toolbar, select the *Open* icon, browse to this textbook's CD, Chapter 01 folder, select the drawing *Overview Prospector*, and click *Open*.

2. If necessary, make sure Prospector is showing by selecting the Toolspace's Prospector tab.

Points

1. In Prospector, select the **Points** heading.

The square with a black dot icon to the left of Points indicates that the drawing has Civil 3D points. When selecting a Prospector object heading, preview displays a corresponding object list (see Figure 1.1). In this case, the preview area lists all of the drawing's points. Any value in preview is editable unless it has a gray background.

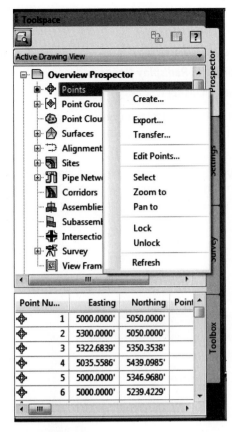

FIGURE 1.1

Each Prospector tree heading has a specific shortcut menu (see Figure 1.1). The number of shortcut menu commands depends on an object's complexity or the user's location in the Prospector tree.

2. With **Points** still highlighted, press the right mouse button and, from the shortcut menu, select CREATE…, displaying the Create Points toolbar.

3. To close the toolbar, click the red **X** in its upper-right corner.

4. In preview, double-click a point's number.

By clicking a preview entry, preview allows edits or style changes.

Point Groups

1. In Prospector, select the **Point Groups** heading and notice that preview lists the drawing's point groups.

Each point group can have a point and/or a label style override. A point and/or a label style override changes the points' display for all point group points.

2. In Prospector, expand the Point Groups branch, viewing the point group list.

3. Select the point group **Existing Ground Points**.

This displays in preview the group's points.

4. With Existing Ground Points still highlighted, press the right mouse button and, from the shortcut menu, select EDIT POINTS….

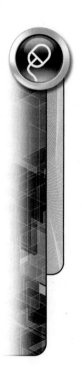

This displays a panorama with the Point Editor vista containing the point group points. The vista allows point value edits.

5. At the top of panorama's mast, click the **X**, closing it.
6. To the left of Prospector's Help icon, click the ***Redisplay Panorama*** icon.
7. At the top of panorama's mast, click the **X**, closing it.

The Point Groups heading's property is the point groups' display order (see Figure 1.2). The drawing's point groups display begins with the list's lowest point group and continues to the top.

This Display Order property "hides" or changes point markers and labels, enabling you to view or display different point data combinations. In the current drawing, the No Show point group hides all of the points, because it is the last drawn point group (top of the list).

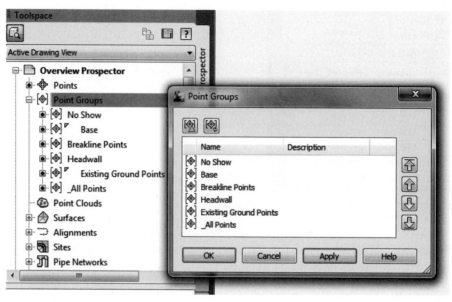

FIGURE 1.2

8. In Prospector, select the heading **Point Groups**, press the right mouse button, and, from the shortcut menu, select PROPERTIES....

The top group is No Show and its overrides suppress all point markers and labels (see Figure 1.2). To view different point groups, change their position relative to the No Show group by using the up and down arrow buttons at the right side of the dialog box. The viewable point groups are above No Show.

If the Event Viewer displays, close it by clicking the green check mark in the upper-right of the panorama.

9. In Point Group Properties, select **Existing Ground Points** and move it to the top position by clicking the up arrow on the dialog box's right.
10. If necessary, select **No Show** and move it to the second position.
11. Click **OK**, viewing the Existing Ground Points point group.

The drawing displays the Existing Ground Points point group points and their assigned point and label styles.

12. Again, in Prospector, select the **Point Groups** heading, press the right mouse button, and, from the shortcut menu, select PROPERTIES....

13. In the Point Group Properties dialog box select the **Breakline Points** point group and move it to the top position.

14. Select the **No Show** point group and move it to the second position.

15. Click **OK**, exiting and displaying the Breakline Points point group. Use the RE-GENALL command to update the display.

This displays the Breakline Points point group (see Figure 1.3). By isolating these points and their surface triangulation you can review the linear objects "successful" triangulation in a surface.

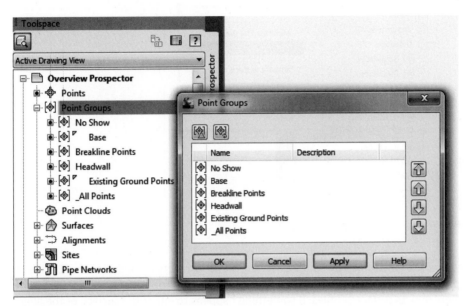

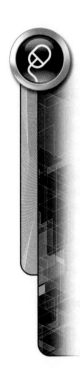

FIGURE 1.3

16. At the displays upper right, select **Close** icon and do not save the changes.

Civil 3D Object Reactions and Dependencies — Alignments and Profiles

All Civil 3D objects react to changes affecting their display or information. When changing the data an object depends on, or changing the object itself, all objects dependent on the changed object react and accommodate its change.

1. At Civil 3D's top left, Quick Access Toolbar, select the **Open** icon, browse to this textbook's CD, Chapter 01 folder, select the drawing *Overview Road*, and click **Open**.

When changing an alignment's endpoint, the profile and profile view react to the change by displaying new elevations and profile length.

2. Use the ZOOM and PAN commands and view the profile view's right side elevation annotation.

3. If necessary, click the **Prospector** tab.

4. In Prospector, expand the Surfaces branch until you view the surface EG (1).

5. Expand the EG (1) branch until you view the Definition heading and its data entries.

6. From the data list, select the heading **Edits**, press the right mouse button, and select RAISE/LOWER SURFACE (see Figure 1.4).

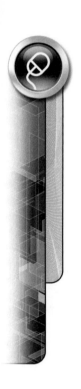

7. In the Command Line, for the amount to add to the surface's elevation, enter **10** and press ENTER.

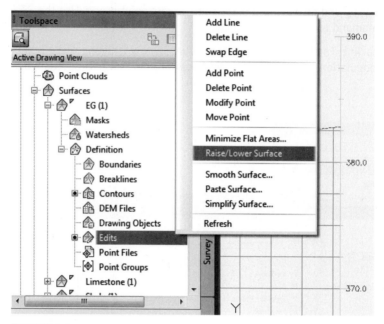

FIGURE 1.4

The edit raises the surface by 10 feet. The profile view reacts by updating the EG(1) surface profile's elevation and annotation. A surface and a profile view are linked by the alignment and its surface profiles. If something happens to either the surface or the alignment, the profile and profile view react and change to correctly show the new situation.

8. At the displays upper right, select the ***Close*** icon and do not save the changes.

Civil 3D Object Reactions and Dependencies — Surface and Contours

When changing surface data, the surface reacts by updating its definition. If the current style displays contours, the contours change, showing the changed surface elevations.

If the Event Viewer displays, close it by clicking the green check mark in the panorama's upper-right corner.

1. At Civil 3D's top left, Quick Access Toolbar, click the ***Open*** icon, browse to this textbook's CD, Chapter 01 folder, select the drawing *Overview Surfaces*, and click ***Open***.

2. In the drawing, select any surface contour, press the right mouse button, and, from the shortcut menu, select SURFACE PROPERTIES....

3. In the Surface Properties dialog box, click the ***Information*** tab. To the right of Surface style (middle left), view the styles list by clicking the drop-list arrow, and change the current surface style by selecting **Contours and Triangles** (see Figure 1.5).

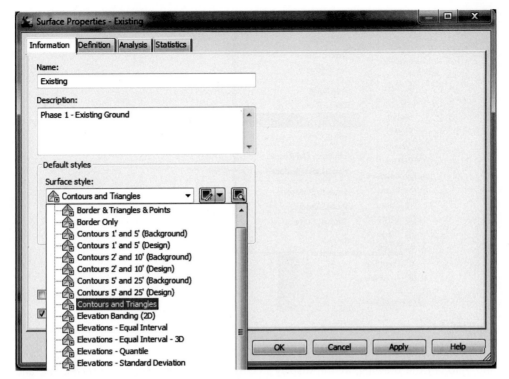

FIGURE 1.5

4. Click **OK**, exiting the dialog box.

The surface now shows contour and triangle surface components. The next step zooms to a point that you are going to modify. After modifying the point, the contours change, reflecting the new surface elevation.

5. In Prospector, at its top, select the **Points** heading, listing all points in preview.

6. In preview, scroll through the point list until you locate point number **71**.

7. In preview, select point number 71's icon, press the right mouse button, and, from the shortcut menu, select ZOOM TO.

The zoom centers the point in the display.

8. In Prospector, expand Surfaces and the Existing surface branches until you view the Definition heading's data list.

9. In the Existing's Definition tree, select **Edits**, press the right mouse button, and, from the shortcut menu, select MODIFY POINT.

Your Prospector should look like Figure 1.6.

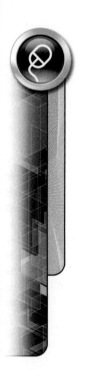

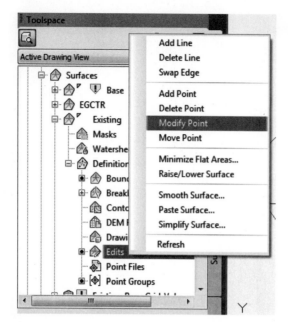

FIGURE 1.6

10. In the drawing, at the display's center, select the triangles' intersection, and press ENTER. In the command line, enter a new point elevation of **740** and press ENTER twice, changing the elevation and exiting the command.

This edit changes the surface data, and the contours change, showing the new surface elevation. Prospector's preview adds the edit to the surface's edits list (see Figure 1.7). The surfaces dependent on Existing become out-of-date because of the change made to Existing.

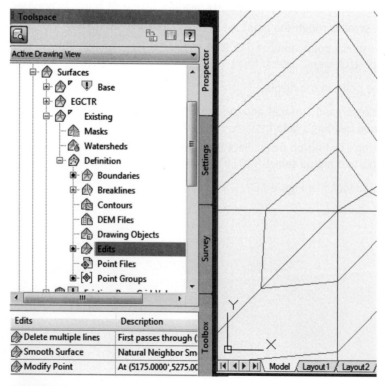

FIGURE 1.7

If the previous surface edit is speculative, you can temporarily remove it from the surface. This is a function of the Surface Properties dialog box's Definition panel. If you want to permanently remove the edit, either you can delete it in the Surface Properties dialog box's Definition panel (Remove From Definition) or you can delete it from preview's edits list.

11. In the drawing, select a contour or triangle leg, press the right mouse button, and, from the shortcut menu, select SURFACE PROPERTIES....

12. In the Surface Properties dialog box, click the ***Definition*** tab; then in Operation Type (at the bottom of the panel) scroll up or down the list until you locate **Modify Point**.

The operation has a check mark, indicating it is an active surface edit.

13. Toggle **OFF** the Modify Point edit and click **OK**, exiting the dialog box.

A warning dialog box displays.

14. In the Warning dialog box, click REBUILD THE SURFACE, rebuilding the surface and exiting the Surface Properties dialog box (see Figure 1.8).

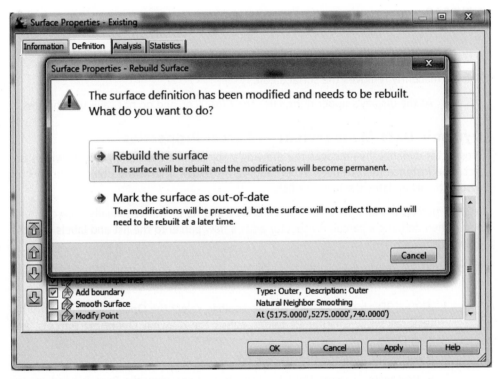

FIGURE 1.8

The surface shows the modified point removed from its data by changing the contours.

15. Return to the Surface Properties dialog box (see Step 11). In the Definition panel, Operation Type, toggle **ON** the Modify Point edit, and click **OK** to exit the dialog box.

16. In the Warning dialog box, click REBUILD THE SURFACE, rebuilding it and exiting the Surface Properties dialog box.

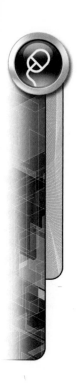

The edit reappears in the surface.

17. If necessary, in Prospector, in the Existing surface branch, from the Definition heading's list, select **Edits**.

The preview area displays the Existing surface edits.

18. In the preview area, select the entry **Modify Point**, press the right mouse button, and, from the shortcut menu, select DELETE... to permanently remove the surface edit.

A Remove From Definition dialog box displays (see Figure 1.9).

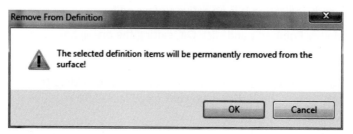

FIGURE 1.9

19. In the Remove From Definition dialog box, click **OK** to accept the deletion.

This permanently removes the surface edit.

20. At the display's upper right, click the **Close** icon and do not save the changes.

Dynamic Data Management — Parcel Properties

Prospector dynamically manages the drawing's objects list. Each entry is a drawing's object type instance (occurrence). When adding or removing drawing objects, Prospector automatically updates the instance list.

A Site's parcels list identifies each parcel's type and its identifier (usually a parcel number). When defining a parcel, Prospector adds a new parcel to the list and labels it in the drawing.

If the Event Viewer displays, close it by clicking the green check mark in the upper-right of the panorama.

1. At Civil 3D's top left, Quick Access Toolbar, select the **Open** icon, browse to this textbook's CD, Chapter 01 folder, select the drawing **Overview Parcels**, and click **Open**.

2. If necessary, click the **Prospector** tab.

3. In Prospector, expand the Sites branch until you view the Parcels heading and its parcels list. Your screen should be similar to Figure 1.10.

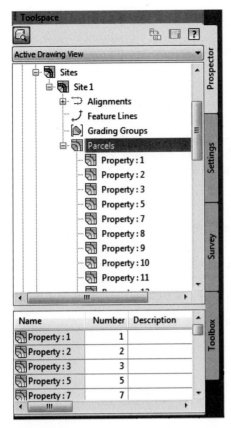

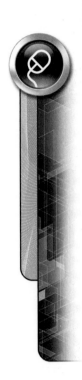

FIGURE 1.10

Each parcel's property includes a boundary analysis (see Figure 1.11).

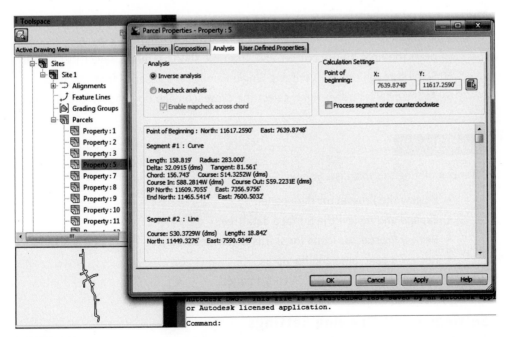

FIGURE 1.11

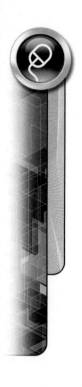

4. In Parcels, select a parcel from the list, press the right mouse button, and, from the shortcut menu, select PROPERTIES....

5. In Parcel Properties, click the **Analysis** tab and review the selected parcel's **Inverse** and **Mapcheck** reports.

6. Click **OK**, exiting Parcel Properties.

Reports Manager — Parcels

1. If the toolbox tab is not displaying, from the Ribbon headings select the View tab. In the View ribbon, Palettes panel, to the Toolspace icon's right, middle row, click the right icon, click **Toolbox** to display Toolspace's Toolbox tab.

2. In the Toolbox panel, expand Reports Manager; then expand Parcel until displaying a parcel reports list.

3. From the parcel reports list, select **Inverse_Report**, press the right mouse button and, from the shortcut menu, select EXECUTE....

4. In the Export to XML Report dialog box, deselect everything except for two parcels and click **OK**.

5. Click **OK** and set the file location and name.

Internet Explorer displays with a Parcel Inverse Report.

6. Review the report and then close Internet Explorer.

Adding and Deleting Parcels

Whenever adding or deleting parcels, Prospector modifies its Parcels list.

1. Click the **Prospector** tab.

2. From the Parcels list, select **Single-Family:17**, press the right mouse button, and, from the shortcut menu, select ZOOM TO.

3. Use the ERASE command, and in the drawing select the north/south side yard line, dividing parcels **4** and **17**. Press ENTER.

The two parcels merge into one.

4. In Prospector, the Parcels list needs refreshing. Select the **Parcels** heading, press the right mouse button, and, from the shortcut menu, select REFRESH.

5. Scroll through the list, verifying Single-Family : 17's absence from the list.

6. At the displays upper right, select the **Close** icon and do not save the drawing.

EXERCISE 1-2

After completing this exercise, you will:

- Review the Prospector data panel.
- Expand and review the Surface data tree.
- Review Prospector icons for status and dependency.
- Review a panorama and its vistas.
- Become familiar with the Settings panel.

Settings - Edit Drawing Settings

This exercise focuses on Settings and its hierarchical tree management of settings and styles.

1. At Civil 3D's top left, Quick Access Toolbar, click the **Open** icon, browse to this textbook's CD, Chapter 01 folder, select the drawing *Overview Surfaces II*, and click **Open**.
2. Click the **Settings** tab.

The Edit Drawing Settings..., Edit Label Style Defaults..., and Edit LandXML Settings... dialog boxes are at the top of Settings' hierarchy.

3. At Settings' top, select the drawing name, press the right mouse button, and, from the shortcut menu, select EDIT DRAWING SETTINGS....
4. If necessary, in Edit Drawing Settings, click the **Units and Zone** tab and review its settings and current Zone.
5. Click the **Transformation** tab.
6. Click the **Object Layers** tab and scroll through the layer list, reviewing layer names and their settings.
7. Click the **Abbreviations** tab and review the abbreviations list.

You can change any of the listed abbreviations.

8. Click the **Ambient Settings** tab and expand different sections to view their settings.

These values set the general drawing tone. Clicking a padlock icon locks the value so no lower style can override or change the locked value.

9. Click **OK**, exiting the Edit Drawing Settings dialog box.

Edit Feature Settings

1. In Settings, select the **Surface** heading, press the right mouse button, and, from the shortcut menu, select EDIT FEATURE SETTINGS....

The Edit Feature Settings dialog box sets the default styles and naming values for Surfaces.

2. Expand the surface sections to review the surface settings, and when done, click **OK**, exiting the dialog box.

Edit Label Style Defaults - Drawing

1. In Settings, select the drawing name, press the right mouse button, and, from the shortcut menu, select EDIT LABEL STYLE DEFAULTS....
2. Expand and review each section.
3. In the Label section, click the Text Style value cell, and then click the ellipses to view the Select Text Style dialog box.

Selecting a text style here sets it for all non-overridden styles.

4. Close the Select Text Style dialog box and click **Cancel**.

Edit Label Style Defaults - Feature

1. In Settings, select the **Surface** heading, press the right mouse button and, from the shortcut menu, select EDIT LABEL STYLE DEFAULTS....

The Edit Label Style Defaults dialog box sets several default values for all surface label styles. The values reflect those set in the drawing name level Edit Label Style Defaults dialog box.

2. Expand each section to review the values and then click **OK**, exiting the dialog box.

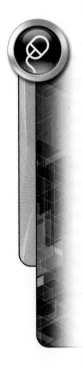

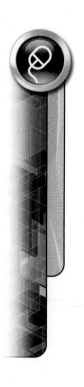

Object Styles, Label Groups, and Their Styles

1. In Settings, click the plus sign to the left of Surface to expand the Surface branch.

The Surface heading has four branches: Surface Styles, Label Styles, Table Styles, and Commands.

2. Click the expand tree icon to the left of Surface Styles, viewing its object styles list.

3. Select the style **Border & Triangles & Points**, press the right mouse button, and, from the shortcut menu, select EDIT.... As an alternative, you can double-click the style to edit it.

You modify a style's values in this dialog box.

4. Click the tabs, expand some of the sections in the different panels, and review the values controlling a style.

5. Click **OK**, exiting the dialog box.

6. In Settings, click the plus sign to the left of Label Styles, displaying the surface label style types list.

7. Expand the Contour branch, viewing its styles.

8. Select a **Contour** style, press the right mouse button and, from the shortcut menu, select EDIT....

9. Click the various tabs and expand some of the sections, reviewing their contents.

10. When you are done, click Cancel to close the label style dialog box.

11. Explore some of the remaining style trees and review their style definitions.

12. CLOSE the drawing and do not save the changes.

EXERCISE 1-3

After completing this exercise, you will:

- View styles from content templates.
- Review the settings and values in Edit Drawing Settings.
- Create a template file.
- Open an object Properties dialog box.
- Review the anatomy of an object style.
- Review the anatomy of a label style.
- Review the anatomy of a table style.

Drawing Templates

Civil 3D has a content template file that you can use and modify. A template is essential to establishing standard layers, labels, and design documentation.

1. If you are not in **Civil 3D**, start the application and close the open drawing. If you are in Civil 3D, close the current drawing and do not save it.

2. If necessary, click the **Prospector** tab.

3. Change Prospector to **Master View**. In Toolspace, top right, under the Help icon, select the drop-list arrow and from the list MASTER VIEW.

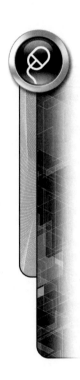

4. Expand the Drawing Templates, AutoCAD node, select and right mouse click over _AutoCAD Civil 3D (Imperial) NCS_, and, from the shortcut menu, select CREATE NEW DRAWING.

5. Click the **Settings** tab.

6. In Settings, expand the Surface branch until you view the Surface Styles heading's styles list.

7. Click the **Surface Styles** heading, press the right mouse button, and from the shortcut menu select NEW….

8. In the Surface Style - New Surface Style dialog box, click the **Display** tab.

A new style does not reference any layers for its components, but it does assign some specific color and linetype properties (see Figure 1.12).

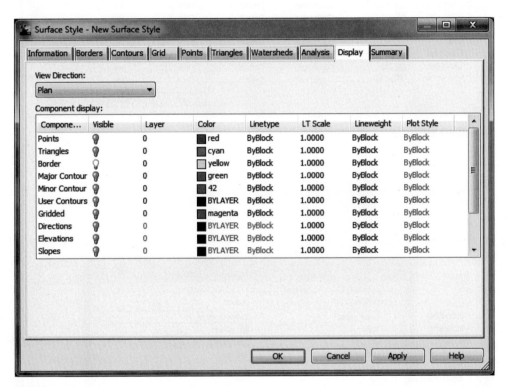

FIGURE 1.12

9. For the Border component, click layer **0**.

The Layer Selection dialog box displays, listing the current drawing's layers. If more than one drawing is open, you can select layers from them by clicking Layer Source's drop-list arrow, or you can create new layers by clicking the upper right New… button (see Figure 1.13 and Figure 1.14).

10. At the dialog box's top right click **New…** to display the Create Layer dialog box.

11. Click **Cancel** until exiting the Surface Style dialog box.

12. In Settings, from the Surface Styles list, select **Contours and Triangles**. Press the right mouse button and, from the shortcut menu, select **Edit…** and select the **Display** tab.

13. Review the layer assignments and their properties and then click **OK** to exit the dialog box.

14. Close the current drawing and do not save the changes.

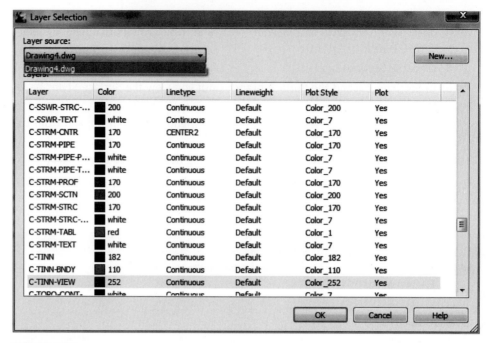

FIGURE 1.13

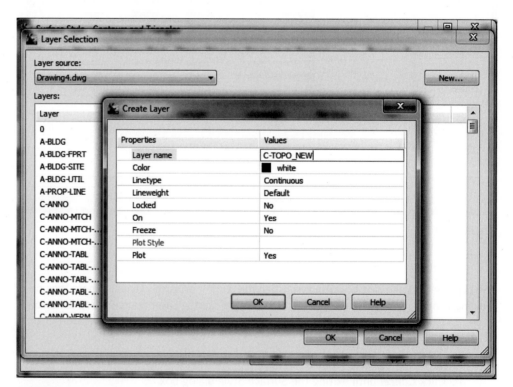

FIGURE 1.14

Edit Drawing Settings

The Edit Drawing Setting, Edit Label Style Defaults, Edit LandXML Settings, and Table Tag Numbering dialog boxes occupy the Settings' tree top. Their values affect all styles below them.

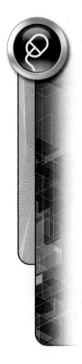

1. At Civil 3D's top left, Quick Access Toolbar, click the **Open** icon, browse to this textbook's CD, Chapter 01 folder, select the drawing *Overview Surfaces II*, and click **Open**.

2. If necessary, click the **Settings** tab.

3. In Settings, at its top, click the drawing name, press the right mouse button, and, from the shortcut menu, select EDIT DRAWING SETTINGS....

4. In Edit Drawing Settings, click the **Units and Zone** tab and review its values.

5. Click the **Transformation** tab and review its values.

6. Click the **Object Layers** tab, review its settings, and scroll through the layer list, noting its modifier settings and layer names.

7. Click the **Abbreviations** tab, viewing Civil 3D's listing and reporting abbreviations. You can change any of them in this panel.

8. Click the **Ambient Settings** tab and review its settings for different sections.

9. Click **OK**, exiting the Edit Drawing Settings dialog box.

Object Style

Each object instance has an assigned object style. Civil 3D has several object styles affecting the display of an object's components and characteristics.

This exercise continues in the Overview Surfaces II drawing.

1. In Settings, expand the Surface branch until you view the Surface Styles heading's styles list.

Each surface style has a function or specific surface characteristic to display. The Border & Triangles & Points style shows a TIN surface structure. This surface view allows a user to review if linear features show correctly or if additional editing or data is necessary for a surface to correctly represent its data.

2. From the Surface Styles list, select **Border & Triangles & Points**, press the right mouse button, and, from the shortcut menu, select EDIT....

3. Click the **Borders**, **Points**, and **Triangles** tabs to review their contents.

4. Click the **Display** tab to review what components this style displays.

5. Exit the dialog box by clicking CANCEL.

6. In Settings, in the Surface Styles list, select **Contours 1' and 5' (Design)**, press the right mouse button, and, from the shortcut menu, select EDIT....

7. Click the **Borders** and **Contours** tabs to review their contents.

8. Click the **Display** tab to review what components this style displays.

9. Exit the dialog box by clicking CANCEL.

10. In Settings, expand the Parcel branch until you can view the Parcel Styles heading's styles list.

11. From the styles list, select **Single-Family**, press the right mouse button, and, from the shortcut menu, select EDIT....

12. Click each tab and review its settings and values.

The Display tab settings define what components a parcel displays

13. Exit the dialog box by clicking CANCEL.

Label Style

1. In Settings, expand the Parcel branch until you view Label Styles' Area heading's styles list.

2. From the list, select **Name Area & Perimeter**, press the right mouse button, and, from the shortcut menu, select EDIT....

3. Click the *Information* tab to review its settings and values.

The Information tab displays the style's name, description, and authorship credits.

4. Click the *General* tab to view its settings and values.

The General tab displays values from the Edit Label Settings dialog box.

5. Click the *Layout* tab, reviewing its settings and values.

6. In the Text section, in Contents, click in the Value cell to display an ellipsis. Click the ellipsis to display the Text Component Editor.

7. Click in the Parcel Area format string on the right side of the dialog box.

This action displays the format string components on the left side of the dialog box. In detail, the left side shows what is in shorthand on the right (i.e., Uacre means the area unit is acre, P2 means two decimal places for areas, etc.). You change values by clicking in their value cell, dropping an options list, and, from the list, selecting a new value.

8. Click in Precision's value cell to display a drop-list, and from the list select a new precision value.

9. Click the **arrow** at the top-center to transfer the new value to the label format.

10. Click CANCEL until exiting all of the dialog boxes.

11. CLOSE the drawing without saving it.

EXERCISE 1-4

After completing this exercise, you will:

- Import a LandXML file from Land Desktop into Civil 3D.
- Read data directly from a Land Desktop project folder.

This exercise uses files found on the textbook's CD. You need to copy the Autodesk Land Desktop Project, Civil 3D, from the CD to the Civil 3D Projects folder on your computer.

Exercise Setup

1. Using Windows Explorer, locate the Civil 3D Projects folder on your hard drive.

2. Copy the *Civil 3D project* from this textbook's CD, and place it in the Civil 3D Projects folder (see Figure 1.15). Close Windows Explorer.

3. Click the *Prospector* tab.

4. If necessary, change Prospector to **Master View**.

5. Expand the Drawing Templates and AutoCAD node, select and right mouse click over **_AutoCAD Civil 3D (Imperial) NCS**, and, from the shortcut menu, select CREATE NEW DRAWING.

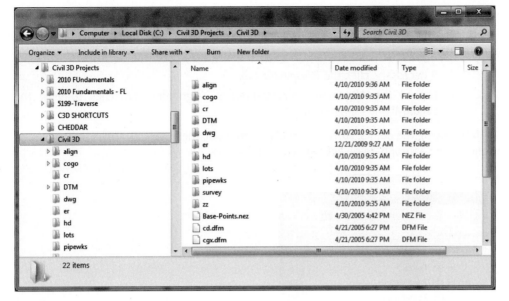

FIGURE 1.15

Importing a LandXML file

A LandXML file contains several data types (surface, alignment, parcel, pipe networks, and profile data). When you view the file in the Import LandXML dialog box, it lists all of the file's data. You select what items to import by toggling them on or off.

1. At Civil 3D's top left click Ribbon's Insert tab.

2. From the Import panel, locate and click the **LandXML** icon.

3. In the file Import LandXML dialog box, browse to the folder C:\Civil 3D Projects\ Civil 3D, select *Overall Proj.xml*, and click **Open**, displaying the Import LandXML dialog box.

The Import LandXML dialog box displays, listing the file's data. It is here that you select the data to import (see Figure 1.16).

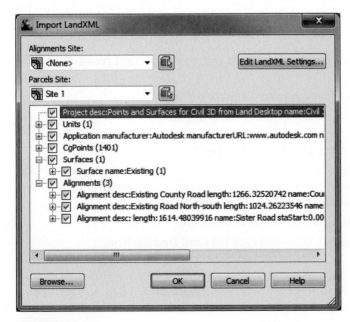

FIGURE 1.16

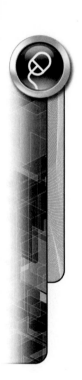

4. Leave all toggles on and click **OK** to import the data.

If the Event Viewer displays, close it by clicking the green check mark in the upper-right of the panorama.

5. If necessary, click the **Prospector** tab.

6. Expand the Alignments and Centerline Alignments branches to view the LandXML file's alignments.

After reading the points, surface, and alignment data, these objects appear in Prospector under their appropriate object type (see Figure 1.17).

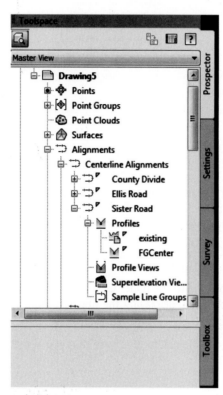

FIGURE 1.17

7. In Prospector, click the heading **Points** and, in Prospector's preview, view the imported points.

8. Expand the Surfaces branch to view the surface list.

9. CLOSE the drawing, and do not save the changes.

Importing from a Land Desktop Project

1. In Prospector's Mater View, expand the Drawing Templates and AutoCAD node, select and right mouse click over **_AutoCAD Civil 3D (Imperial) NCS**, and, from the shortcut menu, select CREATE NEW DRAWING.

2. At Civil 3D's top left click the Ribbon's Insert tab.

3. From the Import panel locate and click the **Land Desktop** icon.

This displays the Import Data from Autodesk Land Desktop Project dialog box, which sets the projects and named project folder (see Figure 1.18). After setting the project folder and identifying the project, the dialog box displays all of the data that you can transfer from the project.

4. In the Import Data from Autodesk Land Desktop Project dialog box, set the project folder to **C:\Civil 3D Projects** and the project name to **Civil 3D**.

5. In the dialog box, click *OK* until you return to the command line to import the project data.

6. If necessary, click the *Prospector* tab.

7. In Prospector, expand the Surfaces branch to view the surface list.

8. CLOSE the drawing and exit Civil 3D.

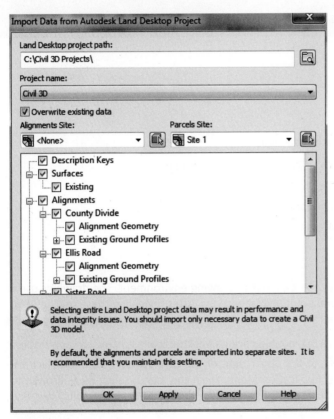

FIGURE 1.18

Each of the following chapters explores in greater detail an object type and its styles.

CHAPTER

2

Points

EXERCISE 2-1

When you complete this exercise, you will:

- Become familiar with Point's Edit Feature Settings.
- Become familiar with the label style defaults.
- Be able to create new point styles.
- Become familiar with the point label style.

Exercise Setup

The first task is setting up the point exercises' drawing environment. The drawings use a template file containing several point and label styles. After reviewing these settings, you will create new point and label styles.

1. If you are not in **Civil 3D**, double-click its desktop icon, starting the application.

2. At Civil 3D's top left, to the upper right of the Civil 3D icon, click the black tri-angle displaying the Application Menu. From the menu, select **New**, browse to this textbook's CD Chapter 02 folder, select the file *Chapter 2 – Unit 1.dwt*, and click **Open**.

3. At Civil 3D's top left, to the upper right of the Civil 3D icon, click the black tri-angle displaying the Application Menu. In the menu highlight **Save as**, in the fly-out menu select AUTOCAD DRAWING, browse to the Civil 3D Projects folder, for the file name, enter **Points-1**, and click **Save**.

Review/Edit Drawing Settings

1. Click the **Settings** tab.

2. In Settings, at the top, select the drawing name, press the right mouse button, and, from the shortcut menu, select EDIT DRAWING SETTINGS....

3. Click the **Units and Zone** tab; if necessary, change the drawing scale to **1″ =40′**, set Imperial to Metric conversion to **US Survey Foot**, and set the zone to **No Datum, No Projection**.

4. Click the ***Object Layers*** tab. Locate the Point Table entry, change its modifier to **Suffix**, and set its value to **-*** (a dash followed by an asterisk).

5. Click the ***Ambient Settings*** tab. Expand the Direction section, click in Format's value cell, and if necessary change the format to **DD.MMSSSS (decimal dms)**.

6. Click **OK** to exit the dialog box.

Edit Feature Settings

Point's Edit Feature Settings dialog box values affect all points. The Default Styles section names point and label styles and the Default Name Format section defines the point groups and point naming format.

1. In Settings, select the **Point** heading, press the right mouse button, and, from the shortcut menu, select EDIT FEATURE SETTINGS....

2. Expand the Default Styles section; if necessary, change the point style to **Basic** and the label style to ***Point#-Elevation-Description***.

3. Expand the Default Name Format section containing the name and sequential counters for both Point Groups and Point Name.

4. Expand the Update Points section and make sure it is set to false.

The Update Points section controls the modification of checked-in points. If this is set to true, you will be able to edit points that should not be changed.

5. Click **OK** to exit the dialog box.

Edit Label Style Defaults

This dialog box's values set the point label styles' general behavior. The Point's Edit Label Style Defaults dialog box can override the drawing's settings. The changes made here only affect the Point branch's labels.

1. In Settings, select the **Point** heading, press the right mouse button, and, from the shortcut menu, select EDIT LABEL STYLE DEFAULTS....

2. Expand each section and review its values.

3. Click **OK** to exit the dialog box.

Point (Marker) Styles

Civil 3D identifies a point's coordinates with a marker. A marker can be a simple Auto-CAD node, a custom Civil 3D object, or a drawing block (symbol) (see Figure 2.1).

1. In Settings, go to the Point branch and expand it until you view the Point Styles list.

2. From the Point Styles list, select **Basic**, press the right mouse button, and, from the shortcut menu, select EDIT....

3. If necessary, select the dialog box's ***Information*** tab.

The Information tab contains the name and other data about the style.

4. Select the ***Marker*** tab.

Currently, the marker is a custom object, an X, whose size is a product of the drawing scale and the value 0.1, and its orientation is relative to the WCS (see Figure 2.1).

5. Change the marker to a plus sign (+) and toggle on the *circle*.

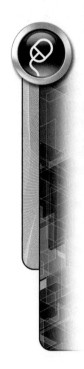

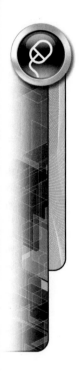

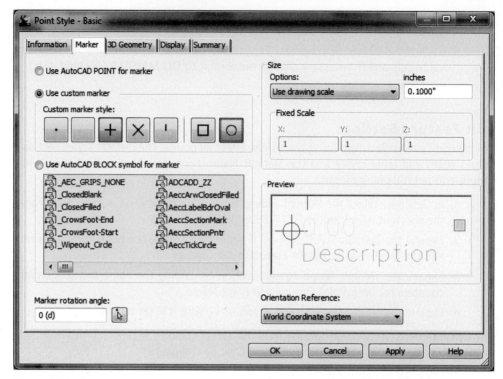

FIGURE 2.1

6. Select the 3D Geometry tab; if necessary, change the Point Display Mode to **Flatten Points to Elevation** and the Point Elevation value to **0** (zero).

These settings produce zero elevation objects when selecting a point with an AutoCAD object snap. When selecting a point with a Civil 3D transparent command (point number or point object), the resulting object will have the point's elevation.

7. Select the ***Display*** tab.

The Display tab indicates that the marker and label are on the default point layer, V-NODE and V-NODE-LBL.

8. Click the ***Summary*** tab; expand each section reviewing its contents.

This tab summarizes all of the style's settings and previews the resulting node.

9. Click ***OK*** to exit the dialog box.

Create Point Styles

The first new point style is Maple (tree). This style's marker is a maple tree symbol. Use Figure 2.2 and the information in Table 2.1 while defining the point style.

1. In Settings, from the point styles list, select **Basic**, press the right mouse button, and, from the shortcut menu, select COPY....

2. Select the ***Information*** tab; for the name, enter **Maple**, enter a short description, and then select the ***Marker*** tab.

3. In the Marker panel, set the marker to **Use AutoCAD BLOCK symbol for marker**, and from the block list, scroll to and select the **Maple** marker.

4. In the Size area, change the inches value to **0.1**, if necessary.

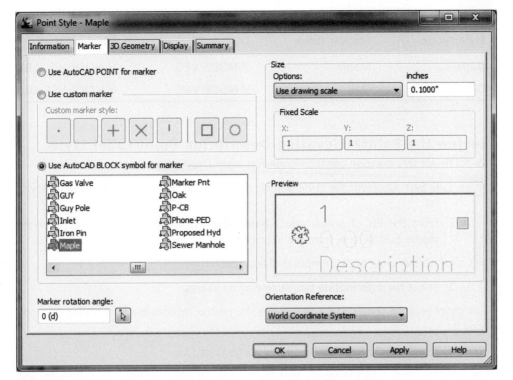

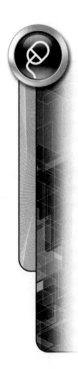

FIGURE 2.2

5. Select the **3D Geometry** tab; if necessary, change Point Display Mode to **Flatten Points to Elevation** and set the Elevation to **0.0**.

6. Select the **Display** tab; click in the cell containing the marker layer name (V-NODE). The Layer Selection dialog box displays.

You select a layer from the current drawing's layer list, from another open drawing (using the Layer Source drop-list at the top-left of the dialog box), or create a new layer (using the New… button at the top right of the dialog box).

7. At the top right of the Layer Selection dialog box, click **New…**. This displays the Create Layer dialog box.

8. In the Create Layer dialog box, click in the Layer name value cell and enter **V-NODE-VEG**.

9. Click in the Color value cell; then click the ellipsis (right side of cell). This displays the Select Color dialog box. In the Select Color dialog box, assign a color, and click **OK** to return to the Create Layer dialog box.

10. Click **OK** to exit the Create Layer dialog box.

11. In the Layer Selection dialog box, if not already selected, locate and, from the layer list, select the layer **V-NODE-VEG**, and click **OK** to assign the selected layer to the marker.

12. Repeat Steps 6 through 11, replacing V-NODE-TEXT layer with **V-NODE-VEG-LBL**.

13. Click **OK**, creating the Maple point style and exiting the dialog box.

14. Repeat Steps 1 through 13 and use the information in Table 2.1 to create an Oak point style. You do not need to create the **V-NODE-VEG** and **V-NODE-VEG-LBL** layers; just select them from the layer list. This style's Point Display Mode is also **Flatten Points to Elevation**, and the Elevation is **0.0**.

TABLE 2.1

Style Name	Marker Block	Display Marker	Display Label
Maple	Maple	V-NODE-VEG	V-NODE-VEG-LBL
Oak	Oak	V-NODE-VEG	V-NODE-VEG-LBL

Review the Point#-Elevation-Description Point Label Style

All points for this exercise use the Point#-Elevation-Description Point Label style.

1. In Settings, from the Point branch, expand Label Styles until you view its styles list.

2. From the list of styles, select **Point#-Elevation-Description**, press the right mouse button, and, from the shortcut menu, select EDIT....

This displays the Label Style Composer dialog box.

3. Select the ***Information*** tab, reviewing its values.

This panel sets the style's name and names the person responsible for creating or modifying it.

4. Select the ***General*** tab, reviewing its settings.

The General panel sets the label's text style, layer, visibility, orientation reference, and plan readability parameters.

5. Select the ***Layout*** tab, reviewing its text component settings.

At the top, the Component name drop-list includes all of the label style's components. The components for this label style are point number, elevation, and description.

6. Click the Component name's drop-list arrow and, from the list, select **Point Elev**.

The Point Elev component's top-left-justified text (Text section's attachment) anchors to the Feature (General section).

7. Click the Component name's drop-list arrow and, from the list, select **Point Number**.

The Point Number component is bottom-left-justified (Text section's attachment value) and is anchored to the top left of the Point Elevation label (General section).

8. Click the Component name's drop-list arrow and, from the list, select **Point Description**.

The Point Description component is top-left-justified (Text section's Attachment value) and is anchored to the Point Elevation's (General section) bottom-left.

Formatting Label Text

Point Number, Point Elev, and Point Description all have their own text format string. The string's values format the label's information (see Figure 2.3).

1. With Point Description as the current component, in the Text section, click in the Contents' value cell.

2. At the Value cell's right, click the ellipsis, displaying the Text Component Editor.

3. On the dialog box's left side, click in the Capitalization's value cell and view the capitalization options list.

4. Click CANCEL to exit the Text Component Editor.

5. In Layout, click Component name's drop-list and, from the list, select **Point Elev**.

6. With Point Elev as the current component, in the Text section, click in the Contents' value cell.

7. At the Value cell's right, click the ellipsis, displaying the Text Component Editor.

8. Click the different value cells, viewing their possible formats and options for this label type.

9. Click CANCEL until exiting the Text Component Editor and the Label Style Composer dialog boxes.

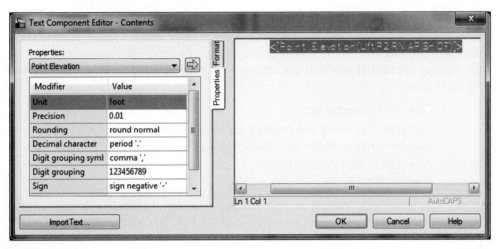

FIGURE 2.3

Table Style

Civil 3D has table styles that list point data in table form.

1. In Settings, expand the Point branch until you can view the Table Styles list.

2. From the table styles list, select **Latitude and Longitude**, press the right mouse button, and, from the shortcut menu, select EDIT....

3. Select the ***Data Properties*** tab and review its values.

4. Select the ***Display*** tab and review its values.

5. Click ***OK***, exiting the dialog box.

6. At Civil 3D's top left, Quick Access Toolbar, click the ***Save*** icon to save the drawing.

Civil 3D has several settings that control visibility and how it works with point data. These values should be set in advance; however, Civil 3D is flexible enough to accommodate most unanticipated situations.

EXERCISE 2-2

When you complete this exercise, you will:

- Create an Import file format.
- Import an ASCII file.
- Import points using LandXML.

- Be familiar with Create Points toolbar's Miscellaneous icon stack commands.
- Be familiar with Create Points toolbar's Intersection icon stack commands.
- Be able to use transparent commands while creating points.

Exercise Setup

1. If you are not in the Points-1 drawing file from the previous unit, open it now. If you are starting with this exercise, from Civil 3D's top left, Quick Access Toolbar, select the **Open** icon, browse to this textbook's CD, Chapter 02 folder, select the drawing *Chapter 02 – Unit 2*, and click **Open**.

Review Point Identity Settings

Before importing points, you need to review the Create Points toolbar's data panel's Point Identity section. This section determines Import points basic values and rules. If the drawing already has points, these settings are critical to not destroy or modify existing point data.

1. Click the **Prospector** tab.
2. In Prospector, select the **Points** heading, press the right mouse button, and, from the shortcut menu, select CREATE....
3. Click the chevron icon (Expand the Create Points dialog) at the toolbar's right to display the point settings data panel.

In an empty drawing the current settings are fine. However, if a drawing contains points, the next point number would be something other than 1. When importing points into a drawing containing points, you may want to change the Sequence Point Numbers From value. Make sure your dialog box matches the values in Figure 2.4.

4. Expand the Point Identity section to change Sequence Point Numbers From value to **10000**.
5. Click the chevron icon (Collapse the Create Points dialog) to close the point settings data panel.

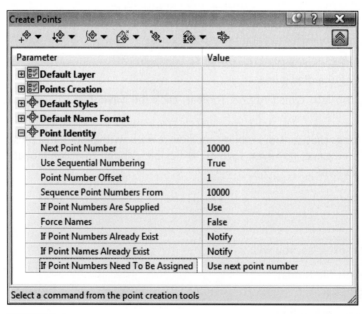

FIGURE 2.4

Defining a File Format

1. Click the various drop-list arrows to review point command icon command stacks.

The Create Points toolbar's Import Points is the rightmost icon. Reading the *Base-Points.nez* file requires a new file format. The file format is point number, northing, easting, elevation, and raw description with a comma delimiter.

2. At the right side of the Create Points toolbar, select the **Import Points** icon.

3. In the Import Points dialog box, at the dialog box's top-right, click the **Point Format** icon.

This displays the Point File Formats dialog box, listing the current file formats.

4. In the Point File Formats dialog box, click **New...**, displaying the Point File Formats – Select Format Type dialog box.

5. In the Point File Format – Select Format Type dialog box, select **User Point File**, and click **OK** to display the Point File Format dialog box.

To create a new file format, enter a format name, set its extension, set its delimiter, and define a record's data types and order. When you are finished setting the values, your dialog box should look like Figure 2.4.

6. At the top left of the Point File Format dialog box, for the name, enter **HC3D - (Comma Delimited)**, and for the extension, click the drop-list arrow, displaying the extension list and, from the list, select **.nez**.

7. In the Point File Format dialog box, at the top-right in Format Options, select **Delimited by** and, in the box to the right of the toggle, type a comma (**,**).

8. In the Point File Format dialog box, in the middle-left, click the first <unused> heading.

This displays the Point File Formats – Select Column Name dialog box.

9. In the Point File Formats – Select Column Name dialog box, click the drop-list arrow, displaying the data types list and, from the list, select **Point Number**.

When you select Point Number, the Point File Formats – Select Column Name dialog box displays with an entry that defines an invalid point number.

10. Click **OK**, assigning Point Number to the first field and returning to the Point File Format dialog box.

11. Click the second column in from the left (<unused>).

12. In the Point File Formats – Select Column Name dialog box, click the drop-list arrow that displays the data types list, and select **Northing** from the list.

After selecting Northing, the Point File Formats – Select Column Name dialog box contains northing entries that identify invalid coordinates and the precision coordinates. The precision value is important with column-formatted data, not comma- or space-delimited data. So, for this exercise, you can ignore precision setting. However, if you are working with a columned file or with invalid indicators, these values need to be set.

13. Click **OK**, returning to the Point File Format dialog box.

14. Repeat Steps 10 through 12 and add **Easting**, **Point Elevation**, and **Raw Description** to the file format. The new headings are to the right of each heading you define.

After adding the headings, your Point File Format dialog box should look like Figure 2.5.

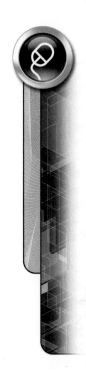

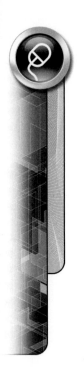

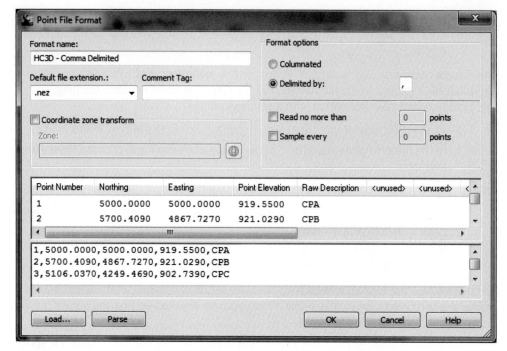

FIGURE 2.5

Testing the Format

Test the new format by loading and parsing *Base-Points.nez*. This file is in this textbook's CD Civil 3D project folder or the Chapter 02 folder.

1. At the bottom-left of the Point File Format dialog box, click **Load...**, and in the Select Source File dialog box browse to the Civil 3D Projects' Civil 3D folder (C:\ Civil 3D Projects\Civil 3D), or browse to this textbook's CD, Chapter 02 folder.

2. In the Select Source File dialog box, at the bottom, change the Files of type to ***.nez**, select the file *Base-Points.nez*, and click **Open**.

3. In Point File Format dialog box, to read the file, at the bottom-left, click **Parse**.

The file reads correctly.

4. In the Point File Format dialog box, click **OK** and notice that the new format is now listed.

5. Click **Close** to return to the Import Points dialog box.

Importing a File

1. If necessary, in the Import Points dialog box, click the Format: drop-list arrow and select the format **HC3D - (Comma Delimited)**.

2. In the Import Points dialog box, at the right of Source File(s):, click the **plus sign** icon, displaying the Select Source File dialog box. If the Civil 3D folder is not the current folder, change the Look in: folder to the local C: drive, double-click Civil 3D Projects, and then the Civil 3D folder to open (C:\Civil 3D Projects\Civil 3D), or navigate to the CD's Chapter 02 folder.

3. Once in the folder, set the Files of type to **.nez*, select the *Base-Points.nez* file, and click Open, returning to the Import Points dialog box.

4. In the Import Points dialog box, do not toggle on any advanced options.

Your Import Points dialog box should look like Figure 2.6.

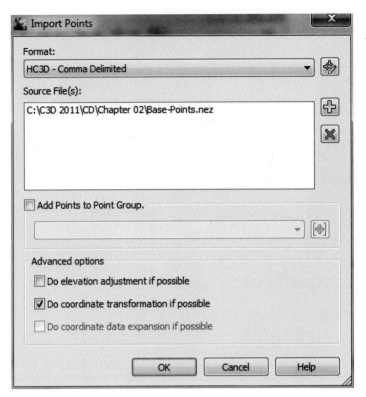

FIGURE 2.6

5. Finally, click **OK**, importing the points.

6. If you cannot see the line work or points, use the ZOOM EXTENTS command to view the entire drawing.

7. In Prospector, expand the Point Groups heading.

8. From the list select **_All Points**, press the right mouse button, and, from the shortcut menu, select **Properties....**

9. Select the Information tab, change the Point label style to **Point#-Elevation-Description**, and click **OK** to exit.

10. In Civil 3D's top left, Quick Access Toolbar, click the **Save** icon, saving the drawing.

LandXML — Settings

Before Importing a LandXML file verify and, if necessary, change the LandXML settings. The Settings, Edit LandXML Settings dialog box contains these settings.

1. Click the **Settings** tab.

2. At the Settings panel's top, select the drawing name, press the right mouse button, and, from the shortcut menu, select EDIT LANDXML SETTINGS....

3. In the LandXML Settings dialog box, click the **Import** tab, expand the Point Import Settings section, and set Point Description to **use "code" then "desc."**

When exporting to LandXML and using **US Survey Foot**, you need to set Export's base unit in the Data Settings section.

4. Click the **Export** tab, expand the Data Settings section, and change the Imperial Units to **Survey Foot** and the Angle/Direction format to **Degrees decimal dms (DDD.MMSSSS)** (see Figure 2.7).

5. Click **OK**, exiting the dialog box.

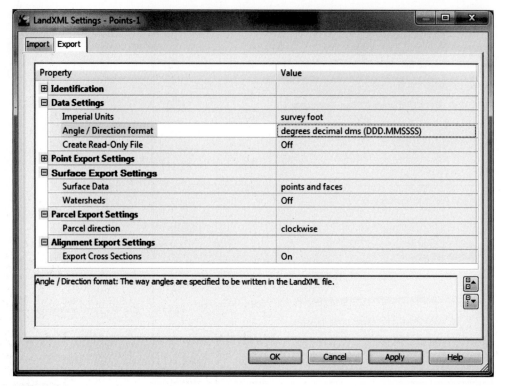

FIGURE 2.7

LandXML — Import

1. In Civil 3D's Ribbon, click the Insert tab. In the Import panel, click the icon *LandXML* to display the Import LandXML dialog box.

2. If you are not in the Civil 3D folder, change the Look in: folder to the Local C: drive, double-click Civil 3D Projects, and then the Civil 3D folder to open it (C:\ Civil 3D Projects\Civil 3D), or navigate to this textbook's CD Chapter 2 folder.

3. From the list, select *Ellis.xml* and click *Open*, reading the file.

4. In the Import LandXML dialog box, toggle **OFF** the Alignments and expand CgPoints to review its contents (see Figure 2.8).

Notice that the dialog box lists the point numbers as point names and point codes and descriptions.

5. Click *OK*, importing the points and closing the dialog box.

6. At Civil 3D's top left, Quick Access Toolbar, click the *Save* icon, saving the drawing.

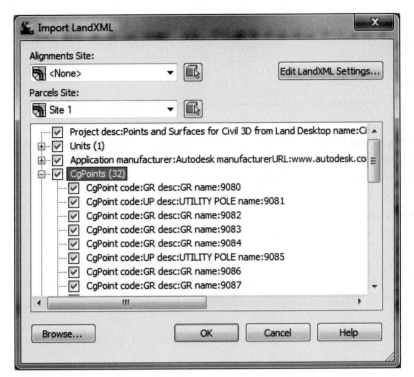

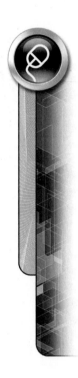

FIGURE 2.8

The new points are to the west of the north–south road, at the western edge of the drawing's line work.

Point Identity — Next Point Number

The Next Point Number is the first unused point number.

1. In the toolbar, click the Expand Create Points chevron, expanding the Create Points toolbar.
2. Expand the Point Identity section, review Next Point Number's value, and if necessary, change it to **10000**.
3. Close the Point Identity section.

Create Points Settings

When setting a point, you need to define how to assign the new points' elevation and description.

1. Expand the Points Creation section, review the values, and make sure Prompt For Elevations and Prompt For Descriptions are set to **Manual**.
2. Close the Points Creation section.
3. Click the Collapse the Create Points dialog chevron, collapsing the Create Points toolbar.
4. Click the **Prospector** tab.

Miscellaneous – Manual — Frame Shed and Residence

This exercise section uses commands from the Create Points toolbar's Miscellaneous icon stack.

There are a few points missing. In Table 2.2, use the point elevations and descriptions, and for the points' locations, refer to Figure 2.9.

TABLE 2.2

Point Number	Elevation	Description
10000	920.07	BLDG
10001	917.33	SWK

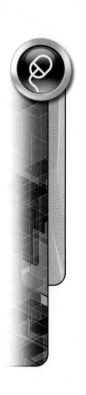

1. Click the Ribbon's View tab, then on the Views panel, select the named view **Frame Shed**.

2. In Create Points toolbar's left side, click the first drop-list arrow and, from the Miscellaneous icon stack, select **Manual**.

3. Use the **Endpoint** object snap and select the shed's southeast corner, setting point 10000. When prompted for the elevation and description, use the values from Table 2.2.

4. Use the **Endpoint** object snap and select the gravel path's intersection east of the Frame Residence. When prompted for the elevation and description, use the values from Table 2.2.

5. Press ENTER to end the command.

6. At the Civil 3D's top left, Quick Access Toolbar, click the **Save** icon, saving the drawing.

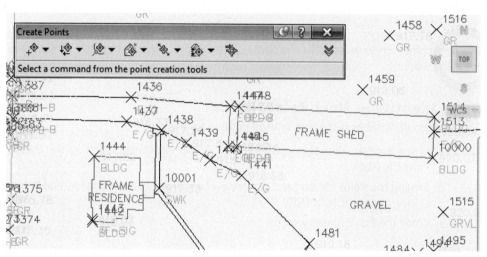

FIGURE 2.9

Miscellaneous – Measure

The next points do not have elevations, but they have the same description. This means changing values in Create Points.

1. From the Ribbon's View tab, Views panel, select the named view **Lot Line**. You will need to click the down arrow to scroll down the list.

2. In the Create Points toolbar, click the Expand the Create Points dialog chevron.

3. Expand the Points Creation section, set Prompt For Elevations to **None**, set Prompt For Descriptions to **Automatic**, and set Default Description to **LLPOINT** (see Figure 2.10).

The Prompt For Elevations and Prompt For Descriptions cells display a drop-list arrow when you click in their value cells.

    4. Click the Collapse the Create Points dialog chevron.

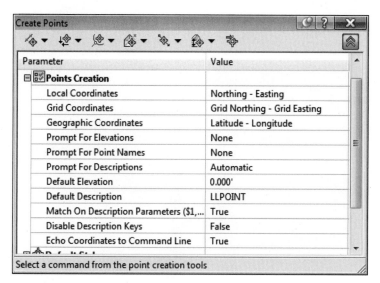

FIGURE 2.10

    5. In the Create Points toolbar, click Miscellaneous' drop-list arrow (first icon on left) and, from the icon stack, select ***Divide Object***.

    6. The routine prompts to select a line, arc, etc. Select lot 16's north lot line, enter **4** as the number of segments, and press ENTER to set the offset to **0.0**.

The routine divides the line into four equal segments with five new points. Each point has no elevation and the description is LLPOINT.

    7. Press ENTER, ending the routine.

    8. In the Create Points toolbar, click the Miscellaneous drop-list arrow and, from the icon stack, select ***Measure Object***.

    9. The routine prompts to select a line, arc, etc. Select lot 16's south lot line, press ENTER twice to accept the beginning and ending stations, set the offset to **0.0**, press ENTER, and for the interval, enter **25** and press ENTER.

The routine creates seven new points with a 25 foot spacing.

    10. Press ENTER, exiting the routine.

    11. At Civil 3D's top left, Quick Access Toolbar, click the ***Save*** icon, saving the drawing.

Your screen should look like Figure 2.11.

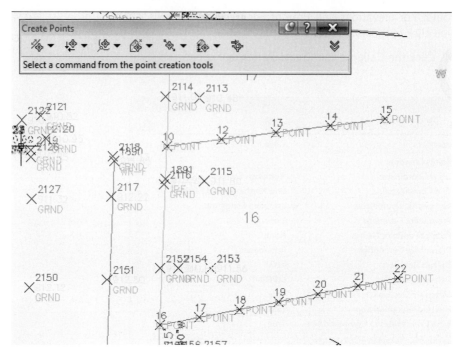

FIGURE 2.11

Polyline Vertices — Automatic

1. In the Create Points toolbar, click the Expand the Create Points dialog chevron. In the dialog box, expand the Points Creation section, set Prompt For Elevations and Prompt For Descriptions to **Automatic**, and set the Default Description to **BERM** (see Figure 2.12).

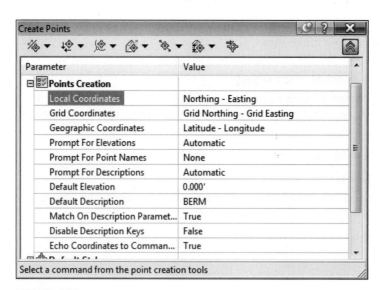

FIGURE 2.12

2. Click the Collapse the Create Points dialog chevron.
3. From the Ribbon's View tab, Views panel, select the named view **Berm**. You will need to click the up arrow to scroll up the list.

4. In the Create Points toolbar, click Miscellaneous' drop-list arrow and, from icon stack, select **Polyline Vertices – Automatic**.

5. The routine prompts for a polyline selection. Select the polyline in the center of the screen.

6. Press ENTER, exiting the routine.

The berm line is a 3D polyline. The routine places a point at each vertex, assigning BERM as the description and the elevation of the vertex to each point (see Figure 2.13).

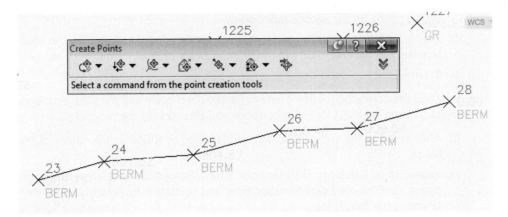

FIGURE 2.13

7. At Civil 3D's top left, Quick Access Toolbar, click the **Save** icon, saving the drawing.

Intersection — House Corners

This exercise section uses commands from the Create Points toolbar's Intersections icon stack.

1. If on, toggle **OFF** Dynamic Input mode.

2. In the Create Points toolbar, click the Expand the Create Points dialog chevron.

3. If necessary, expand the Points Creation section and set Prompt For Elevations and Prompt For Descriptions to **Manual**.

4. Expand the Point Identity section, set the Next Point Number to **10050**, and click the Collapse the Create Points dialog chevron.

5. From the Ribbon's View tab, Views panel, select the named view **Lot 11**.

This point series references a point number as the starting point. A field crew measured distances from two points, locating new points that represent the Lot 11 house's south corners. Use the values in Table 2.3 to answer the command prompts.

TABLE 2.3

From Point	Distance	New Point	Elevation	Description
1001	88.94	10050	920.54	BLDG
1002	113.14			
1001	116.21	10051	920.52	BLDG
1002	141.06			

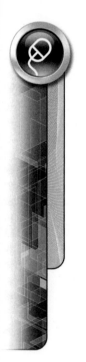

6. In the Create Points toolbar, click the Intersections drop-list arrow (second icon in from left) and, from the icon stack, select ***Distance/Distance***.

7. In the command line, the routine prompts you to Please specify a location for the radial point location. Click the Transparent Commands' ***Point Number*** filter icon, overriding the current prompt with Enter point number.

8. In the command line, as the first radial point, enter **1001**, press ENTER, enter the distance of **88.94**, and press ENTER, defining the first radial point.

9. The prompt changes, asking for the second point's radial location. Enter as the point number **1002**, press ENTER, enter the distance of **113.14**, and press ENTER, defining the second radial point.

10. The routine then prompts again for a point number. Press ESC, exiting the Point Number prompt and returning to the Point or All prompt.

11. In Lot 11, select a point near the green **X** at the house's southern corner.

After selecting the point, the routine prompts for the description and then the elevation. Use the values from Table 2.3 for the description and elevation of the new point.

12. After entering the two values, routine places point **10050** at the corner of the house.

13. Repeat Steps 8 through 13 to set point **10051**. Remember to reselect the Transparent Command's **Point Number** filter and to stop it by pressing ESC when selecting the point's location.

14. Exit the *Distance/Distance* command by pressing ESC.

15. In Civil 3D's upper left, Quick Access Toolbar, click the ***Save*** icon, saving the drawing.

House Corners by Side Shots

The next point set represents the current building's northern corners and two additional adjacent parcel (Lot 12) building corners. Table 2.4 contains the setup and backsight points, and each new point's side shot angles, distances, elevations, and descriptions.

The Transparent Commands Side Shot override uses point 2341 as the setup point and point 1001 as the backsight point.

Make sure AutoCAD's Dynamic Input mode is off while entering the point values.

TABLE 2.4

Setup	Backsight	Angle	Distance	Description	Elevation
2341	1001	318.2034	56.55	BLDG	920.54
		294.0043	67.05	BLDG	920.53
		252.5419	128.29	BLDG	920.55
		243.5822	96.75	BLDG	920.53

1. In the Create Points toolbar, click Miscellaneous' drop-list arrow and, from the icon stack, select ***Manual***.

2. From the Transparent Commands toolbar, select the override **Side Shot**.

Side Shot changes the prompting to selecting a line or points, establishing a setup and backsight (turned angle and distance from this setup).

3. To establish a setup using points 2341 and 1001, type the letter 'P' (for points mode) and press ENTER.

In points mode, .p changes the prompting to point number and .g changes it to select a point object.

4. In the command line, at the Select line or [Points]: prompt, type '**P**', and press ENTER.

5. In the command line, at the Specify starting point or [.P/.N/.G]: prompt, type '**.p**' and press ENTER.

6. The prompt changes to Enter Point Number; enter point **2341** and press ENTER.

7. The next prompt is Specify ending point or [.P/.N/.G]: again, type '**.p**' and press ENTER.

8. The prompt changes to Enter Point Number; enter point **1001** and press ENTER.

9. In the command line, the prompt is for an angle; for the angle, enter **318.2034** and press ENTER.

10. In the command line, the prompt is for a distance; for the distance, enter **56.55** and press ENTER.

11. The next two command prompts are for the new point's description and elevation. For their respective prompts, enter **BLDG** and **920.54**.

12. Repeat Steps 9 through 11, creating the three remaining building corner points.

13. Press ESC, exiting the Side Shot override, and press ENTER, ending the command.

14. In the Create Points toolbar, select the red **X** at the right to close it.

15. In Civil 3D's upper left, Quick Access Toolbar, click the **Save** icon, saving the drawing.

This completes this exercise. After points are created they may need evaluation and possibly editing.

EXERCISE 2-3

When you complete this exercise, you will:

- Generate a points report.
- Be familiar with the LandXML Report Application.

Generating a Point Report

1. If you are not in the *Points-1* drawing file from the previous unit, open it now. If you are starting with this exercise, in Civil 3D's upper left, Quick Access Toolbar, click the **Open** icon, browse to this textbook's CD Chapter 2 folder, select the *Chapter 02 - Unit 3.dwg*, and click **Open**.

2. If necessary, click the Ribbon's View tab. To the Toolspace icon's right and in the bottom row, click the **TOOLBOX** icon, and click the Toolbox tab.

3. Expand the Reports Manager and Points sections.

4. From the list of Points reports, select **Points_List**, press the right mouse button, and select EXECUTE..., displaying the Export to XML Report dialog box (see Figure 2.14).

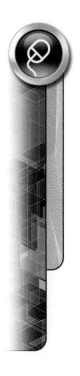

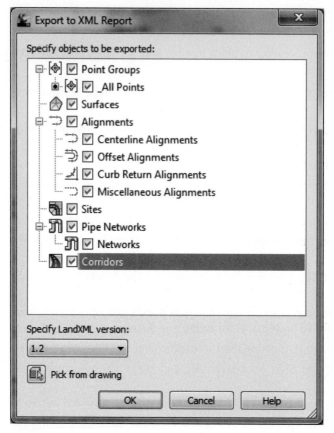

FIGURE 2.14

    5. If not selected, toggle **ON** the point group _All Points and click **OK**.

Internet Explorer displays a point list report. You can print or save this report.

    6. Review the report and close Internet Explorer.

    7. If you want to export an Excel or CSV file, from the Points report list, select **Points_in_CSV**, press the right mouse button, and select **Execute....** The Export to XML Report dialog box displays.

    8. If not selected, toggle **ON** the point group _All Points and click **OK**.

    9. A File Download dialog box displays, asking to Open or Save the file. Click SAVE, browse to the Civil 3D Projects folder, click SAVE, and CLOSE.

    10. If you have Excel, start Excel and view the file.

    11. After reviewing the file, exit Excel and return to Civil 3D.

Zoom to and Pan to

Locating individual points can be difficult. The Points heading and preview area's shortcut menus have the commands **ZOOM TO** and **PAN TO**.

    1. Click the **Prospector** tab.

    2. In Prospector, select the **Points** heading.

The preview area lists all points.

    3. In preview, scroll through the points list, select a point, press the right mouse button, and, from the shortcut menu, select **ZOOM TO**.

The display centers on the point's location.

4. In preview, scroll through the list, select another point, press the right mouse button, and, from the shortcut menu, select **PAN TO**.

5. Select a few additional points from preview and use the commands **ZOOM TO** or **PAN TO**.

6. In Civil 3D's upper left, Quick Access Toolbar, click the **Save** icon, saving the drawing.

Available Point Numbers

When setting points it is always good to know what point numbers are available (unused) *List Available Point Numbers*.

1. On the Ribbon, select the Modify tab. In the Modify tab's left side, click the Points icon to display the Points ribbon.

2. On the Points ribbon's left click the Cogo Point Tools panel's drop list arrow.

3. From the Cogo Point Tools panel's drop list select List Available Point Numbers to view the list in the command line.

4. In Civil 3D's upper left, Quick Access Toolbar, click the **Save** icon, saving the drawing.

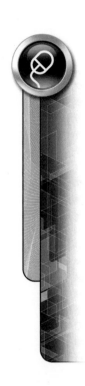

EXERCISE 2-4

When you complete this exercise, you will:

- Be able to edit points from various locations in the Civil 3D interface.
- Be able to edit points in a vista.
- Renumber selected points.
- Adjust the datum and the elevation of selected points.
- Export points to an ASCII file.

Exercise Setup

This exercise continues with the drawing from the previous exercise.

1. If you are starting with this exercise and have not done the previous exercises, from the CD that accompanies this textbook, browse to the Chapter 02 folder and open the *Chapter 02 - Unit 4.dwg*.

Prospector Preview - Edit Points

1. If necessary, click the **Prospector** tab.

2. In Prospector, click the **Points** heading, listing the points in preview.

3. In preview, use the scroll bars to view the points.

4. In preview, click the Point Elevation heading, sorting the point elevations.

5. In preview, scroll to and locate point **3001**.

6. In preview, scroll through point 3001's record until locating its elevation (0.0). Click the Point Elevation cell, change the elevation to **898.76**, and accept the new value by pressing ENTER.

7. In Prospector, click the **Points** heading, press the right mouse button, and, from the shortcut menu, select EDIT POINTS... to display the Point Editor vista.

8. In the Point Editor vista, click the Point Number heading, sorting the list.

9. For point 10001, click the Point Elevation cell and change its elevation to **914.15**.

10. For point 10002, click the Point Elevation cell and change its elevation to **914.90**.

11. After editing the points, at the top of Panorama's mast, click the **X**, closing it.

12. In Prospector, select the Points heading, listing the points in preview.

13. In preview, click the Point Number column heading, sorting the points.

14. In preview, scroll thorough the point list, locating point **10003**.

15. In preview, select 10003's point icon, press the right mouse button, and, from the shortcut menu, select *ZOOM TO*.

16. In the drawing, select point **10003**, press the right mouse button, and, from the shortcut menu, select EDIT POINTS....

17. In the Point Editor vista, click in the Point Elevation cell, enter the value of **916.78**, click the green check in the upper-right to close the vista.

18. In preview, scroll through the point list, locating 3002, select 3002's point icon, press the right mouse button, and, from the shortcut menu, select *ZOOM TO* .

19. In the drawing, select point **3002**, press the right mouse button, and, from the shortcut menu, select EDIT POINTS....

20. In the Point Editor vista, click in the Point Elevation cell, enter the value of **914.96**, and close the vista.

21. In preview, scroll through the point list, locating number 3003, select 3003's point icon, press the right mouse button, and, from the shortcut menu, select *ZOOM TO*.

22. In preview, select point **3003**, press the right mouse button, and, from the shortcut menu, select EDIT POINTS....

23. In the Point Editor vista, click in the Point Elevation cell, enter the value of **912.40**, and close the panorama.

24. At Civil 3D's upper left, Quick Access Toolbar, click the **Save** icon, saving the drawing.

Renumber Points

There will be times when you will need to renumber points to new numbers.

1. In preview, if necessary, click the Point Number column heading, sorting the points.

2. In preview, scroll through the point list, locating point **3001,** and select the point.

3. While holding down SHIFT, select point number **3003**, selecting points 3001 to 3003.

4. With the cursor over the highlighted points, press the right mouse button, and, from the shortcut menu, select RENUMBER....

5. In the command line, the routine prompts for an additive factor. Enter **2000** and press ENTER to renumber the points to 5001, 5002, and 5003.

The points change to the new point numbers.

6. In preview, scroll until you locate points 5001 through 5003.

7. At Civil 3D's upper left, Quick Access Toolbar, click the **Save** icon, saving the drawing.

Edit Datum

There will be times when points have an assumed or incorrect setup elevation. You adjust these points with the Datum routine. The Datum routine prompts for an old and new elevation or an amount (positive or negative) to add to the point's elevation.

1. If necessary, in Prospector, select the Points heading, listing the points in preview.

2. In preview, scroll through the point list, locating point 2025.

3. Select point **2025**, press CTRL, and select point **2024**.

4. With both points highlighted, press the right mouse button, and, from the shortcut menu, select DATUM....

5. In the command line, the routine prompts for an amount of change, enter **5**, and press ENTER.

This adds 5 feet to the two points' elevations.

6. In preview, with the two points highlighted, press the right mouse button, and, from the shortcut menu, select DATUM.... You may have to reselect the two points (2024 and 2025).

7. In the command line, the routine prompts for a change in elevation; for reference, enter the letter '**R**'; as the reference elevation, enter **100**; and for the "new" elevation, enter **95**.

This subtracts 5 feet of elevation from the two points.

8. At Civil 3D's upper left, Quick Access Toolbar, click the ***Save*** icon.

Export Points

There will be times when you will need to export points to an ASCII file.

1. In Prospector, select the **Points** heading, press the right mouse button, and, from the shortcut menu, select EXPORT....

2. In the Export Points dialog box, select the **HC3D - (Comma Delimited)** format. If you didn't make the format, use the **PNEZD (comma delimited)** format.

3. To the right of Destination File, click the ***folder*** icon, browse to the Civil 3D Projects folder, and name the file Points-1.

4. Click ***Open*** and ***OK***, exporting the points and dismissing the dialog boxes.

5. At Civil 3D's upper left, Quick Access Toolbar, click the ***Save*** icon, saving the drawing and close the file.

This ends the edit point exercise. The next unit covers layer and point and label styles assignments using a Description Key Set.

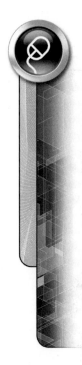

EXERCISE 2-5

When you complete this exercise, you will:

- Import a Description Key Set.
- Add entries to a Description Key Set.
- Create point groups.
- Assign point group style overrides.
- Adjust the point group display properties.
- Create a point table style.
- Create a point table.

Exercise Setup

This exercise uses a new drawing template file, Chapter 02 – Unit 5.dwt file. The file is on this textbook's CD, in the Chapter 2 folder.

1. At Civil 3D's top left, click the Civil 3D's icon drop-list arrow, from the Application Menu, select **New**, browse to the textbook's CD, Chapter 02 folder, select the template file *Chapter 02 – Unit 5.dwt*, and click OPEN.

2. At Civil 3D's top left, click Civil 3D's icon drop-list arrow, from the Application menu, select SAVE AS, in the flyout select AutoCAD Drawing, and save it as *Points-2* in the Civil 3D Projects folder.

Add Point Styles

There are three missing point styles: Inlet, Maple, and Oak. This exercise creates these point styles, defining a point label layer. The Marker component layer is 0 (zero) for all three point styles because Description Key Set entries assign the point style's layer (see Table 2.5).

TABLE 2.5

Point Style	Marker	3D Geometry	2D and 3D Label Layer
Inlet	INLET	Flatten Point to Elevation	V-ESTM-LBL
Maple	Maple	Flatten Point to Elevation	V-VEG-LBL
Oak	Oak	Flatten Point to Elevation	V-VEG-LBL

1. Click the **Settings** tab.
2. In Settings, expand the Point branch until you can view the Point Styles heading and its styles list.
3. Select the **Point Styles** heading, press the right mouse button, and, from the shortcut menu, select NEW....
4. Click the **Information** tab, and for the name, enter **Inlet**.
5. Click the **Marker** tab, making it current.
6. Click the **Use AutoCAD BLOCK symbol for marker**, scroll through the blocks, and select **Inlet**.
7. Click the **3D Geometry** tab and change the Point Display Mode to **Flatten Points to Elevation**.
8. Click the **Display** tab, and leave layer 0 (zero) for the Marker.
9. Click Label layer 0 (zero); the Layer Selection dialog box displays. Scroll down the list, select the layer **V-ESTM-LBL**, and click OK to return to the Point Style dialog box.
10. If necessary, for both the Marker and Label entries, change the Color and Linetype properties to **ByLayer**.
11. Click **OK**, closing the dialog box.
12. Repeat Steps 3 through 11, making the point styles **Maple** and **Oak** by using the information in Table 2.5.
13. At Civil 3D's top left, Quick Access Toolbar, click the **Save** icon, saving the drawing.

Importing a Description Key Set

Civil 3D imports a Land Desktop project's description keys, creating a Description Key Set. If you did not copy it from this textbook's CD Civil 3D project to the Civil 3D Projects folder, please do so now.

1. In Settings, click the **Point** heading, press the right mouse button, and, from the shortcut menu, select EDIT FEATURE SETTINGS....

2. Expand the Default Styles section, reviewing the default Point and Label style names.

Each Description Key Set entry will have these as their initial styles.

3. Click **OK**, exiting the dialog box.

The Ribbon's LAND DESKTOP command imports the Description Key file from a LDT project.

4. On the Ribbon click the Insert tab. In the Import panel select LAND DESKTOP.

5. Set the Land Desktop Project Path to *Civil 3D Projects*, set the Project Name to *Civil 3D*, and in the data list, except for Description Keys, toggle them **OFF** (see Figure 2.15).

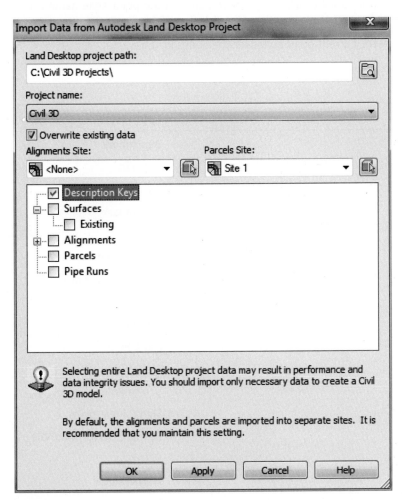

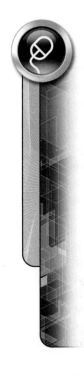

FIGURE 2.15

6. Click **OK**, migrating the description keys.

7. The Description Keys Migration Completed dialog box displays. Click **OK**, exiting the dialog box and completing the description key migration.

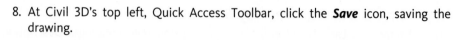

8. At Civil 3D's top left, Quick Access Toolbar, click the **Save** icon, saving the drawing.

Review Imported Description Keys

1. In Settings, expand the Point branch until you can view the Description Key Sets list.

The LAND DESKTOP command imported three description key files: **Civil 3D**, **DEFAULT**, and **HNC3D-DEFAULT**.

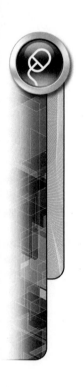

2. In Settings, in the Point branch, select the **Civil 3D** description key set, press the right mouse button, and, from the shortcut menu, select DELETE.... The "Are you sure?" dialog box displays, and click **YES** to delete the description key set.

3. Repeat step 2 and delete the **Default** description key set.

Change Codes Point Styles

When importing a Land Desktop Description Key file, each code has the default point style. However, several codes need a point styles with a symbol.

Changing the point style is a two-step process. First, toggle on point style and then select the new point style from a list of styles. When you click in the point style name cell, a Select Point Style dialog box displays with a drop-list of available point styles.

You need to change the current point styles for the codes in Table 2.6.

TABLE 2.6

Code	Point Style
BM###	Benchmark
BP	Iron Pin
CB*	Catch Basin
CP@	Benchmark
DATUM	STA
DMH	Storm Sewer Manhole
EFYD	Hydrant (existing)
I[PR]F	Iron Pin
LC	Iron Pin
LP*	Light Pole
MH*	Manhole
PP*	Utility Pole
SIGN*	Sign (single pole)
SMH	Sanitary Sewer Manhole
STA#	STA
TF-SIG	Traffic Signal
TR*	Tree
VV	Valve Vault
WELL	Well
WV	Water Valve

1. In Settings, in the Description Key Set list, select **HNC3D-DEFAULT**, press the right mouse button, and, from the shortcut menu, select EDIT KEYS....

2. In the DescKey Editor vista, for each entry, change the Use Drawing Scale toggle to **NO**. Right-click the column heading, select EDIT..., and a drop-list below the heading will appear, allowing you to select **NO** from the list.

3. In the DescKey Editor vista, in the Point Style column for the BM### entry, indicate the intent to override the default point style by togging it **ON**.

4. In the DescKey Editor vista, in the Point Style column for the **BM###** entry, click <default>, displaying the Point Style dialog box. Click the drop-list arrow, from the list select **Benchmark**, and click **OK**, selecting the style and closing the dialog box.

The BM### code has the Benchmark point style.

5. Repeat Steps 3 and 4, assigning the remaining keys in Table 2.6 to the appropriate point styles.

Create New Description Key Set Entries

The Description Key Set has three missing entries: Inlet, Maple, and Oak. The next exercise section creates these keys.

The new keys use related point styles (Inlet, Maple, and Oak). The Maple and Oak codes (MP 2 and OK 6) have a parameter (the tree's trunk diameter). The format entry for these description keys uses the $1 parameter and is the full description's first item. The MP 2's resulting full description is 2″ MAPLE and OK 6's is 6″ OAK. The format coding is in Table 2.7.

The first parameter is also a symbol multiplier. This displays a 6-inch oak tree as a larger symbol than an oak tree that is only 2 inches in diameter.

The three keys' scaling settings are in Table 2.8. All three keys use the drawing scale to make them the correct size for the drawing. The Maple and Oak entries use Parameter 1 as a secondary scaling factor, which is listed in Table 2.8. This entry makes an MP 4 code's symbol larger than an MP 2's symbol.

TABLE 2.7

Code	Point Style	Format	Layer
IN@	Inlet	$*	V-ESTM-PNT
MP*	Maple	$1″ MAPLE	V-VEG-PNT
OK*	Oak	$1″ OAK	V-VEG-PNT

TABLE 2.8

Code	Scale Parameter1	Drawing Scale	Apply to X-Y
IN@	ON	OFF	OFF
MP*	ON	OFF	ON
OK*	ON	OFF	ON

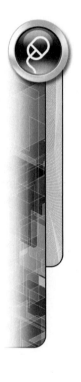

Inlet Key

When creating a new Description Key Set entry, you have to select an existing code entry. You should review and adjust all key settings before going on to the next key.

1. In the DescKey Editor vista, select the entry **MH***, press the right mouse button, and, from the shortcut menu, select NEW....

This makes a new key entry, New DescKey, which is at the bottom of the list.

2. Locate the New DescKey entry, click in the Code cell, and for the new code, enter **IN@**.

The list resorts and you must locate IN@ in the sorted code list.

3. Click the Point Style override toggle to **ON** and click in the point style cell, displaying the Point Style dialog box. Click the drop-list arrow, from the list select **Inlet**, and click *OK*, returning to the DescKey Editor vista.
4. Make sure the Format cell value is **$***.
5. Click the Layer override toggle, turning it **ON,** and click in the cell, displaying the Layer Selection dialog box. From its list, select the layer of **V-ESTM-PNT**, and click *OK*, returning to the DescKey Editor Vista.
6. If the Scale Parameter is on, toggle it **OFF**.
7. If the Use Drawing Scale is on, toggle it **OFF**.

Maple Key

1. While still in the DescKey Editor, select the entry **ALLEY**, press the right mouse button, and, from the shortcut menu, select NEW....

This makes a new key entry, New DescKey, and sorts it to the bottom of the code list.

2. Locate the New DescKey entry, click in its Code cell and, as the new code, enter **MP***.

The code is again sorted and you must look for the MP* code entry.

3. Click the Point Style override toggle to turn it **ON** and, as in step 8 of inlet (above), select the style **Maple**.
4. Change the Format to **$1" MAPLE**.
5. Click the Layer override toggle to turn it **ON** and, as in step 10 of inlet (above), select the layer **V-VEG-PNT**.
6. Make sure the Scale Parameter is **ON** and is set to **Parameter 1**.
7. If the Use Drawing Scale is on, toggle it **OFF**.
8. Toggle **ON Apply to X-Y**.

Oak Key

1. Repeat the Maple steps 1 through 8, creating an entry for the **OK*** code, but use the **Oak** point style and for the format use **$1" OAK**.
2. After creating the new codes, close the DescKey Editor vista by clicking the **X** at the top of Panorama's mast.
3. At Civil 3D's top left, Quick Access Toolbar, click the *Save* icon to save the drawing.

Importing Points

Everything is set to import points using the current Description Key Set. The file Base-Points.nez is in this textbook's CD, Chapter 02 folder, or in Civil 3D Projects folder, the Civil 3D project.

1. In the Ribbon click the Insert tab. On the Import panel click POINTS FROM FILE to display the Import points dialog box.

2. In the Import Points dialog box, at its top, click the drop-list arrow and, from the format list, select **PNEZD (comma delimited)**.

3. To select the Source File(s), on the dialog box's right side, click the plus (+) icon.

4. In the Select Source File dialog box, at the bottom, to the right of Files of Type, click the drop-list arrow, and, from the list of file extensions, select ***.nez**.

5. Navigate to the location of the Base-Points.nez file (C:\Civil 3D Projects\Civil 3D), select the file, and click **Open**, returning to the Import Points dialog box.

6. In the Import Points dialog box, do not set any Advanced Options, and click **OK**, importing the points.

7. Click the **prospector** tab, making it current.

8. In Prospector, expand Point Groups, select **_All Points**, press the right mouse button, and, from the shortcut menu, select PROPERTIES....

9. In _All Points Properties, change the Label Style to **Point #-Elevation-Description** and click **OK** to exit.

10. At the drawing's bottom right, the Status Bar, set the Annotation Scale to 1" =30'.

11. In Prospector, select the Points heading, listing the points in preview.

12. In preview, scroll the point list to the right until you can view the Raw Description heading, and click the heading, sorting the list.

13. In preview, scroll through the list until you can view a **WELL** description, and then scroll the listing back until you can view its point number.

14. In preview, select the point number's point icon, press the right mouse button, and, from the shortcut menu, select ZOOM TO.

You should see the Well marker style in the drawing specified by the Description Key Set.

15. At Civil 3D's top left, Quick Access Toolbar, click the **Save** icon, saving the drawing.

You can zoom to any point from the preview area (Points or Point Group) and the Edit Points vista.

Creating Point Groups

The easiest way to create a point group is by using raw descriptions. The Description Key Set's raw description entries create the point groups' selection list.

Use Table 2.9's description keys list to define three point groups.

TABLE 2.9

Name	Key(s)
Well	WELL
Iron	BP, I[PR]F, LC, SIP
Site Control	BM###, CP@, STA#

1. If necessary, click the ***Prospector*** tab, making it current.
2. In Prospector, click the **Point Groups** heading, press the right mouse button, and, from the shortcut menu, select NEW....
3. Select the Information tab and as the name of the point group, enter **Well**.
4. Select the Raw Desc Matching tab, scroll through the list, and toggle **ON** the description **WELL**.
5. To view the selected points, click the ***Point List*** tab.
6. Click **OK**, exiting the Point Group Properties dialog box.
7. Repeat Steps 2 through 6 and, using key entries from Table 2.9, create the Iron and Site Control groups.

No Point or Label Point Group — Point and Label Style Overrides

A point group that includes all points, with its point and label styles overridden, set to None, and that is at the point group display's list top, hides all points.

1. If necessary, click ***Prospector's*** tab, making it current.
2. In Prospector, select the **Point Groups** heading, press the right mouse button, and, from the shortcut menu, select NEW....
3. Click the ***Information*** tab and for the name, enter **No Point or Label**.
4. In the Information panel, for the point and label styles, assign **<none>**.
5. Select the ***Include*** tab, and at the panel's bottom-left toggle **ON** include all points.
6. Select the ***Overrides*** tab, toggle **ON** the Style and Point Label Style overrides, and set their Override to <none>
7. Click **OK**, creating the point group.

All points disappear.

8. In Prospector, select the ***Point Groups*** heading, press the right mouse button, and, from the shortcut menu, select PROPERTIES....
9. In the Point Group Properties dialog box, select the point group **Well**, and move it to the list's top. If necessary, select the point group **No Point or Label**, move it to the second position, and click **OK**. You will need to REGEN (RE) the drawing.

Only the Well points display on the screen.

10. In Prospector, select the **Point Groups** heading, press the right mouse button, and, from the shortcut menu, select PROPERTIES....
11. Finally, in the Point Groups Properties dialog box, select the point group **_All Points**, move it to the list's top, and click **OK**.
12. At Civil 3D's top left, Quick Access Toolbar, click the ***Save*** icon, saving the drawing.

Defining a Point Table Style

Many times a site includes a benchmarks and/or survey control points list. A convenient method of documenting them is creating a table. In this exercise you identify the point group Well and create a table containing their coordinates.

The current drawing does not have a table style for the well data.

1. Click the ***Settings*** tab.

2. In Settings, expand the Point branch until you can view the Table Styles' styles list.

3. From the list select the style **PNEZD format**, press ENTER, and from the shortcut menu select EDIT....

4. In the style dialog box, select the tab ***Data Properties***.

This dialog box's panel defines the sorting order, text styles and heights, data options, and the table's structure (see Figure 2.16).

FIGURE 2.16

5. Select the Northing heading and drag it to the second column.

6. Select the Easting heading and drag it to the third column.

7. In the Description column, double-click in the Column Value cell displaying the Text Component Editor.

8. On the right side of the dialog box, select Raw Description and delete the entry.

9. In the Text Component Editor's top left, click the Properties drop-list arrow, select **Full Description**, click the arrow (in the top center of the dialog box), and click ***OK***, returning to the Table Style dialog box and setting the cell's value.

Your dialog box should look like Figure 2.17.

10. Click ***OK***, creating the table style.

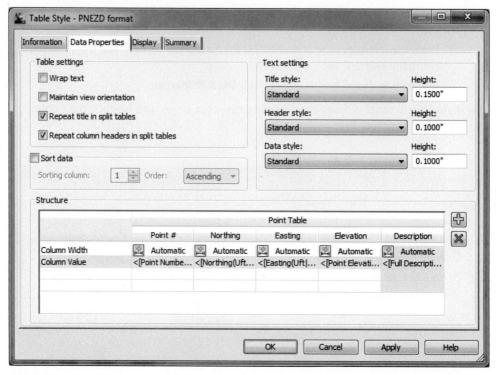

FIGURE 2.17

Creating a Point Table

1. Click the ***Prospector*** tab, making it current.

2. In Prospector, select the **Point Groups** heading, press the right mouse button, and, from the shortcut menu, select PROPERTIES....

3. In the Point Groups Properties dialog box, select the point group **Well**, move it to the list's top, select and move to the second position the **No Point or Label** group, and click **OK**, dismissing the Point Groups Properties dialog box.

4. Use the PAN and ZOOM commands to display a clear area to the east of the site.

5. From the Ribbon, click the Annotate tab. On the Labels & tables panel's left, click the Add Tables drop-list arrow, and from the shortcut menu select ADD POINT TABLE.

The Point Table Creation dialog box displays (see Figure 2.18).

6. If necessary, at the Point Table Creation dialog box's top, change the Table Style to **PNEZD format**.

The dialog box's middle selects the table's points. You select points by Point Label Style Name, Point Groups, or by selecting points from those showing on the screen. The points must be showing on the screen to be a part of a table.

7. At the center-left of the dialog box, click the ***Select Point Groups*** icon.

A list of point groups displays.

8. In the Point Groups dialog box, select the point group **Well** and click OK, returning to the Point Table Creation dialog box.

The dialog box's bottom half defines how many rows are in one table column and how to stack a split table.

9. The current values are correct, including the relationship between the points and their table entry, i.e., Dynamic.

10. Click **OK**, closing the dialog box.

11. In the drawing, select a point, locating the table.

12. At Civil 3D's top left, Quick Access Toolbar, click the **Save** icon, saving the drawing.

13. Use the ZOOM command to view the table better.

If you are editing the point data, the table updates to the new values.

FIGURE 2.18

Dynamic Table Entries

1. In Prospector, click the **Points** heading, previewing the points.

2. In preview, click the Point Number heading to sort the list.

3. In preview, scroll through the points list until you locate point **1473**.

4. Click in the Point Number cell for **1473** and change it to **3050**.

A number of reactions occur. The Well point group goes out of date, displaying an out-of-date icon.

5. In Prospector, expand Point Groups, select the entry Well, press the right mouse button, and, from the shortcut menu, select SHOW CHANGES....

The Point Group Changes dialog box displays that point number 3050 is a new member, while 1473 is to be removed.

6. In the dialog box's upper-left, click **UPDATE POINT GROUP**, updating the point group and click Close, exiting the dialog box.

The Well group updates, the out-of-date icon disappears, and the new point number (3050) appears in the table.

In Prospector, a reference (orange triangle) icon appears to the left of the point group's name Well. This icon indicates that another object references the point group. Because there is a reference to the group, Civil 3D will not display in the shortcut menu for this group the Delete command.

7. In Prospector, click the point group Well, press the right mouse button, and review the shortcut menu commands; it does not include DELETE....

8. At Civil 3D's top left, Quick Access Toolbar, click the **Save** icon, saving the drawing.

9. Exit Civil 3D.

This concludes the point discussion. There are many more point commands in Civil 3D that were not discussed in this chapter. The point commands used for creating points from surface, alignment, profile, and corridor data will be discussed in later chapters.

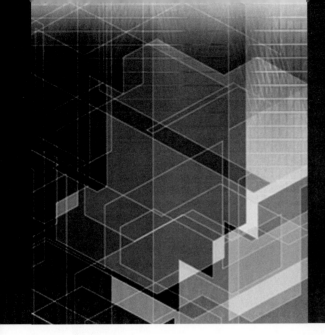

Site and Parcels

EXERCISE 3-1

After completing this exercise, you will:

- Be familiar with parcel settings.
- Be familiar with parcel styles.

Exercise Setup

This exercise's drawing uses a template file, Chapter 03 - Unit 1.dwt. It is found in the Chapter 3 folder of the CD that accompanies this textbook.

1. Start Civil 3D by double-clicking its **Desktop** icon.

2. Close the startup drawing and do not save it.

3. At Civil 3D's top left, click the Application Menu's drop-list arrow, from the shortcut menu select the **New**, browse to the Chapter 03 folder of the CD that accompanies this textbook, and select and open the file *Chapter 03 – Unit 1.dwt*.

4. At Civil 3D's top left, click the Application Menu's drop-list icon. From the menu place your cursor over SAVE AS..., in the flyout menu select AutoCAD Drawing, save the file to your machine's Civil 3D Projects folder, and name it **Parcels-1**.

Edit Drawing Settings

1. Click the **Settings** tab.

2. In Settings, at the top, select the drawing name, press the right mouse button, and from the shortcut menu, select EDIT DRAWING SETTINGS....

3. Select the **Object Layers** tab and scroll down the list until you are viewing the Parcel layer entries.

4. For Parcel, Parcel-Labeling, Parcel Segment, Parcel Segment-Labeling, and Parcel Table, change the modifier to **Suffix** and for each Suffix, enter the value of **-*** (dash asterisk) (see Figure 3.1).

5. Select the **Ambient Settings** tab, and expand the Direction section.

6. Click in the Format value cell and, if necessary, set its format to **DD.MMSSSS (decimal dms)**.

7. Click **OK** to create the layer suffixes and close the dialog box.

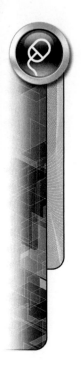

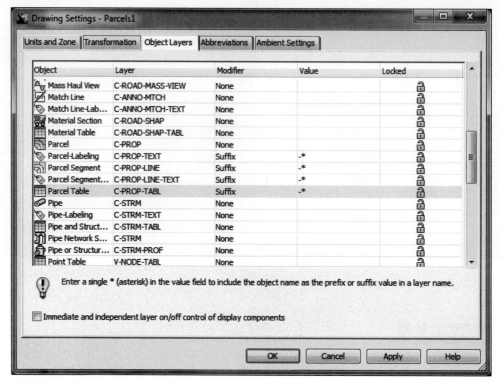

FIGURE 3.1

Edit Feature Settings

The Edit Feature Settings dialog box sets initial parcel and label styles.

1. In Settings, select the Parcel heading, press the right mouse button, and from the shortcut menu, select EDIT FEATURE SETTINGS....

2. Expand the **Default Styles** section, and if necessary, change styles to match those listed in Figure 3.2.

3. Click **OK** to close the dialog box.

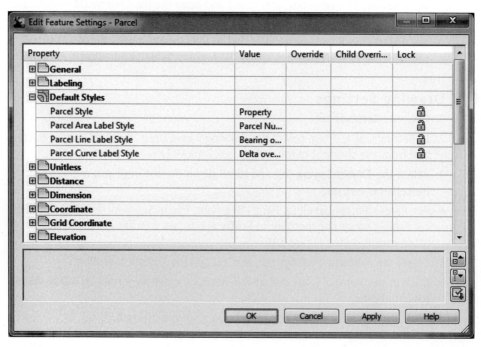

FIGURE 3.2

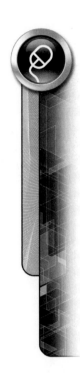

Create Site

The CreateSite command creates a site, parcel, tag counters, and increment values.

1. In Settings, expand the Parcel branch until you are viewing the Commands heading's command list.

2. From the Commands list, select CreateSite, press the right mouse button, and from the shortcut menu, select EDIT COMMAND SETTINGS....

3. Review the CreateSite settings and, if necessary, match them to the settings in Figure 3.3.

4. After making any needed changes, click **OK** to close the dialog box.

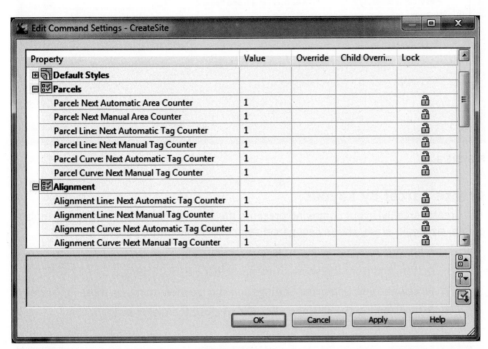

FIGURE 3.3

Create a Parcel from Objects

When converting the parcel's lines, curves, and polylines, Civil 3D wants to know what to do with the selected line work. The choices are to keep or to erase the line work. When defining the parcel, the routine assigns default parcel, area, line, and curve styles.

1. From the Parcel Commands list, select CreateParcelFromObjects, press the right mouse button, and from the shortcut menu, select EDIT COMMAND SETTINGS....

2. Review the Edit Command Settings and, if necessary, match them to those in Figure 3.4.

3. After making any needed changes, click **OK** to close the dialog box.

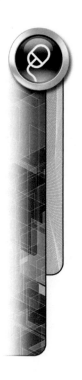

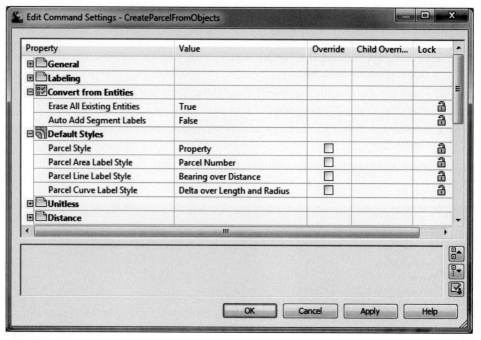

FIGURE 3.4

Create Parcel By Layout

The CreateParcelByLayout command presents the user with a toolbar containing tools to define, size, edit, erase, and review parcel segments or an entire parcel.

1. From the Commands list, select CreateParcelByLayout, press the right mouse button, and from the shortcut menu, select EDIT COMMAND SETTINGS....

2. Review the Edit Command Settings and change them to match those in Figure 3.5.

3. After changing the settings, click **OK** to close the dialog box.

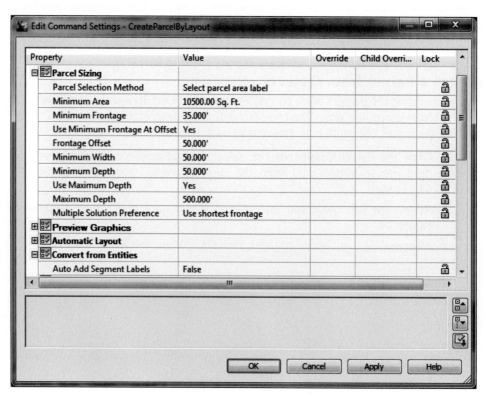

FIGURE 3.5

Parcel Styles

Parcel object styles set the layers, hatch pattern, and parcel area label style.

1. In Settings, expand the Parcel branch until you are viewing the Parcel Styles heading and its styles list.

2. From the styles list, select the Property style, press the right mouse button, and from the shortcut menu, select EDIT....

3. Select the **Information** tab and review its settings.

4. Select the **Design** tab and review its settings.

5. Select the **Display** tab and review its settings.

6. Click **OK** to close the dialog box.

7. From the list of styles, select the Single-Family style, press the right mouse button, and from the shortcut menu, select EDIT....

8. Select the **Information** tab and review its settings.

9. Select the **Design** tab and review its settings.

10. Select the **Display** tab, review its settings, and, if necessary, match them to those in Figure 3.6.

11. After making any needed changes, click **OK** to close the dialog box.

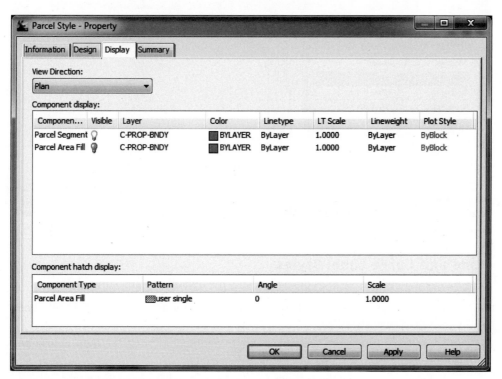

FIGURE 3.6

Area Label Styles

When defining a parcel, all commands creating parcels assign area and segment labels. The segment styles include labels for lines (bearing and distance) and curves (radius, length, chord, etc.). Civil 3D content template files provide basic area label styles.

1. In Settings, expand the Parcel branch until you are viewing the Label Styles' Area label styles list.
2. From the list, select Name Area & Perimeter, press the right mouse button, and from the shortcut menu, select EDIT....
3. Select the *Information* tab and review its settings.
4. Select the *General* tab and review its settings.
5. Select the *Layout* tab and review its settings.

The Name Area & Perimeter label has one named Component: Area. This component has three parcel properties: Parcel Name, Area, and Perimeter (see Figure 3.7).

6. Click in the Component's Text Contents value cell, and then click the ellipsis to display the Text Component Editor – Contents dialog box. This dialog box defines the parcel name, area, and perimeter format.
7. In the dialog box's right side, click a property and review its values on the left side of the dialog box.
8. After reviewing the properties and their format values, click **OK** until the dialog boxes are closed and you return to the command prompt.

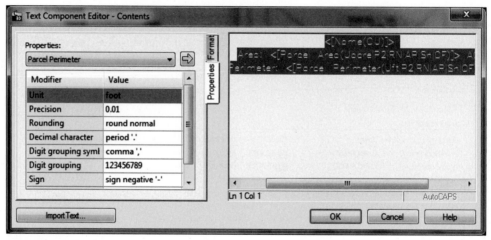

FIGURE 3.7

Line and Curve Label Styles

The Parcel Edit Feature Settings dialog box sets the default line and curve styles.

1. In Settings, expand the Parcel branch until you are viewing the Label Styles' Curve label styles list.
2. From the list, select Delta over Length and Radius, press the right mouse button, and from the shortcut menu, select EDIT....
3. Select and review the settings for the *Layout* and *Dragged State* tabs.
4. Click **OK** to exit the dialog box.
5. Under Label Styles expand Line to view its styles list.
6. From the list, select Bearing over Distance, press the right mouse button, and from the shortcut menu, select EDIT....
7. Select the *Layout* tab to review its contents (see Figure 3.8).

8. Click **OK** to exit the dialog box.

9. At Civil 3D's upper left, Quick Access Toolbar, click the **Save** icon to save the current drawing.

This completes the parcel drawing setup and styles review. The next unit reviews site and parcel drafting and defining. The settings reviewed in this unit directly affect the results of the exercises in the next unit.

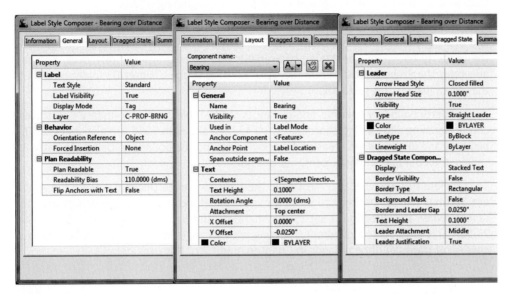

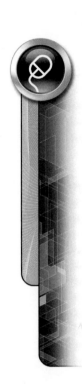

FIGURE 3.8

EXERCISE 3-2

After completing this exercise, you will:

- Be able to draw a boundary using transparent commands.
- Be able to define a site and its parcels using the Create Parcel From Objects command.
- Be able to subdivide parcel blocks using the Create Parcel By Layout command.

Exercise Drawing Setup

This exercise continues using the Unit 1 drawing, Parcels-1. If you did not do the Unit 1 exercise, you can open the drawing in the Chapter 03 folder of the CD that accompanies this textbook: *Chapter 03 - Unit 2.*

1. If the previous exercise's drawing is not open, open it now, or browse to the Chapter 03 folder of the CD that accompanies this textbook and open the *Chapter 03 - Unit 2.dwg* file.

2. In the Layer Properties Manager, freeze the following layers: **BLDG, BNDY, BNDYTXT, CONT, CONT2,** and **EX-ADJLOTS.**

3. Create a New layer, name the layer **BNDY-INPUT,** assign it a color, and make it the current layer.

4. Close Layer Properties Manager.

5. At Civil 3D's top left, Quick Access Toolbar, click the **Save** icon to save the drawing. Or if working in Chapter 03 – Unit 2, from the Menu Browser, select Save as and save the drawing as Parcels-1.

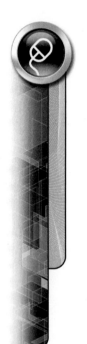

Drafting a Property Boundary

When drafting a boundary by bearings and distances, a small gap will exist between the last boundary segment's end and the first segment's beginning. This gap represents the closure error. You close this gap using the drafting command's Close option. If you are using a transparent command override, you press ESC once to exit the override and return to the drafting command, and then close the boundary. If you exit before completing the boundary, you will have to pick up where you left off and use an Endpoint object snap to draw the last small segment to the point-of-beginning (POB).

This exercise uses the Line command with the transparent commands Northing/Easting and Bearing and Distance overrides. The *Boundary Calls.txt* file is a Notepad file that has the property boundary's beginning coordinates and its bearings and distances. You can also print the *Property Outline.PDF* file. Both files are in the Chapter 03 folder of the CD that accompanies this textbook.

The property boundary bearings and distances are in Table 3.1. The boundary's northwest corner northing and easting coordinates, 14742.1740 and 6716.9054, are also its POB. The bearings and distances are measured clockwise around the boundary. Using the Table 3.1 values, the boundary almost closes, but not quite.

To complete this exercise, you must have the direction set to Decimal dms (in the Ambient Settings panel). Decimal dms is a surveyor's method of entering bearings and distances. When the direction is set to Decimal dms, the first bearing in the table, N 88d 25'50" E, is entered as a quadrant 1 bearing with an angle value of 88.2550. See the discussion of transparent commands in Chapter 2, unit 2 of this textbook.

```
Point 1:
Northing=14742.1740
Easting=6716.9054
```

TABLE 3.1

Segment	From/To Point	Quadrant	Bearing	Distance
1	1 to 2	1	N 88d25'50" E	504.9805
2	2 to 3	2	S 1d34'10" E	1001.1900
3	3 to 4	1	N 87d59'17" E	333.0686
4	4 to 5	2	S 1d37'46" E	1322.0609
5	5 to 6	3	S 88d03'56" W	1.8090
6	6 to 7	2	S 1d31'46" E	1028.4872
7	7 to 8	4	N 59d22'31" W	392.2954
8	8 to 9	4	N 1d33'54" W	817.3547
9	9 to 10	4	N 1d34'10" W	614.2499
10	10 to 11	3	S 87d56'21" W	504.9986
11	11 to 1	4	N 1d34'10" W	1712.8900

1. If necessary, click the **Settings** tab.
2. In Settings, at the top, select the drawing name, press the right mouse button, and from the shortcut menu, select EDIT DRAWING SETTINGS....
3. Select the **Ambient Settings** tab and expand the **Direction** section.

4. If necessary, change the Format to **DD.MMSSSS (decimal dms)**.

5. Click **OK** to set the format and exit the dialog box.

6. Make sure the Civil 3D Transparent Commands toolbar is visible.

7. Start the L I N E command. From the Transparent Commands toolbar, select the **Northing Easting** icon, and the command line prompt changes to a starting northing coordinate.

After entering the northing coordinate value, press ENTER for the easting prompt. After entering the easting value, press ENTER to enter both coordinates.

8. At the Northing prompt, enter **14742.1740** and press ENTER.

9. At the Easting prompt, enter **6716.9054** and press ENTER to set the starting point of the line.

10. Press ESC once to end the Northing Easting transparent override and resume the L I N E command.

11. Without exiting the L I N E command, from the Transparent Commands toolbar select the **BEARING DISTANCE** icon and answer the prompts using the information in Table 3.1 for quadrant and bearing and the distance information for each remaining boundary segment.

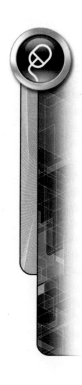

Civil 3D displays a tripod and prism representing the starting coordinates and each segment's direction. You can graphically (by using selections in the drawing) define a quadrant, direction, and distance while in this override mode.

You will not return to the exact starting coordinates. This is because the distance and angle values contain rounding errors even at four decimal places. If you are looking at the bearings and distances on a plat of survey, the precision is a whole second for bearings and only two decimal places for distances. This precision is less than what you have just entered for this boundary.

12. After entering the last bearing and distance, press ESC only once to exit the Bearing Distance override.

13. At the command line, enter the letter '**C**' and press ENTER to close the boundary by drawing the last line segment to the beginning.

14. At Civil 3D's top left, Quick Access Toolbar, click the **Save** icon to save the drawing.

Create Parcel From Objects

The current line work represents the site definition.

1. In the Ribbon Home tab, Layers panel, click the **Layer Isolate** icon, select a line on the BNDY-INPUT layer, and press ENTER to isolate the layer's line work.

2. From the Ribbon's Home tab, Create Design panel, click the **Parcel** icon, and from the menu select CREATE PARCEL FROM OBJECTS.

3. Select all of the lines you just drew and press ENTER to display the Create Parcels - From objects dialog box.

4. Match your settings to those in the dialog box in Figure 3.9 and when done, click **OK** to exit the dialog box and create the site, parcel, and the parcel area label.

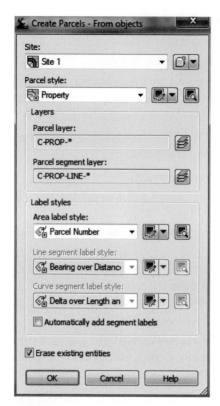

FIGURE 3.9

5. In the Ribbon's Home tab, Layers panel, at its middle right, click the UNISO-LATE icon to restore the previous layer state.

6. If necessary, open the Layer Properties Manager, toggle **ON** and **Unlock** the **C-PROP-Site1** and **C-PROP-LINE-Site1** layers, and click **X** to exit the Layer Properties Manager palette.

7. If necessary, type REGENALL (REA) at the command prompt and press ENTER.

Defining the Right-of-Way and the Wetlands Parcels

The ROW and wetlands boundary are polylines on the PR-ROW and Wetlands layers.

1. Isolate the two layers **PR-ROW** (magenta) and **Wetlands** (rust).

2. From the Ribbon's Home tab, Create Design panel, click the ***Parcel*** icon, and from the menu, select CREATE PARCEL FROM OBJECTS, select the line work on the screen, and press ENTER.

3. The Create Parcels - From Objects dialog box opens. Adjust the settings to match those in Figure 3.9.

4. After adjusting the settings, click ***OK*** to exit and define the parcels.

5. From the Ribbon's Home tab, Layers panel, click the ***Unisolate*** icon to restore the previous layer state.

6. If necessary, type REGENALL (REA) at the command prompt and press ENTER.

7. At Civil 3D's top left, Quick Access Toolbar, click the ***Save*** icon to save the drawing.

Creating BackYard Parcel Lines

The previous steps create parcels with uniform land-use types: blocks. These "block" parcels are further subdivided into individual parcels. Some blocks need further subdividing to create front yard and backyard lines. These particular blocks have two frontage lines and no backyard line. Having front yard and backyard lines makes creating individual parcels much easier; all that is left to create is the side yard segment, which, when done, means the final parcel is created. To define the backyard line, you select polylines on the Interior Boundary layer.

1. Isolate the **Interior Boundary** layer.

2. From the Ribbon's Home tab, Create Design panel, click the **Parcel** icon, and from the menu, select CREATE PARCEL FROM OBJECTS, select the line work on the screen, and press ENTER.

3. The Create Parcels - From Objects dialog box opens. Adjust the settings to match those in Figure 3.9.

4. After adjusting the settings, click **OK** to define the parcels.

5. From the Ribbon's Home tab, Layers panel, click the **Unisolate** icon to restore the previous layer state.

6. If necessary, type REGENALL (REA) at the command prompt and press ENTER.

7. Click the **Prospector** tab.

8. In Prospector, expand the Sites branch until you can see the Site 1 parcel list.

9. At Civil 3D's top left, Quick Access Toolbar, click the **Save** icon to save the drawing.

Creating Single-Family Parcels

The final step is dividing the parcel blocks into individual resalable parcels. You do this using the Parcel Layout Tools toolbar's sizing tools.

1. Click Ribbon's **View** tab. In the Views panel's left side select the named view **Residential Parcels 1**.

2. In the Layer Properties Manager (Ribbon's **Home** tab, Layers panel), turn on the **Area** layer and click the **X** on the mast to close the palette.

3. From the Ribbon's Home tab, Create Design panel, click the **Parcel** icon, and from the shortcut menu, select PARCEL CREATION TOOLS to display the Parcel Layout Tools toolbar.

4. Click the **chevron** at the right side of the toolbar to expand the Parcel Sizing section.

5. If necessary, adjust your settings to match those in Figure 3.10.

6. Go to the toolbar's fifth icon in from the left, and then click the drop-list arrow to the right of the Parcel Sizing Tools icon stack. From the list select SLIDE LINE - CREATE to display the Create Parcels - Layout dialog box.

7. In the dialog box, set the Parcel Style to **Single-Family**, the Area Label Style to **Name Area & Perimeter**, and click **OK** to start sizing parcels (see Figure 3.11).

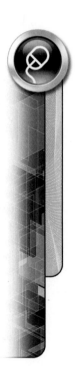

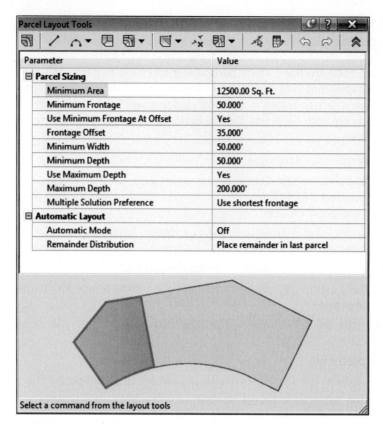

FIGURE 3.10

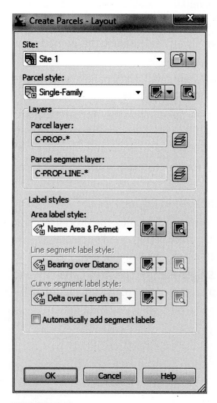

FIGURE 3.11

The command prompts you to select a parcel to subdivide.

8. Click the **chevron** at the right side of the toolbar, to collapse the toolbar.
9. If Object Snap is on, toggle it **OFF**.
10. Select the parcel's parcel number.

Next, you are prompted to define the parcel's frontage start and ending points.

11. In the drawing, using the Endpoint object snap, start the frontage by selecting the parcels' northeast end and again using the Endpoint object snap, end the frontage by selecting the parcels' southwestern end (see Figure 3.12).

While doing this, the routine displays a jig following the frontage geometry. The jig recognizes changes in the frontage's geometry.

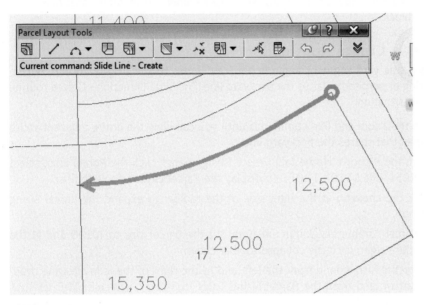

FIGURE 3.12

Next, you are prompted to define the frontage line's relative turned angle.

12. At the angle from frontage prompt, enter **90** degrees as the angle.
13. A parcel with 12,500 square feet displays on the screen. The command line prompts to accept the parcel; press ENTER to accept the new parcel and to continue to the next parcel.
14. A parcel with 12,500 square feet displays on the screen. The command line prompts to accept the parcel; press ENTER to accept the new parcel and to continue to the next parcel.
15. A warning dialog box appears stating that the minimum frontage is less than 100 ', click OK.
16. Press ESC to stop the routine. The command line prompts to select a command or to exit.

The area is now three parcels with one still having the property style and the west most parcel having the block's remaining area.

17. In the command line, enter **'X'** and press ENTER to exit the command.
18. At Civil 3D's top left, Quick Access Toolbar, click the *Save* icon to save the drawing.

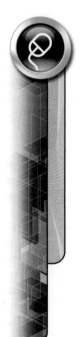

Changing a Parcel's Type and Label

When you need to change a parcel's type, change it in the Parcel Properties dialog box, Information panel. When you need to change the area label, change it in the Parcel Properties dialog box, Composition panel.

1. In the drawing, select the central parcel's label, press the right mouse button, and from the shortcut menu, select PARCEL PROPERTIES....
2. Select the **Information** tab, click the Object Style drop-list arrow, and from the list, select **Single-Family**.
3. Select the **Composition** tab, click the Area Section Label Style drop-list arrow, and select **Name Area & Perimeter**.
4. Click **OK** to change the parcel type and label.

The remaining parcel is now single-family with an area label and segments that match the others in the block.

Slide Line — Create Routine

Many times a side yard segment angle is not 90 degrees to the front yard line, but instead a bearing or perpendicular to the backyard line. The Slide Direction - Create routine is best for this situation.

1. Use the Zoom and Pan commands until you can view the entire adjacent south parcel that shares the backyard line.
2. From the Ribbon's Home tab, Create Design panel, click the **Parcel** icon, select PARCEL CREATION TOOLS to display the Parcel Layout Tools toolbar.
3. Click the **chevron** at the right side of the toolbar to expand the Parcel Sizing section.
4. In Parcel Parameters' Minimum Area, set the parcel size to 10500 and at the toolbar's right, click the collapse toolbar chevron.
5. Go to the fifth icon in from the left, and to the right of the icon click the drop-list arrow, and from the Parcel Sizing Tools icon stack, select SLIDE LINE - CREATE.
6. The Create Parcels - Layout dialog box opens. Match the settings to those in Figure 3.11, and when set, click **OK** to start the parcel-sizing process.
7. If necessary, close Parcel Parameters by clicking the toolbar's right side chevron.

The command line prompts you to select a label or point within the parcel to subdivide.

8. In the drawing, select the parcel number (in the parcel south center).

Next, you select the parcel frontage's beginning and ending points.

9. In the drawing, at the frontage's southeast end, select its endpoint (Endpoint object snap). Next, select the frontage's northeastern end (Endpoint object snap) to define the parcel's frontage geometry.

While defining the frontage, the routine displays a jig following the frontage geometry. The jig understands the changing frontage geometry (see Figure 3.13).

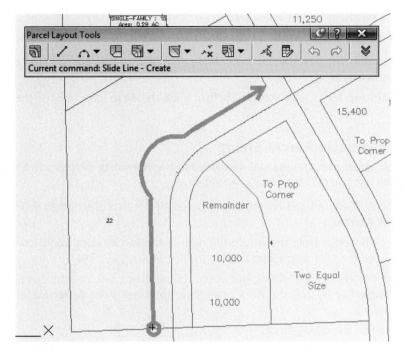

FIGURE 3.13

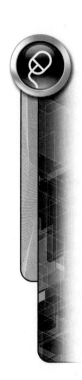

Next, you are prompted for a side yard direction.

10. Using the **Nearest** object snap, select a frontage line point and a second point, using the **Perpendicular** object snap, to the backyard line.

A parcel displays between the front and backyard segments.

11. The command line prompts to accept the parcel; press ENTER twice to accept two parcels with 10,500 square feet.

The next parcel is 12,000 square feet.

12. While the command line prompts to accept the parcel, click the toolbar's Expand Toolbar chevron to display the parcel parameters.

13. In parcel parameters, change the parcel size to 12000, close parcel parameters by clicking on the chevron and press ENTER.

The next parcel is 14,000 square feet.

14. While the command line prompts to accept the parcel, click the toolbar's Expand Toolbar chevron to display the parcel parameters.

15. In parcel parameters, change the parcel size to 14000, close parcel parameters by clicking on the chevron and press ENTER.

The next parcel is 10,800 square feet.

16. While the command line prompts to accept the parcel, click the toolbar's Expand Toolbar chevron to display the parcel parameters.

17. In parcel parameters, change the parcel size to 10800, close parcel parameters by clicking on the chevron and press ENTER.

18. A warning dialog box appears stating that the minimum frontage is less than 100 ', click OK.

The last parcel is the remainder of the block area.

19. Press ESC to exit the Parcel Sizing command.

20. Change the last parcel in the block to **Single-Family** with the label of **Name Area & Perimeter**.

21. At Civil 3D's top left, Quick Access Toolbar, click the **Save** icon to save the drawing.

Manual and Automatic Parcel Sizing

The next area's first parcel uses the manual method, and the remaining parcels use an automatic mode. The last parcel is the block's remainder.

1. Click Ribbon's **View** tab and on the Views panel's left, select the named view **Residential Parcels 2**.

2. Go to the fifth icon in from the left, to the right of the Parcel Sizing Tools icon stack, click the drop-list arrow, and select SLIDE LINE - CREATE.

3. The Create Parcels - Layout dialog box opens. In the dialog box, set the Parcel Style to **Single-Family** and the Area Label Style to **Name Area & Perimeter** (see Figure 3.11).

4. Click **OK** to exit the dialog box.

5. In the parcel parameters sizing section, set the parcel size to **12500**, press ENTER, and at the toolbar's right side, click the **chevron** to collapse the toolbar.

6. The routine prompts you to select the parcel to subdivide, select the parcel's label.

Next, you are prompted to define the frontage.

7. Using the Endpoint object snap, define the block's eastern frontage by selecting its southern and northern endpoints (see Figure 3.14).

A jig appears, tracing the frontage line and recognizing its geometry.

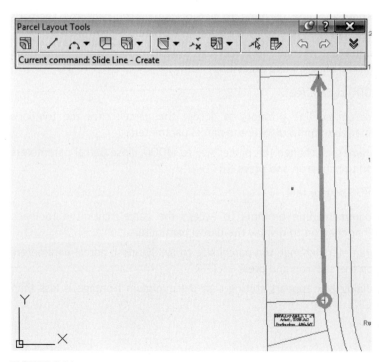

FIGURE 3.14

Next, you are prompted for an angle.

8. For the angle enter **90** and press ENTER.

9. A 12,500 square feet displays and the command line prompts to accept the parcel; press ENTER to accept.

The next parcel displays with 12,500 square feet.

10. On the right side of the Parcel Layout Tools toolbar, click the **chevron** to expand the toolbar.

11. In the Parcel Sizing, set the parcel size to 10,500 square feet.

12. In the Automatic Layout section, change Automatic Mode to **ON**, and then set Remainder Distribution to **Redistribute remainder**.

13. Click the **chevron** to collapse the toolbar.

14. The 10,500 square feet parcels display and the command line prompts to accept the resulting parcels; press ENTER to accept.

The parcel block is divided into 10,500-square-foot parcels with the remaining area distributed evenly to every xparcel.

The routine prompts you to select the next parcel to subdivide.

15. Press ESC once, then enter an '**X**' and press ENTER to exit the Parcel Layout Tools toolbar.

16. At Civil 3D's top left, Quick Access Toolbar, click the *Save* icon to save the drawing.

The remaining parcel (block's top) needs to have its type and label changed to Single-Family and Name Area & Perimeter.

17. Select the central parcel's label, press the right mouse button, and, from the shortcut menu, select PARCEL PROPERTIES....

18. Select the **Information** tab and change the Object Style to **Single-Family**.

19. Select the **Composition** tab, assign the **Name Area & Perimeter** to the Area selection label style, and click *OK* to exit.

20. At Civil 3D's top left, Quick Access Toolbar, click the *Save* icon to save the drawing.

The next phase reviews the new residential parcels' areas and boundaries.

EXERCISE 3-3

After completing this exercise, you will:

- Be able to review site properties.
- Be able to review parcel properties.
- Be able to create parcel reports with Toolbox.
- Be able to create a parcel data LandXML file.

Exercise Drawing Setup

This exercise continues with Unit 2's drawing. If you did not complete that exercise, browse to the Chapter 03 folder of the CD that accompanies this textbook and open the *Chapter 03 - Unit 3.dwg* file.

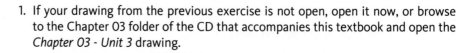

1. If your drawing from the previous exercise is not open, open it now, or browse to the Chapter 03 folder of the CD that accompanies this textbook and open the *Chapter 03 - Unit 3* drawing.

View Site Properties

Site properties affect a site's layers and the initial parcel's numbers.

1. If necessary, click the ***Prospector*** tab.

2. In Prospector, expand the Sites branch until you are viewing Site 1.

3. Click **Site 1**, press the right mouse button, and from the shortcut menu, select PROPERTIES....

4. Click the ***Information***, ***3D Geometry***, and ***Numbering*** tabs to review their current values.

5. Click **OK** to close the dialog box.

Preview Parcels

When selecting an item, named site, or individual parcel, if on, Prospector previews the item's geometry.

1. At Prospector's top, select the **magnifying glass** until it displays a border around the icon.

2. In Prospector, expand the **Site 1** branch until you are viewing the Parcels heading and its parcels list.

3. Select the **Parcels** heading, press the right mouse button, and if not on, select SHOW PREVIEW.

4. From the list of parcels, select a parcel to display its vector geometry in the preview area.

Parcel Properties

Parcel properties include a parcel's name, parcel type, parcel label, and area and perimeter statistics.

1. In Prospector, from the parcels list, select a **Single-Family** parcel, press the right mouse button, and from the shortcut menu, select ***ZOOM TO***.

2. In Prospector, with the parcel entry still highlighted, press the right mouse button, and from the shortcut menu, select PROPERTIES....

3. Click the ***Information*** tab to view its contents.

The parcel name is grayed out and is not editable.

4. Click the ***Composition*** tab to view its contents.

Composition lists the area, perimeter, and currently assigned parcel area selection label style.

5. Click the ***Analysis*** tab to view its contents.

Analysis displays a parcel's inverse or map check values. These reports summarize the boundary's quality based on the line work defining the parcel.

6. Click INVERSE ANALYSIS and review its values.

7. Click MAPCHECK ANALYSIS and review its contents.

8. Click the PICK NEW POB icon to the right of its current coordinates.

The dialog box closes and a glyph indicating the current POB position is displayed on the parcel's perimeter. The default is to move it to the next parcel corner.

9. Press ENTER to move the POB clockwise around the boundary.

10. Type '**S**' and press ENTER to select a new POB.

The dialog box is displayed again with new POB coordinates and a new parcel analysis.

11. Click **OK** to exit the Parcel Properties dialog box.

12. At Civil 3D's top left, Quick Access Toolbar, click the **Save** icon to save the drawing.

Toolbox Parcel Report

Toolbox creates printable and customizable parcel reports.

1. If the Toolbox tab is not displaying on the Toolspace, from the Ribbon, click the View tab, and at the Palettes panel click the **Toolbox** icon.

2. In the Toolbox, expand Reports Manager until you are viewing the Parcel report section.

3. Select the METES_AND_BOUNDS report, press the right mouse button, and from the shortcut menu, select EXECUTE....

4. In the Export To LandXML Report dialog box, click **OK** to create reports for the selected parcels in Internet Explorer.

5. Scroll through the report and, when finished, from the File menu select SAVE AS ... and save the report.

6. Close **Internet Explorer**.

7. At Civil 3D's top left, Quick Access Toolbar, click the **Save** icon to save the drawing.

This completes the analysis of parcels. The next unit reviews the various parcel-editing tools.

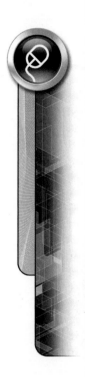

EXERCISE 3-4

After completing this exercise, you will:

- Be able to edit parcels by manipulating their segments.
- Be able to edit with Parcel Layout Tools toolbar tools.

Exercise Setup

The exercise continues with the previous exercise's drawing. If you did not complete the previous exercise, start this exercise by browsing to the Chapter 03 folder of the CD that accompanies this textbook and open the *Chapter 03 - Unit 4* drawing file.

1. If the previous exercise's drawing is not open, open it now, or open the *Chapter 03 – Unit 4* drawing file.

Grip-Editing Parcel Segments

Grips can slide a side yard segment along the frontage line to adjust the parcel area. You can even transfer the segment to an adjacent parcel block by selecting a new point on the block's frontage (see Figure 3.15).

1. On the Ribbon click the *View* tab. At the Views panel's left, select the named view **Residential Parcels 1**.

2. Click a side yard line between two adjacent parcels. A special grip appears, indicating that the side yard is "tied" to the two lots' frontage. Select the **grip**.

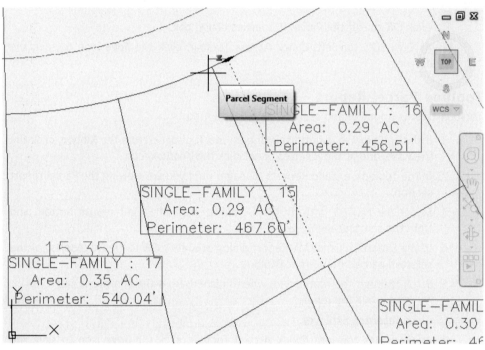

FIGURE 3.15

3. If on, toggle **OFF** Osnap and slide the segment along the frontage line.

4. When the side yard line is in the desired position, press the left mouse button to set its location, and press ESC to deselect the object.

If you are dragging a parcel segment into an adjacent parcel block, the segment divides the block. The original parcels sharing the parcel line merge. If moving the segment back to its original location, the newly merged parcels divide and the newly divided block returns to its original form.

Parcel Layout Tools Toolbar — Delete Parcel Sub-Entity

There are reasons to delete parcel segments and redraw them, to change the segment's angle, or to further divide the parcels. The Parcel Layout Tools toolbar has the tools to perform these actions.

1. In the Ribbon select the Modify tab. At the Design panel's left click the Parcel icon. The Ribbon changes to a Parcels tab. In the Modify panel select PARCEL LAYOUT TOOLS, to display the Parcel Layout Tools toolbar.

2. Near the toolbar's middle, click the ***DELETE SUB-ENTITY*** icon (see Figure 3.16).

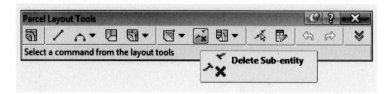

FIGURE 3.16

This command deletes parcel segments. If using Delete Sub-entity on frontage, backyard, or boundary segments, this command destroys the parcel definition results in parcel segments. If deleting a side yard segment, the two parcels merge into a single parcel.

3. In the current view, delete one or both side yard segments.

Parcel Layout Tools Toolbar — Add New Parcel Segment

You can use any of the following parcel-sizing methods to redraft the side yard segment: Slide Line or Swing Line.

1. Redraft the segment(s) with one of the just-mentioned Parcels Layout Tools toolbar commands. Set Parcel Style to **Single-Family** and use the Area Label Style of **Name Area & Perimeter**.

Parcel Layout Tools Toolbar — Edit Parcel Segment Angle/ Direction

There are times when users want to change the side yard segment's angle or direction. The Slide Line - Edit routine redefines the side yard's angle relative to the frontage geometry or its direction or bearing (Direction).

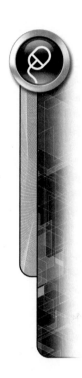

1. From the Parcel Layout Tools toolbar, Parcel Sizing Tools icon stack, select the command *SLIDE LINE - EDIT*.

2. If the Create Parcel From Layout dialog box opens, set the parcel type to **Single-Family**, the label to **Name Area & Perimeter**, and click *OK*.

3. Click the collapse toolbar chevron.

4. You are prompted to select a parcel line to adjust. Select a side yard segment between two parcels.

5. Next, select the parcel to edit by placing the highlight over the parcel and clicking in the parcel area.

6. Next, you are prompted to define the parcel's frontage start and ending points.

7. Next, you are prompted for a new angle. For the new side yard line, enter a new angle.

8. After entering a new angle, the command line prompts to accept the adjustment. Press ENTER to accept the new parcel definition.

9. The command line prompts for the next parcel to adjust. Edit a few more parcels in the area.

10. After editing a few parcels, press ESC twice to exit the Parcel Layout Toolbar.

11. At Civil 3D's top left click, Quick Access Toolbar, the **Save** icon to save the drawing.

Renumbering Parcels

After subdividing a site into parcels, the parcel numbers may not represent the desired numbering. The Renumber/Rename Parcels command renumbers selected parcels.

1. If you are not in the Residential Parcels 1 view area, click the Ribbon's View tab. In the Views panel, select and restore the named view **Residential Parcels 1**.

2. In the Ribbon's Parcel tab, Modify panel select RENUMBER/RENAME to display the Renumber/Rename Parcels dialog box (see Figure 3.17).

3. Set the Starting Number to **100** and click *OK* to close the dialog box.

4. Next, select a point in the short block's westernmost parcel, and then select a second point in the easternmost parcel.

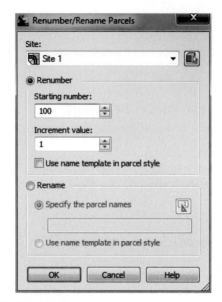

FIGURE 3.17

This selects the parcels for renumbering.

>   5. Press ENTER twice to end the selection and renumber the parcels west to east, from **100** to **102**.

Renaming Parcels

There are times when it is desirable to rename the parcels. It may be that the parcel number needs to be a Parcel Identification Number (PIN) or a more general name, such as Parcel #.

>   1. In Ribbon's Parcel tab, from the Modify panel, select RENUMBER/RENAME to display the Renumber/Rename Parcels dialog box (see Figure 3.17).
>   2. Toggle **ON** Rename and Specify the Parcel Names and click ***Edit Name Template*** (the icon to the toggle's right).
>   3. In the Name Template dialog box, Name Formatting Template section, set the Property Fields to **Next Counter**, in the Name area type **Parcel #**, and click **Insert** to put the next counter after "Parcel #."
>   4. In the Incremental Number Format area, set the Number Style to **1, 2, 3...**, and set the Starting Number to **300**. Your settings should match Figure 3.18.
>   5. Click ***OK*** to exit the Name Template dialog box and return to the Renumber/Rename Parcels dialog box.
>   6. Click ***OK*** to close the dialog box.
>   7. Select the lots for renumbering by first selecting a point in the short block's westernmost parcel and then selecting a point in the easternmost parcel.
>   8. Press ENTER twice to rename the parcels.
>   9. Close the Parcel tab by selecting Close on the right.
>   10. Click the ***Prospector*** tab to make it current.
>   11. In Prospector, expand the Sites branch until you are able to view the Parcels heading and the list of parcels with new names and numbers in Preview.
>   12. At Civil 3D's top left, Quick Access Toolbar, click the ***Save*** icon to save the drawing.

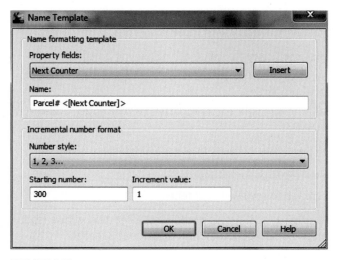

FIGURE 3.18

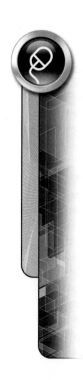

This completes the parcels editing exercise. The next section reviews parcel segment labels and their annotation.

EXERCISE 3-5

After completing this exercise, you will:

- Be familiar with parcel area label styles.
- Be familiar with parcel segment label styles.
- Be familiar with parcel table styles.
- Be able to create parcel segment labels.
- Be able to create parcel area tables.
- Be able to create parcel segment tables.
- Be able to create a Map Check analysis.

Exercise Setup

This exercise continues with the previous unit's exercise drawing. If starting with this exercise, browse to the Chapter 03 folder of the CD that accompanies this textbook and open the Chapter 03 - Unit 5 drawing.

1. If the previous exercise's drawing is not open, open it now, or if not completing the previous unit, open the *Chapter 03 - Unit 5* drawing file.

Parcel Area Labels

Labeling a parcel area or number occurs when creating a parcel.

1. Click the ***Settings*** tab.
2. In Settings, expand the **Parcel** branch until you view the Area label styles list.
3. Click the **Name Area & Perimeter** area label style, press the right mouse button, and from the shortcut menu, select EDIT....
4. If necessary, click the ***Layout*** tab to review its contents. To do this, at the top left click the drop-list arrow and select a component's name from the list.

The label component contains three parcel properties. These properties are not individual components, but are one text component entry with three properties.

5. In the Text section, click the **Contents'** value cell displaying an ellipsis.

6. At the cell's right side, click the **ellipsis** to display the Text Component Editor dialog box with the label properties.

7. On the right side, click and highlight each property's format string, and review its settings on the left side.

A border encompasses all three properties because they are a single label component.

8. Click **Cancel** to exit the Text Component Editor.

9. Click the **Dragged State** tab to review its contents.

If this label is dragged from its original position, it becomes stacked text with a leader. Again, you can change it to remain composed either with or without a leader.

10. Click **Cancel** to exit the Label Style Composer.

Segment Labels — Line Segments

As with area labels, each segment label has a layout and dragged state definition. A line segment label annotates a line's distance, bearing, or both.

1. In Settings, in the **Parcel** branch, expand **Line** label styles until you view its styles list.

2. Select **Bearing over Distance**, press the right mouse button, and from the shortcut menu, select EDIT....

3. Click the **Layout** tab and view the style's components. You do this at the top left by clicking the drop-down list arrow and selecting a component's name from the list.

This style has a table tag and two components: bearing and distance.

4. Click the **Dragged State** tab to review its contents.

5. Click CANCEL to exit the style.

6. From the list, select **(Span) Bearing over Distance with Crows Feet**, press the right mouse button, and from the shortcut menu, select EDIT....

7. Click the **Layout** tab and review its components. To do this, at the top left, click the drop-down list arrow and select a component's name from the list. This style has a table tag and four components; bearing and distance, direction arrow, and crow's feet (start and end).

You use this label on segments that extend over individual line segments (backyard line).

8. Click the **Dragged State** tab to review its contents.

9. Click **Cancel** to exit the style.

Segment Labels — Curve Segments

A curve segment label annotates a curve's radius and length. A curve label can annotate additional curve properties.

1. In Settings, expand the **Parcel** branch until you are viewing the Curve styles list.

2. From the list, select **Delta over Length and Radius** style, press the right mouse button, and from the shortcut menu, select EDIT....

3. Click the **Layout** tab and change the Component name to **Distance & Radius**.

This label component anchors to the curve's bottom center (the Text Attachment value).

4. Change the Component name to **Delta**.

This label's component anchors to the curve's top center (the Text Attachment value).

5. Click the ***Dragged State*** tab to review its settings.

6. Click ***Cancel*** to exit the dialog box.

Adding Segment Labels — Multiple Segment

When defining parcels, you can label their segments. If you are not labeling parcels at this time, the Add Labels routine creates the segment labels. Add Labels has two parcel segment labeling methods: multiple and single segment.

1. If necessary, use PAN and ZOOM to view a few parcels.

2. On the Ribbon, click the Annotate tab, on the Labels & Tables panel's left, click the Add Labels drop-list arrow to display a shortcut menu. Place the cursor over the Parcel entry and in its flyout menu select ADD PARCEL LABELS.

3. In the Add Labels dialog box, set the Feature to **Parcel**, the Label type to **Multiple Segment**, the Line label style to **Bearing over Distance**, and the Curve label style to **Delta over Length and Radius** (see Figure 3.19).

4. Click **Add**.

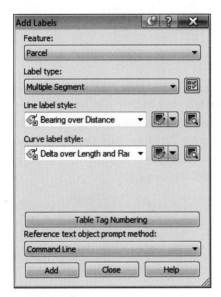
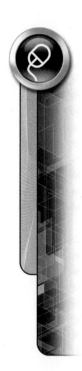

FIGURE 3.19

You are prompted to select a parcel label and to identify what segments to label.

5. Select a parcel label to create the segment labels.

6. Press ENTER to specify the clockwise direction for the labels.

There is only one label at each common side yard segment.

7. Undo the labels. You may have to use the ERASE command to remove all segment labels.

Adding Segment Labels — Single Segment

Labeling individual backyard segments is a problem with multiple-segment labeling. It may be necessary to have a label spanning the backyard with a bearing and overall distance, as well as individual parcel backyard distances (see Figure 3.20). In Figure 3.20, the left side uses multiple-segment labeling, and the right side shows single-segment labeling with the spanning label style.

When using single-segment labeling, the segment's selection point is the label's anchoring point. If it is not in the correct location, you can graphically position the label.

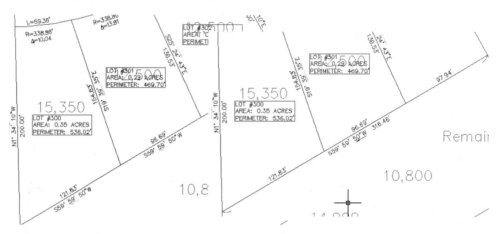

FIGURE 3.20

8. If you are not working in the Residential Parcel 1 view area, from Ribbon's **View** tab, Views panel, select and restore the named view **Residential Parcel 1**.

9. In the Add Labels dialog box, set the Feature to **Parcel**, the Label type to **Single Segment**, the Line label style to **Bearing over Distance**, and the Curve label style to **Delta over Length and Radius** (see Figure 3.21).

10. Click **Add** and select the side and front parcel segments.

11. In the Add Labels dialog box, change the Line label style to **Distance**.

12. Click **Add** and label the three parcels' individual backyard distances.

13. Click **Close** to exit the Add Labels dialog box.

Your parcel labels should look similar to Figure 3.20.

FIGURE 3.21

Changing Segment Labels

There will be times when a label shows the wrong bearing (north instead of south), or the label is on the segment's wrong side. When you select a label and press the right mouse button, a shortcut menu is displayed with several editing options (see Figure 3.22).

FIGURE 3.22

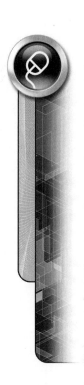

Reverse Label

If a label's direction is incorrect, use Reverse Label to change its direction.

1. Zoom in to view a few segment labels.

2. In the drawing, select a label, press the right mouse button, and from the shortcut menu, select REVERSE LABEL.

Flip Label

If a label is on the segment's wrong side, use Flip Label to exchange the label element's location.

1. In the drawing, select a label, press the right mouse button, and from the shortcut menu, select FLIP LABEL.

Drag Label

If selecting and activating a label's grip, you can drag it from its original position. Depending on the label's dragged state settings, the label may dramatically change. Depending on where the label is located, it may use right- or left-justification.

1. In the drawing, select a label, and then select its square grip and move it to a new location.

2. Try different locations and watch the label switch from left- to right-justified.

Reset Label

If you want to return a label to its original position and composition, use Reset Label.

1. In the drawing, select the dragged label, press the right mouse button, and from the shortcut menu, select RESET LABEL.

2. In the drawing, select a label, and then select its square grip and move it to a new location.

3. In the drawing, select the dragged label and reset the label by clicking its round grip.

Change Label

Settings in a label's Label Properties dialog box can change its label style.

1. In the drawing, select a label, press the right mouse button, and from the shortcut menu, select LABEL PROPERTIES....

2. In the Properties panel, click in the **Line Label Style** value cell. Click the drop-list arrow to display a list of styles; from the list, select a different label style (see Figure 3.22). Press ESC to deselect the label.

3. Click the **X** on the top of the Properties mast to close the palette.

When you have exited, the new style appears on the segment.

4. At Civil 3D's top left, Quick Access Toolbar, click the ***Save*** icon to save the drawing.

Label Plan Readability

All parcel labels are sensitive to view rotation and maintain plan readability (reading from left to right). Whether rotating model space or a layout viewport, the labels react to the rotation and read correctly.

1. Start DVIEW command and select the parcels and their line work. Use the **TWist** option and twist the view until the backyard line is nearly horizontal and select a point.

While rotating the view, the labels remain plan-readable. No matter what the rotation angle is, the area labels and segment annotation are plan-oriented.

2. Press ENTER to exit DVIEW.

3. On the Ribbon click the View tab, in the Views panel, click NAMED VIEWS, click ***New...***, and save the current view as **Parcels Twisted**. Click **OK** until you return to the command line.

4. Use the Plan command and press ENTER twice to return to the World view.

Scale Sensitivity of Labels

All labels and annotation are scale sensitive. Labels and annotation react to scale's value and size themselves for that scale. Use the Regenall (REA) command to resize the label.

1. Click the **Layout 1** to enter the paper space.

2. Double-click inside the viewport to enter its model space.

3. In Ribbon's Views panel's left, select and restore the named view **Parcels Twisted**.

4. Double-click outside the viewport to return to the paper space.

5. Select the viewport border, press the right mouse button, and from the shortcut menu, select PROPERTIES....

6. In the Properties dialog box, the Misc section, set the Standard Scale to **1″=50′**.

7. Use the Regenall (REA) command to resize the text (if necessary).

The label and annotation text is now correct for the viewport's scale.

8. In the Properties palette, set the viewport's scale to **1″=20′**.

9. Use the Regenall (REA) command to resize the text.

10. Click **X** to close the Properties palette.

11. Click the **Model** tab to reenter the model space.

Parcel Segment Table

Tables list the segments' geometric values. A table is created from existing segment labeling. The Table routine reads the values, creates the table, and toggles the label's visibility to tags (L1, L2..., C1, C2...).

1. On the Ribbon click the **Annotate** tab. In the Labels & Tables panel, click the Add Tables drop-list arrow to display a shortcut menu. Place the cursor over Parcel to display its flyout menu and from the flyout, select ADD SEGMENT.

The Table Creation dialog box opens. The Table Style heading at the top lists the current table style. The Select by label or style area lists all label styles.

2. Scroll through the Select by label or style's list, toggling on the styles: Parcel curve; Delta over Length and Radius and Parcel line; Bearing over Distance.

Next, you will set the table type.

3. If necessary, set the Table type to **Dynamic**.

Your dialog box should look like Figure 3.23.

4. Click **OK** to close the dialog box.

5. Next, you are prompted to locate the table's upper-left corner. In the drawing, select a point to create the table.

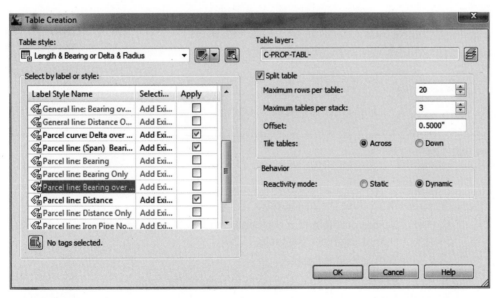

FIGURE 3.23

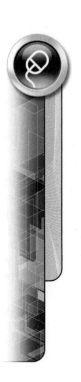

The routine reads the label's values, creates the table, and toggles the styles to tag mode (L1, L2..., C1, C2...).

6. Zoom in to the table and parcels to view their tags.

7. Use undo to remove the tables and return the annotation to verbose mode.

8. At Civil 3D's upper left, Quick Access Toolbar, click the **Save** icon to save the drawing.

Parcel Area Table

When submitting documents, you may need to list the parcel areas as a table.

1. From the Labels & Tables panel, click the Add Tables drop-list arrow to display a shortcut menu. Place your cursor over Parcel to display its flyout menu, select ADD AREA.

The Table Creation dialog box opens.

2. In the Select by Label or Style area, toggle **ON** each style.

3. Change Selection Rule to **Add Existing and New**. To do this, double-click in the Selection Rule cell, and select **Add Existing and New** from the list.

4. If necessary, leave Split Table **ON** and set the table to **Dynamic**. Your Table Creation dialog box should look like Figure 3.24.

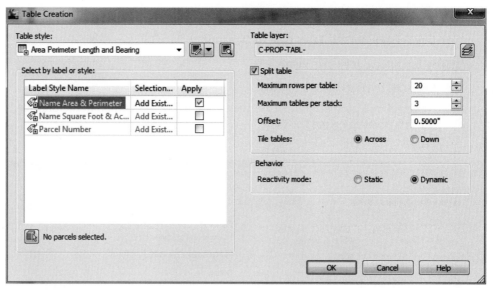

FIGURE 3.24

5. Click **OK** to close the dialog box.

6. Next, you are prompted for the upper-left location of the table. In the drawing, select a point to create the table.

7. Zoom in to review the table.

8. After reviewing the table, use undo to remove the table.

9. At Civil 3D's top left, Quick Access Toolbar, click the icon to **Save** the drawing.

Map Check

All parcels need their closure documented; the Map Check tool creates this legal documentation.

1. If necessary, click the Prospector tab.
2. In Prospector, click Site 1's parcel heading to list the parcels in the preview area.
3. In preview from the parcel's list select **Parcel #302**, press the right mouse button, and from the shortcut menu, select ZOOM TO.
4. Use the ERASE command and erase all of the parcel segment labels.
5. If necessary, on the Ribbon click the **Annotate** tab. On the Labels & Tables panel's left, click the Add Labels drop-list icon and from the shortcut menu's Parcel flyout, select ADD PARCEL LABELS.

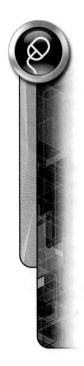

Create a New Curve Label

Mapcheck must have a curve label with a bearing to correctly follow the brokenback curve, i.e., non-tangential curve on the parcel's west frontage.

1. Set the Label Type to **Multiple Segment** and Line Label Style to **Bearing over Distance**.
2. Click the **Edit Style** icon to the left of the Line Label Style.
3. If necessary, click the style's General tab and change the Display mode to **Label**.
4. Click OK to return to the Add Labels dialog box.
5. To the right of the curve label style, click the drop-list arrow and from the shortcut menu select COPY CURRENT SELECTION.
6. In the Information panel, change the style's name to **Delta and Bearing over Radius and Length**.
7. Click the General tab and change the Display mode to Label.
8. Click the Layout tab.
9. In the Layout tab's upper left, click the Component Name drop-list arrow and from the list select **Delta**.
10. In Delta's Text section, click on the Content's value cell to display an ellipsis. Click the ellipsis to display the Text Component Editor.
11. At the top right, place the cursor after the Segment Delta Angle format string and press ENTER to create a new line.
12. At the Text Component Editor's top left, click the Properties' drop-list arrow and from the properties list select **Segment Chord Direction**.
13. At the Text Component Editor's top center, click the transfer **arrow** to place the property on the editor's right side.
14. Click **OK** until returning to the Add Labels dialog box.

The new style is the current curve label style.

15. Click the **Add** button and in the drawing select the area labels for Parcels #300, #301, and #302.
16. The command line prompts for clockwise or counterclockwise, press ENTER for clockwise.
17. Close the Add Labels dialog box.
18. Zoom in on Parcel #302 to view its labeling.

Define a Mapcheck

1. Click Ribbon's **Analyze** tab. On the Analyze tab's left, Ground Data panel, click the Survey drop-list icon and from the shortcut menu, select MAPCHECK.

2. Click the New Mapcheck icon (fifth in from the left) and in the command line for the mapcheck name, enter **Parcel #302** and press ENTER.

3. The command line prompts for a Northing/Easting starting point. In the drawing, using an Endpoint object snap, select the parcel's northeast front yard corner.

The corner's coordinates appear in the Mapcheck Analysis panel.

4. If necessary, toggle off object snaps.

5. The command line prompts for a segment label selection. In the drawing, select parcel #302's east property line label.

An entry for the line appears in the palette and Civil 3D draws a line over the segment.

6. In the drawing select the next two backyard segment labels and then select parcel #302's west side yard line label.

The west side yard was drawn from front to back and the mapcheck bearing arrow points away from the front yard line.

7. In the command line, type the letter '**R**' and press ENTER to reverse the line's bearing to point to the front yard.

8. In the drawing, select the short curve segment label on the front yard's west side.

An arrow should point to parcel's last front yard line segment.

9. In the drawing, select the closing front yard segment label.

10. Press ENTER to end the routine.

11. Review each segment's values for the current mapcheck.

12. At Mapcheck Analysis' top center, click the **Output View** icon to display the mapcheck results and review the mapcheck results.

13. At Mapcheck Analysis' top center, click the **Insert Mtext** icon and select a point to place the mapcheck values in parcel #302.

14. Close the Mapcheck Analysis palette.

15. At Civil 3D's top left, click the **Save** icon to save the drawing.

16. Exit Civil 3D.

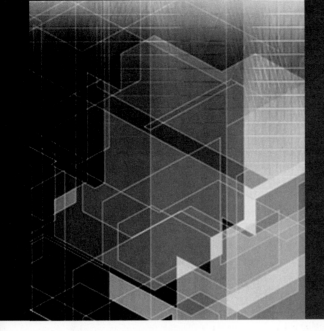

EXERCISE 4-1

After completing this exercise, you will:

- Be familiar with the settings in Edit Drawing Settings.
- Be familiar with the settings in Edit Feature Settings.
- Be familiar with the settings in Edit Label Default Settings.
- Be familiar with the settings in Surface Object Styles.
- Be familiar with the settings in Surface Label Styles.

Exercise Preparation

1. If not in Civil 3D, start it by double-clicking the **Civil 3D** desktop icon.
2. Close and do not save **Drawing1**.
3. At Civil 3D's top left, click the Civil 3D drop-list icon and from the Application Menu select NEW.
4. Browse to the Chapter 04 folder of the CD that accompanies this textbook and select the *Chapter 04 – Unit 1.dwt* template file and click **Open**. Click **OK** to create the new file.
5. At Civil 3D's top left, click the Civil 3D drop-list icon and in the Application Menu, place your cursor over SAVE AS, in the flyout menu select AutoCAD Drawing, browse to the Civil 3D Projects folder, name it **Surfaces**, and click **Save** to return to Civil 3D.

Edit Drawing Settings — Review and Set Values

The following steps set a suffix modifier, use the surface's name as part of the layer name, and place a dash separating the surface name from the root layer name, e.g., C-TOPO-EG.

1. Click the **Settings** tab.
2. At the top, click the drawing name, **Surfaces**, press the right mouse button, and from the shortcut menu, select EDIT DRAWING SETTINGS....
3. Click the **Units and Zone** tab to make it current.

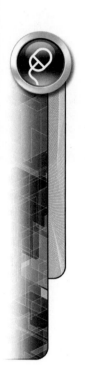

4. If necessary, set the scale to **1" = 40'**, set the Imperial to Metric conversion to **US Survey Foot,** and set the Zone Category to **No Datum, No Projection** (see Figure 4.1).

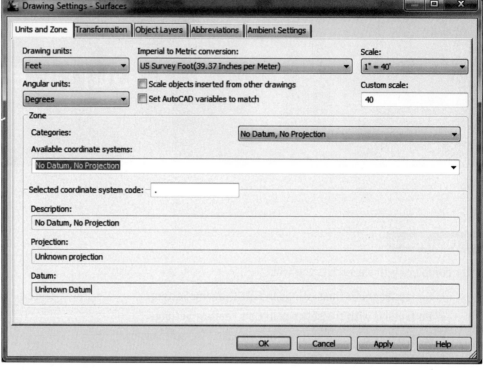

FIGURE 4.1

5. Click the ***Object Layers*** tab to view its values.

6. Scroll down the layer list until you are viewing the **Grid Surface** and **Grid Surface-Labeling** entries. Keep the default layer names, but change the Modifier by clicking in the cell and changing its value from None to **Suffix**.

7. Set the Modifier value to **-*** (a dash followed by an asterisk).

8. Continue scrolling down the layer list until you are viewing the **TIN Surface** and **TIN Surface-Labeling** entries. Keep the default layer names, but change the Modifier by clicking in the cell and changing the value from None to **Suffix**.

9. Set the Value of the Modifier to **-*** (dash followed by an asterisk) (see Figure 4.2).

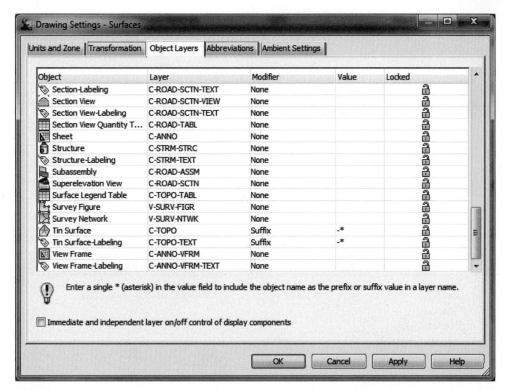

FIGURE 4.2

10. Click the **Ambient Settings** tab.
11. Expand the **Coordinate** and **Elevation** sections, review their settings, and if necessary adjust them (see Figure 4.3).

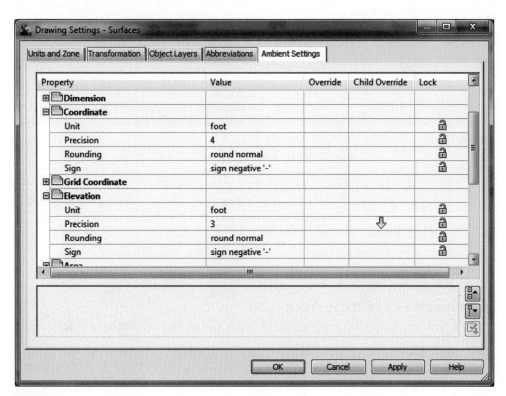

FIGURE 4.3

The settings affect the coordinate and elevation display and reporting. The Child Override column's downward-pointing arrow indicates the elevation precision is overridden by a lower style.

12. Click **OK** to close the Edit Drawing Settings dialog box.

Edit Feature Settings Dialog Box

1. In Settings, expand the **Surface** branch by clicking the surface heading's expand tree icon (plus sign).

2. Click the heading **Surface**, press the right mouse button, and from the shortcut menu, select EDIT FEATURE SETTINGS....

3. In Edit Feature Settings, expand **Default Styles** to view the assigned object and Label Styles (see Figure 4.4).

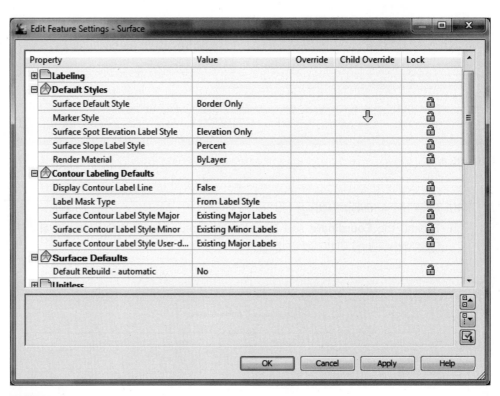

FIGURE 4.4

4. Change the Surface Default Style to **Border Only**. This is done by clicking in the Surface Default Style's value cell to display an ellipsis. Clicking the ellipsis displays the Surface Default Style dialog box. Click the drop-list arrow, from the list select **Border Only**, and click **OK** to exit the dialog box and assign the style.

5. Change in Contour Label Lines, Display Contour Label Line to False.

6. Click **OK** to exit the dialog box.

Command — CreateSurface

Create Surface's command settings control new surface default values and set surface build options.

1. In Settings, expand the **Surface** branch, until you are viewing its **Commands** branch and commands list.

2. From the list, select **CreateSurface**, press the right mouse button, and from the shortcut menu, select EDIT COMMAND SETTINGS....

3. Expand the **Surface Creation** section to view its values (see Figure 4.5).

The Default Surface Type is TIN surface. The grid spacing and rotation values apply a grid surface type. The default naming convention is Surface1, Surface2, etc.

4. Close the **Surface Creation** section.

5. Expand the **Build Options** section to view its values (see Figure 4.5).

This section contains critical surface-building parameters. The settings and their values are discussed in a later unit's exercise.

6. Click *OK* to exit the Edit Command Settings dialog box.

7. Collapse Surface's **Commands** branch.

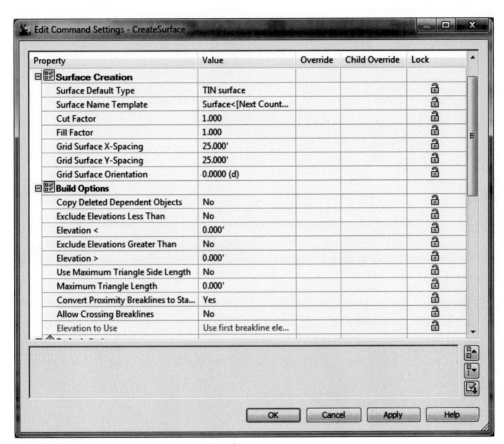

Property	Value	Override	Child Override	Lock
Surface Creation				
Surface Default Type	TIN surface			🔒
Surface Name Template	Surface<[Next Count...			🔒
Cut Factor	1.000			🔒
Fill Factor	1.000			🔒
Grid Surface X-Spacing	25.000'			🔒
Grid Surface Y-Spacing	25.000'			🔒
Grid Surface Orientation	0.0000 (d)			🔒
Build Options				
Copy Deleted Dependent Objects	No			🔒
Exclude Elevations Less Than	No			🔒
Elevation <	0.000'			🔒
Exclude Elevations Greater Than	No			🔒
Elevation >	0.000'			🔒
Use Maximum Triangle Side Length	No			🔒
Maximum Triangle Length	0.000'			🔒
Convert Proximity Breaklines to Sta...	Yes			🔒
Allow Crossing Breaklines	No			🔒
Elevation to Use	Use first breakline ele...			🔒

Edit Command Settings - CreateSurface

OK   Cancel   Apply   Help

FIGURE 4.5

Surface Style — Border Triangles Points

The surface style, Border Triangles Points, displays surface components important to building the surface and reviewing the surface quality. Viewing the triangulation allows you to decide whether additional data and/or breaklines are necessary, or if editing the surface suffices.

1. In the Settings, expand the **Surface** branch until you are viewing Surface Styles' list of styles.

2. From the list of Surface Styles, select the style **Contours and Triangles**, press the right mouse button, and from the shortcut menu, select COPY....

3. Click the *Information* tab. Change the style name to **Border Triangles Points** and give the style a short description.

The information panel identifies the Name, Description, and the created and modified dates for the style (see Figure 4.6).

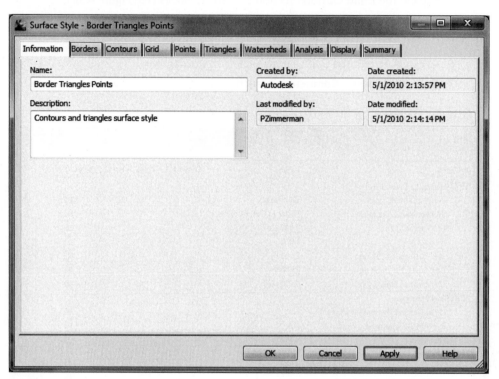

FIGURE 4.6

4. Click the *Borders* tab.

The 3D Geometry section defines how the surface displays its border (at its true Z-elevation), if it shows an exterior and any interior borders, and if those borders use a datum (see Figure 4.7).

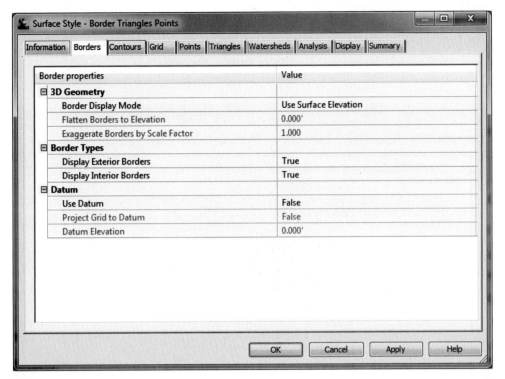

FIGURE 4.7

5. Click the ***Points*** tab.

The 3D Geometry section defines how surface points represent surface elevations (they will be at their true Z-elevation). The Point Size section sets the marker's size (this size is set by the scale of the drawing). The Point Display section defines the three surface point types' shape and color: Point Data (actual points from the drawing); Derived Points (smoothing points); and Non-Destructive Points (points on the surface from boundaries) (see Figure 4.8). These settings do not affect cogo points.

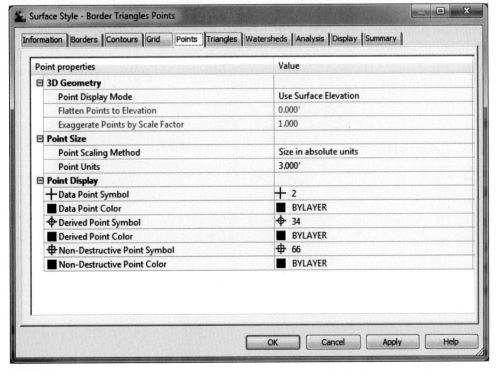

FIGURE 4.8

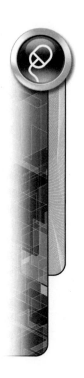

6. Click the ***Triangles*** tab to make it current.

The setting's value affects the surface triangle elevations.

7. If necessary, set the value to **Use Surface Elevation** (see Figure 4.9).

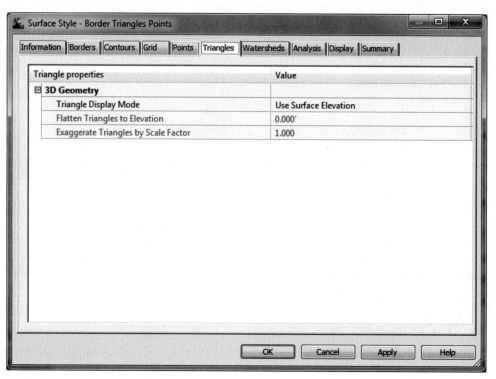

FIGURE 4.9

8. Click the ***Display*** tab to make it current.

The Display panel sets surface components and characteristics layers and their properties. The initial settings for plan views (directly over the surface) are surface border, triangles, and points. In the top left, clicking the View Direction drop-list arrow changes the settings to Plan, Model, or Section view (see Figure 4.10).

This surface style's plan view displays a border, triangles, and points, and its Model view shows only triangles.

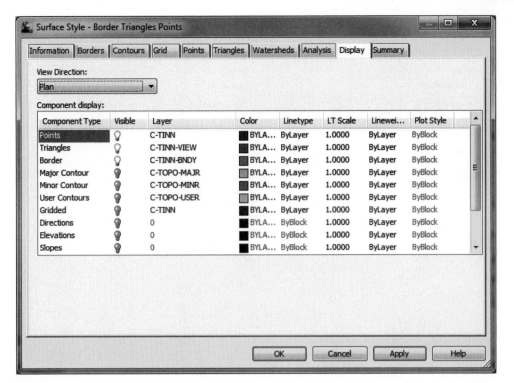

FIGURE 4.10

9. In the Display panel for Plan View Direction, toggle **ON Points** and toggle **OFF Contours Minor** and **Major**.

10. In the Display panel, click the **View Direction** drop-list arrow and change the View Direction to **Model**. Only Triangles are visible.

11. In the Display panel, click the **View Direction** drop-list arrow and change the View Direction to **Section**. Only Surface Section is visible.

12. Click *OK* to create the Border Triangles Points style.

13. Collapse the **Surface Styles** branch.

14. At Civil 3D's top left, Quick Access Toolbar, click the *Save* icon to save the drawing.

Spot Elevation Label Style – Elevation Only

The Spot Elevation, Elevation Only label style defines surface elevation labels. The drawing contains several additional labeling styles for contours, slopes, watersheds, and spot elevations.

1. While still in the Settings' Surface branch, expand **Label Styles** to view its style's list.

2. Expand the label type **Spot Elevation** to view its label styles list.

3. Select **Elevation Only**, press the right mouse button, and from the shortcut menu, select EDIT....

The Label Style Composer dialog box opens.

4. Click the *Information* tab to make it current.

This panel contains the label style's Name and Description. The panel's right side displays the style's creation and modified dates (see Figure 4.11).

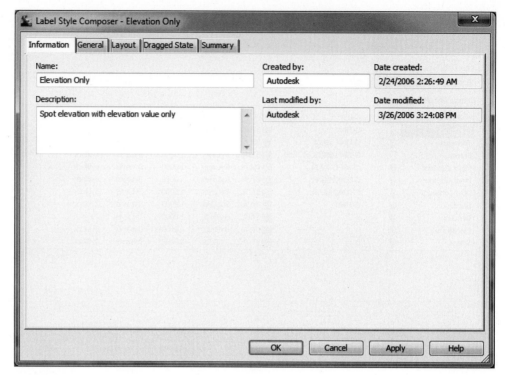

FIGURE 4.11

5. Click the *General* tab to make it current.

6. If necessary, at the panel's upper right, click the drop-list arrow and set the preview to **Surface Spot Elevation Label Style**.

This panel sets the label style's overall visibility. If you change the Label section's Visibility to false, all of this style's labels are not displayed. If the Label Layer is set to 0, the label will use the label layer defined in Edit Drawing Setting, Object Layer list (Surface Labels — C-TOPO-TEXT). The Orientation Reference sets how the label attaches itself to the surface. The label's initial orientation is the object's orientation. Optional settings include View (displays the label horizontally in all views), or WCS (label reads toward the World Coordinate System's Zero direction). The Plan Readability section defines how the Elevation Only label is displayed in any view (see Figure 4.12).

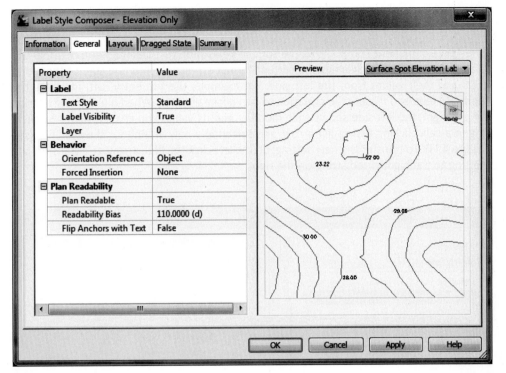

FIGURE 4.12

7. Click the **Layout** tab to make it current.

The Layout panel defines how the label anchors itself in the object. This label is anchored on the <Feature> (surface) at the selected point's middle right. If toggled on, the Border section defines an enclosing box around the text. The Text section defines the label's text height, rotation, and color. The Attachment setting attaches the text's middle left to the feature's anchor point (see Figure 4.13).

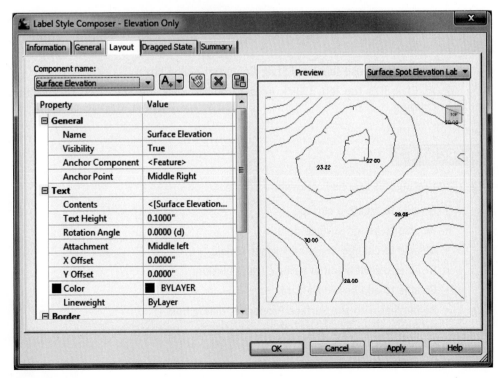

FIGURE 4.13

8. In the Text section, in Contents, click in its **Value** cell to display an ellipsis, and click the ellipsis to display the **Text Component Editor**.

The Text Component Editor shows the surface's elevation text labeling format string. The dialog box's right side displays the current format string, which is the sum of the modifiers and their values from the left side. For example, the top entry, Uft, indicates the units are feet, and P2 sets the precision to two decimal places. When highlighting a string on the right, the left side shows each modifier and its value. You change values by clicking in a cell, clicking a drop-list arrow, and selecting a new setting for the Modifier (see Figure 4.14). To transfer the new setting, on the left click the blue arrow; this transfers the setting to the highlighted string on the right.

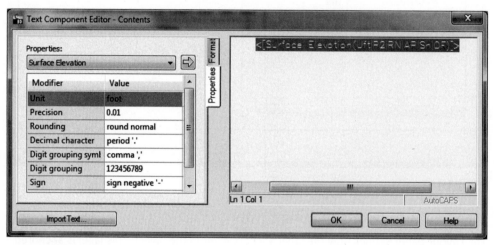

FIGURE 4.14

9. Click **OK** to exit the Text Component Editor.
10. Click the ***Dragged State*** tab to make it current.

If you are dragging a label from its original location, these settings define what happens to it. In this label style, the dragged label will have a leader and the text will be stacked (left-justified).

11. Click **OK** to close the Label Style Composer dialog box.
12. At Civil 3D's top left, Quick Access Toolbar, click the ***Save*** icon to save the drawing.

EXERCISE 4-2

After completing this exercise, you will:

- Be able to create a surface.
- Be able to assign surface data.
- Be able to change assigned surface styles.
- Be able to review initial surface triangulation.
- Be able to add breakline data.
- Be familiar with the Surface Properties dialog box.
- Be familiar with the Surface Properties' Surface Statistics panel.

Exercise Setup

This exercise continues with the previous unit's exercise drawing, Surfaces. If you did not complete the previous exercise or starting with this exercise, browse to the Chapter 04 folder of the CD that accompanies this textbook and open the drawing, *Chapter 04 - Unit 2*.

1. If in the previous unit's exercise drawing, Surfaces is not open, open it now, or browse to the Chapter 04 folder of the CD that accompanies this textbook and open the *Chapter 04 - Unit 2.dwg* file.

2. Use the ZOOM EXTENTS command to view the drawing.

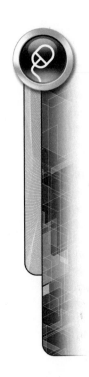

Creating a Surface

The first step is to create a surface from the Ribbon's Home tab, Create Ground Data panel, Surfaces drop-list menu or from Prospector's Surfaces shortcut menu.

1. If necessary, click the **Prospector** tab.

2. In the Ribbon's Home tab, Create Ground Data panel, click the Surfaces drop-list arrow, from the menu select CREATE SURFACE or in Prospector, click the **Surfaces** heading, press the right mouse button, and from the shortcut menu, select CREATE SURFACE.... Either action displays the Create Surface dialog box (see Figure 4.15).

3. In the Create Surface dialog box, for the surface name enter **Existing**, for the description enter **Phase 1 – Existing Ground**, and make sure the surface type is **TIN surface**.

4. Set the surface style to **Border Triangles Points**. Click in the Style value cell, and at the cell's right side, click the ellipsis (...) to display the Select Surface Style dialog box, click the droplist arrow, select **Border Triangles Points** from the list, and click **OK**.

5. Using the same method as the previous step, set the render material to **Sitework.Planting.Grass.Short** and after setting the values, click **OK** to create the surface.

The Surfaces branch updates the Existing listing.

6. If necessary, select the **Prospector** tab.

7. In Prospector, expand **Surfaces** to view the Existing instance, its surface masks, watersheds, and definition headings.

8. Expand the **Definition** branch to view the surface data tree. You may have to scroll Prospector to view all of the entries.

The surface's Definition branch lists a surface's data types. When assigning a type, Prospector indicates the assignment with an icon (square with a black dot).

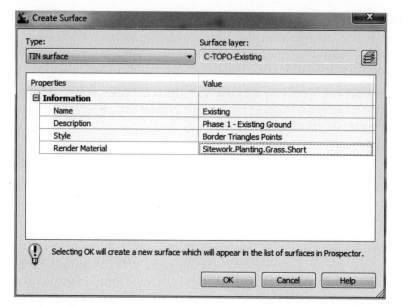

FIGURE 4.15

Surface Point Group — Existing Ground Point Group

Next, you will organize point data into point groups. This exercise creates three groups. The first is the surface data. The second consists of points representing possible break-lines. The last hides all point markers and labels. The last group controls point visibility.

The Existing Ground point group contains Existing's surface point data. The group contains all drawing points (point numbers 1–228), except for existing fire hydrants. Fire hydrant elevations are their tops. This creates an incorrect surface elevation at each hydrant. Remove these points from a group by using the group's exclude panel's By Raw descriptions matching option (EFYD).

1. Expand **Point Groups** to view the point group list.
2. Select **_All Points**, and in preview, scroll through its point list (Prospector's lower panel).

The group contains point numbers 1–228, and existing fire hydrants are point numbers 149, 181, 198, and 216.

3. In Prospector, click the **Point Groups** heading, press the right mouse button, and from the shortcut menu, select NEW....
4. In the *Information* tab for the point group's name, enter **Existing Ground Points**.
5. Click the *Include* tab, toggle **ON With Numbers Matching**, and for the point number range, enter **1–228** (see Figure 4.16).

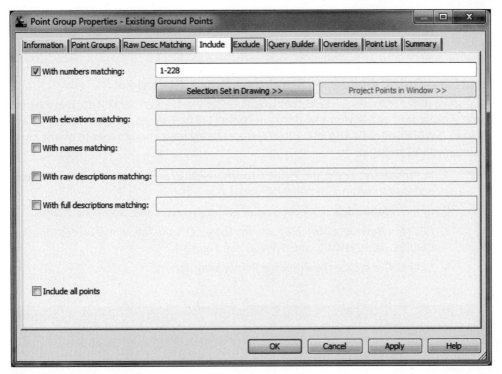

FIGURE 4.16

6. Click the *Exclude* tab, toggle **ON With Raw Descriptions Matching**, and enter **EFYD** (see Figure 4.17).

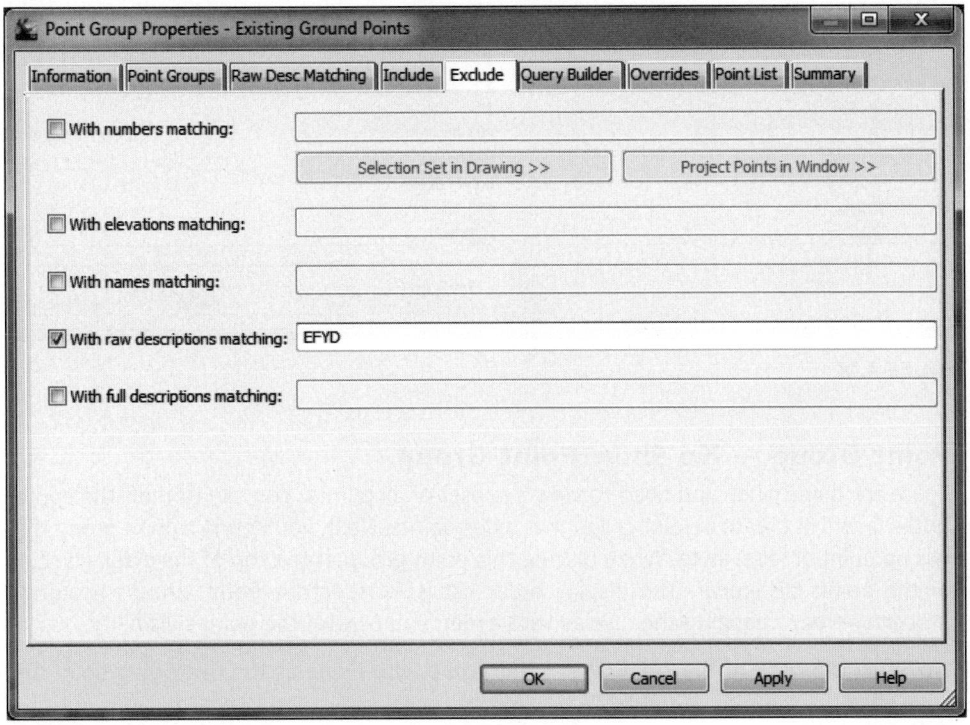

FIGURE 4.17

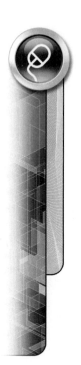

7. Click the ***Point List*** tab and sort the Raw Description list by clicking its heading. Notice the absence of existing fire hydrants (EFYD).

8. Click **OK** to create the Existing Ground Points point group.

Surface Point Group — Breakline Point Group

For some linear features to correctly triangulate, they may need breaklines. These points may include tops and bottoms of slopes or banks (T/S or T/B), edges-of-travelway (EOP), swales (SWALE), ditches (DL), etc. These features rarely have enough data to clearly show their linear relationship.

1. In Prospector, select the **Point Groups** heading, press the right mouse button, and from the shortcut menu, select NEW....

2. In the ***Information*** tab for the point group's name, enter **Breakline Points**.

3. Click the ***Raw Desc Matching*** tab and toggle **ON** the following raw descriptions: **CL***, **DL**, **EOP**, **HDWL**, and **SWALE** (see Figure 4.18).

4. Click **OK** to create the Breakline Points point group.

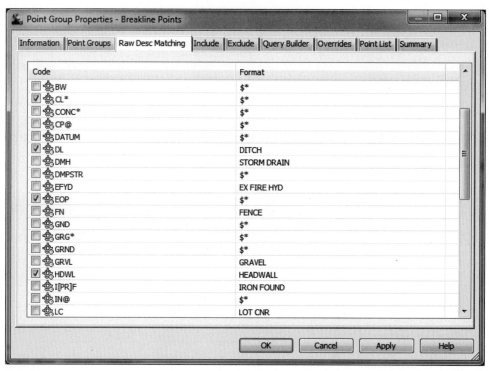

FIGURE 4.18

Point Group — No Show Point Group

There are times when you need to view a subset of all points. You can turn off the appropriate layers or create a point group that hides points. Next, you define a point group that has no point or label style. When placing this point group at the top of the display list, all of the points disappear. The display order list is Prospector's Point Groups heading's property. When changing the display list's order, you control the point's visibility.

No Show is a copy of the _All Points point group with None as the overriding point and label style.

1. In Prospector, from the list of point groups, select the **_All Points**, press the right mouse button, and from the shortcut menu, select COPY....

2. From the list of point groups, select **Copy of _All Points**, press the right mouse button, and from the shortcut menu, select PROPERTIES....

3. In the *Information* tab change the point group's name, **No Show**.

4. On the panel's middle left, change the Point Style to **<none>** and the Point Label Style to **<none>**.

5. Click the *Overrides* tab and toggle **ON** the overrides for point **Style** and **Point Label Style**.

6. Click **OK** to create the No Show point group.

7. In Prospector, Point Groups, select the **Point Groups** heading, press the right mouse button, and from the shortcut menu, select PROPERTIES....

8. In Properties, select **Existing Ground Points** and click the up arrow to move the group to the top of the list.

9. Click **OK** to exit and display the points.

10. In Civil 3D's top left, Quick Access Toolbar, click the *Save* icon to save the drawing.

Assigning Surface Data

To assign surface data, you select a data type from the surface definition branch, press the right mouse button, and from the shortcut menu select ADD....

1. If necessary, in Prospector, expand **Existing** until you are viewing its Definition branch.

2. In the Definition branch, select the **Point Groups** heading, press the right mouse button, and from the shortcut menu, select ADD....

3. From the point group list select **Existing Ground Points** and click **OK** to close the dialog box and add the group (see Figure 4.19).

The surface builds from the newly assigned data.

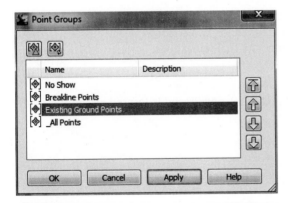

FIGURE 4.19

Review Initial Surface

1. Use the ZOOM and PAN commands to navigate the site to review the surface triangulation.

2. In Prospector, select the **Point Groups** heading, press the right mouse button, and from the shortcut menu, select PROPERTIES....

3. In the Point Groups Properties dialog box, select **Breakline Points**, and move it to the list's top by clicking the up arrow on the dialog box's right side.

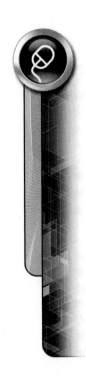

4. Select the **No Show** point group and using the up arrow icon on the right, move it to the list's second position (see Figure 4.20).

5. Click **OK** to close the dialog box and change the point group display. You may need to use Regenall (REA) to see the update.

The current point group display order hides all points except for Breakline Points.

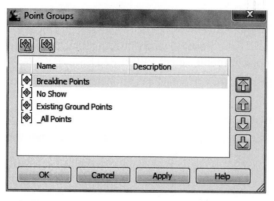

FIGURE 4.20

6. In Prospector, click the **Existing** surface name, press the right mouse button, and from the shortcut menu, select REBUILD – AUTOMATIC.

7. Open the **Layer Properties Manager** (Ribbon's Home panel, Layers panel), create a new layer, for its name enter **Existing-Breakline**, assign it a color, make it the current layer, and click **X** to close the palette.

8. At Civil 3D's top left, Quick Access Toolbar, click the **Save** icon to save the drawing.

9. Use the ZOOM command to review the site's southern portion and its triangulation for the points EOP, DITCH, CL, and SWALE.

The initial triangulation does not correctly connect the points representing the north and south edges-of-pavement, the centerline, ditch, or swale (south of the roadway). These points need breaklines to correctly triangulate their elevations.

Ditch Breakline

The ditch breakline consists of points 21, 22, 180, and 23 and is located just to the road's northwest side (see Figure 4.21).

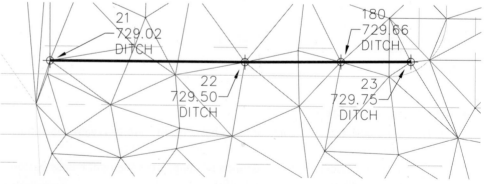

FIGURE 4.21

1. Zoom into the area shown in Figure 4.21.

2. Check and, if necessary, display the **Transparent Commands** toolbar.

3. Start the 3DPOLY command, from the Transparent Commands toolbar select either filter, Point Number, or Point Object, and enter or select points **21, 22, 180,** and **23**.

The following are the 3DPOLY command's responses when using the point number transparent command.

```
Command: 3dpoly
Specify start point of polyline: '_PN
>>Enter point number: 21
Resuming 3DPOLY command.
Specify start point of polyline: (4962.18 5043.75 729.02)
Specify endpoint of line or [Undo]:
>>Enter point number: 22
Resuming 3DPOLY command.
Specify endpoint of line or [Undo]: (5048.7 5043.75 729.5)
Specify endpoint of line or [Undo]:
>>Enter point number: 180
Resuming 3DPOLY command.
Specify endpoint of line or [Undo]: (5090.95 5044.15 729.66)
Specify endpoint of line or [Close/Undo]:
>>Enter point number: 23
Resuming 3DPOLY command.
Specify endpoint of line or [Close/Undo]: (5121.48 5044.44
   729.75)
Specify endpoint of line or [Close/Undo]:
>>Enter point number: *Cancel* <Press ESC to exit transparent
   command>
Specify endpoint of line or [Close/Undo]:
Resuming 3DPOLY command.
Specify endpoint of line or [Close/Undo]: <Press ENTER to
   exit>
Command:
```

The command echoes the selected point's coordinates and elevations.

4. After drafting the 3D polyline, press ESC to exit the transparent command, and press ENTER to exit the 3DPOLY command.

5. In the Existing surface Definition branch, select the **Breaklines** heading, press the right mouse button, and from the shortcut menu, select ADD....

6. In the Add Breaklines dialog box, for the description, enter **Ditch**, set its type to **Standard**, and click **OK** to select the polyline (see Figure 4.22).

7. Select the 3D polyline representing the ditch and press ENTER.

8. If Civil 3D issues a duplicate points warning, close the warning panorama and review the new triangulation.

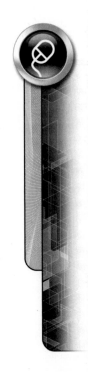

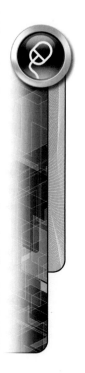

Add Breaklines

Description:
Ditch

Type:
Standard

File link options:
Break link to file

☐ Weeding factors

Distance:
15.000'

Angle:
4.0000 (d)

Supplementing factors

☐ Distance:
100.000'

Mid-ordinate distance:
1.000'

OK   Cancel   Help

FIGURE 4.22

After adding the breakline, the surface updates its definition list and rebuilds its triangulation to resolve the ditch.

9. In the Existing surface Definition branch, expand **Breaklines** to view the Ditch entry.

10. Select the list's **Ditch** entry, press the right mouse button, from the shortcut menu, select PROPERTIES..., and review its coordinate and elevation values.

11. From the list select a vertex, press the right mouse button, and from the shortcut menu select ZOOM TO.

12. Close the Breakline Properties vista by selecting the Panorama's **X**.

Whenever using Zoom to, the object is put at the display's center.

Swale Breakline

A swale breakline just south of the roadway is next. The swale starts near the road's eastern end and extends all the way to the western edge of the surface. The swale point numbers are (east to west) 190, 135, 191, 185, 184, 134, 183, 182, and 133 (see Figure 4.23).

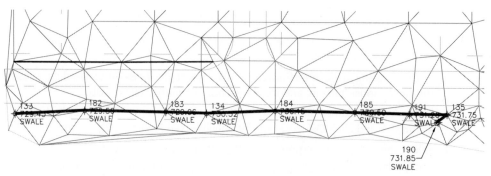

FIGURE 4.23

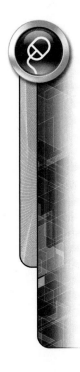

1. Use the ZOOM command and navigate to the area shown in Figure 4.23 to view the swale points.

2. To better view the points, change the Annotation Scale to 1″ = 20′.

3. Start the 3DPOLY command, from the Transparent Commands toolbar select either filter, Point Number or Point Object, and enter or select points **190**, **135**, **191**, **185**, **184**, **134**, **183**, **182**, and **133**.

4. When you are done, press ESC to exit the transparent command, and press ENTER to exit the 3DPOLY command.

5. In the Existing surface Definition branch, select **Breaklines**, press the right mouse button, and from the shortcut menu, select ADD....

6. In the Add Breaklines dialog box, for the breakline description enter **Swale**, set the type to **Standard**, and click **OK** to select the polyline.

7. Select the 3D polyline representing the Swale and press ENTER.

8. If you are warned about duplicate surface points, close the Event Viewer and view the new triangulation.

The surface updates its triangulation, correctly resolving the swale.

9. If necessary, in the Existing surface Definition area of Prospector, expand **Breaklines**, to view its list, and notice the Swale breakline is on it.

10. Select the **Swale** entry, press the right mouse button, and from the shortcut menu, select PROPERTIES... and review its coordinate and elevation values.

11. Close the Breakline Properties vista by selecting the Panorama's **X**.

12. At Civil 3D's top left, Quick Access Toolbar, click the **Save** icon to save the drawing.

Reviewing a surface's triangulation does not necessarily give you a good idea of what the surface looks like. A Quick Profile is an additional tool to review a surface and its breaklines and edit effects.

Evaluate Surface with Quick Profile

Reviewing a surface includes looking at its triangulation along the road's southern edge. Few, if any, triangles correctly defined the north ditch and the south swale. As a result of this initial review, two breaklines were added to the surface. Even with these two breaklines, the triangulation representing the roadway is wrong. One triangle even connects a ditch point (180) north of the road to a swale point (183) south of the road (see Figure 4.24).

It is one thing to see the triangulation, but another to understand what the triangles mean as surface elevations or relief. Quick Profile displays elevations along an object as a profile. This exercise portion creates a Quick Profile view that reviews surface elevations and the effects of breaklines.

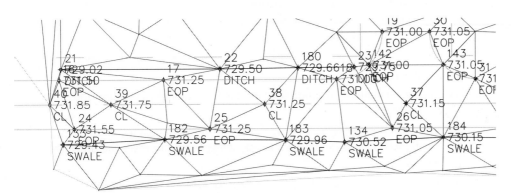

FIGURE 4.24

1. From the Ribbon's View tab, Viewports panel, click the Named Viewports icon.

2. In the Viewports dialog box, click the **New Viewports** tab and for the name enter, **A and P**. From the Standard Viewports list, select Two: Horizontal and at the bottom right set the view to Current and the Visual Style to 2D Wireframe.

3. Click **OK** to exit the dialog box.

4. Click the top viewport to make it current, and use the PAN command to move the site to the left until the screen is clear.

5. Click in the lower viewport to make it current and use the PAN and ZOOM commands to view the roadway's western half (see Figure 4.25).

6. In the lower viewport, start the PLINE command and draw a polyline from south of the swale (near point number 133) to a point north of the ditch (past point 21) at the western end of the roadway (see Figure 4.25).

7. In the Ribbon, click the Analyze tab, Ground Data panel, select the **Quick Profile** icon.

8. Select the polyline just drawn, and the Create Quick Profiles dialog box opens. (You can also select the polyline, right-click, and from the shortcut menu, select QUICK PROFILE...).

9. Click **OK** to accept the default values and continue drawing the profile.

10. Click in the top viewport to make it current and select a point in its lower left to locate the profile view.

A Quick Profile appears in the upper viewport and an Event Viewer is displayed to remind you that the profile is only temporary.

11. Click the **X** on Panorama's mast to close the Events vista.

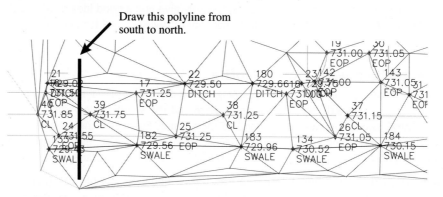

Draw this polyline from south to north.

FIGURE 4.25

Relocating the polyline segment causes it to update the profile. What should be displayed in the profile is a ditch or swale with a poorly defined road section. Because triangulated points rarely triangulate linear features correctly, you must use breaklines to make them appear correctly in a surface.

12. Use the MOVE command or grips to relocate the polyline segment reviewing various road locations.

13. Finally, relocate the polyline segment back to its original position and press ESC to deselect it.

14. With the lower viewport active from the Viewports panel, click the Viewport Configuration icon, and from the list, select **Single** to reset the drawing to one viewport.

15. At Civil 3D's top left click, Quick Access Toolbar, the **Save** icon to save the drawing.

16. If necessary, use the ZOOM and PAN commands to view the roadway from its western end to just east of the intersection.

Roadway Breaklines

Next, you add four roadway breaklines. The first two will be feature lines and the remaining two repeat the steps you used to create and define the ditch and swale breaklines. The reason for feature lines is the north edges-of-travelway (NEOP-W and NEOP-E) have curve return segments. When defining feature lines containing arcs as breaklines, you process the curves with a mid-ordinate value.

Table 4.1 contains the breakline names, their point numbers, and their breaklines type.

TABLE 4.1

Name	Point Numbers	Type of Breakline
NEOP-W	16, 17, 18, 142, and 19	Standard from Feature Line object
NEOP-E	176, 34, 33, 31, 143, and 34	Standard from Feature Line object
CL	177, 35, 36, 37, 38, 39, and 40	Standard
SEOP	24, 25, 26, 27, 28, 29, and 178	Standard

1. Click the Ribbon's Home tab. On the Create Design panel's left click the Feature Line icon, and from the menu select the CREATE FEATURE LINE to display the Create Feature Lines dialog box.
2. Match your values to those in Figure 4.26 and click **OK** to continue.

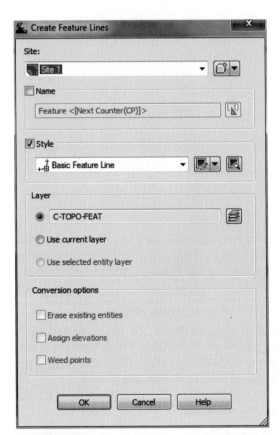

FIGURE 4.26

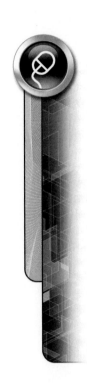

3. From the Transparent Commands toolbar, select the **Point Object** override ('PO), and identify the points listed in Table 4.1 for the breakline NEOP-W. When you select the first point object, the routine echoes back its elevation (point 16), and you press ENTER to accept the elevation.

When selecting the second point, the prompt may not be for elevation.

4. Select point 17. Enter the letter '**E**', press ENTER twice using the point's elevation, and the prompt returns to Select point object.

The following segment shows the prompt for the start of the command:

```
Command:
Specify start point: '_PO Start override
>>
Select point object: Select Point 16
Resuming DRAWFEATURELINE command.
Specify start point: (4961.15 5038.0 731.5)
Specify elevation or [Surface] <731.50>: (Press ENTER)
Specify the next point or [Arc]:
>>
Select point object: Select Point 17
Resuming DRAWFEATURELINE command.
Specify the next point or [Arc]: (5018.14 5038.0 731.25)
Distance 56.991', Grade -0.44, Slope -227.96:1, Elevation
731.250'
Specify grade or [SLope/Elevation/Difference/SUrface]
<0.00>: e
   (To set feature line to Elevation)
Specify elevation or [Grade/SLope/Difference/SUrface]
<731.250>:
   (Press ENTER)
Specify the next point or [Arc/Length/Undo]:
>>
Select point object: Select Point 18
```

There needs to be a vertex at each point on the arc. Make sure you draw an arc from point 18 to 142 and from 142 to 19. You will want to zoom in to better view the points. Press ESC to toggle off the transparent command, enter A for arc, and restart the point object override.

5. Select point number **18** and press ENTER to accept its elevation.

6. Press ESC to exit the Point Object override, enter the letter '**A**' and press ENTER to set the Arc option. Toggle **ON** the **Point Object** filter from the Transparent Commands toolbar, select point **142**, and press ENTER to accept the point's elevation. You are now ready to select point **19**, and press ENTER to accept the point's elevation to finish the arc. Press ESC to exit the Point Object override and press ENTER to exit the command.

The following is the command sequence to create the arc segment of the feature line:

```
Select point object:
Resuming DRAWFEATURELINE command.
```

Specify the next point or [Arc/Length/Undo]: (5112.0 5038.0
  731.0)

Distance 93.857', Grade -0.27, Slope -375.43:1, Elevation
  731.000'

Specify elevation or [Grade/SLope/Difference/SUrface]
<731.000>:

  (Press ENTER)

Specify the next point or [Arc/Length/Close/Undo]:

>>

Select point object: *Cancel* (Press ESC)

>>

Specify the next point or [Arc/Length/Close/Undo]:

Resuming DRAWFEATURELINE command.

Specify the next point or [Arc/Length/Close/Undo]: **a** (for
Arc)

Specify arc end point or [Radius/Secondpnt/Line/Close/
Undo]: '_PO

>>

Select point object: Select Point **142**

Resuming DRAWFEATURELINE command.

Specify arc end point or [Radius/Secondpnt/Line/Close/
Undo]:

  (5129.68 5045.32731.0)

Distance 19.635', Grade 0.00, Slope Horizontal, Elevation
  731.000'

Specify elevation or [Grade/SLope/Difference/SUrface]
<731.000>:

  (Press ENTER)

Specify arc end point or [Radius/Secondpnt/Line/Close/
Undo]:

>>

Select point object: Select Point **19**

Resuming DRAWFEATURELINE command.

Specify arc end point or [Radius/Secondpnt/Line/Close/
Undo]:

  (5137.0 5063.0731.0)

Distance 19.635', Grade 0.00, Slope Horizontal, Elevation
  731.000'

Specify elevation or [Grade/SLope/Difference/SUrface]
<731.000>:

  (Press ENTER)

Specify arc end point or [Radius/Secondpnt/Line/Close/
Undo]:

>>

Select point object: *Cancel* (Press ESC)

>>

Specify arc end point or [Radius/Secondpnt/Line/Close/
Undo]:

Resuming DRAWFEATURELINE command.

Specify arc end point or [Radius/Secondpnt/Line/Close/
Undo]:

  (Press ENTER)

Command:

7. Use the PAN command to view the eastern portion of the north edge-of-travelway.
8. Repeat Steps 1–6 and draft the NEOP-E feature line. Start the feature line at point 176. Make sure you draw an arc from point 31 to 143 and from 143 to 30. There must be a vertex at each point on the arc segment.

Review Feature Lines Elevations

1. Select the **NEOP-E** feature line, press the right mouse button, and from the shortcut menu, select ELEVATION EDITOR....
2. Compare the elevations of the feature to those in Table 4.2, and, if necessary, correct their elevations.

TABLE 4.2

Point	Elevation
176	733.75
34	733.45
33	732.35
31	731.05
143	731.05
30	731.05

3. Click the Select icon in the upper left of the vista.
4. Repeat Step 1 and select the **NEOP-W** feature line. If necessary, correct its elevations using those listed in Table 4.3.

TABLE 4.3

Point	Elevation
16	731.50
17	731.25
18	731.00
142	731.00
19	731.00

5. Click Panorama's **X** to close it.

Add Feature Lines as Breaklines

1. In the Existing surface's Definition branch, select **Breaklines**, press the right mouse button, and from the shortcut menu, select ADD....

2. In the Add Breaklines dialog box, name the breakline **NEOP-W**, set its type to **Standard**, set the Mid-ordinate value to **0.1**, click **OK** to exit the dialog box, select the westernmost feature line, and press Enter.

3. Repeat Steps 1–2 for the eastern feature line using the same values found in Step 2, but describe the breakline as **NEOP-E**.

4. If a Panorama is displayed and informs you about duplicate points, close the Events vista by clicking the **X** and view the new triangulation.

Each time you add a new breakline, the surface rebuilds and the triangulation contains additional data around the curve return arc segments.

Draw and Define the CL and SEOP Breaklines

1. Draw 3D polyines for the **CL** and **SEOP** breaklines using the transparent command override of Point Object to reference the points and their elevations.

2. After drafting the 3D polylines, define them as standard breaklines, and assign each one an appropriate name.

3. If a Panorama is displayed and informs you about duplicate points, close the Events vista by clicking the **X** and view the new triangulation.

View Roadway Quick Profiles

1. Click the Ribbon's **View** tab. In the Viewports panel, click NAMED.

2. In the Viewports dialog box's, Named Viewports tab, select the named viewport **A and P**, and click **OK** to exit the dialog box.

3. Click the Ribbon's **Modify** tab. In the Design panel, click the **Feature Line** icon. At the Launch Pad panel's left, click QUICK PROFILE.

4. Select the quick profile polyline drawn earlier, the Create Quick Profiles dialog box displays, click **OK** to close it, and in the upper viewport select the profile's insertion point.

5. Close the Panorama to view the quick profile.

6. Review the road by moving the polyline in the lower viewport.

7. After reviewing the road, move the quick profile polyline's southern end to a point just south of point 188 and its northern end just south of the lot line north of the roadway.

A headwall is just south of the road between points 186 and 189.

8. Click the Ribbon's **View** tab. In the Viewports panel, click the Set VIEWPORT icon, and from the menu, select SINGLE.

9. At Civil 3D's top left, Quick Access Toolbar, click the **Save** icon to save the drawing.

Wall Breakline

The last breakline is a wall breakline. It starts as a 3D polyline representing the top of the wall. The breakline's offset side (bottom of the wall) is north toward the road.

A wall breakline in a surface greatly changes the surface's slope statistics report. Before adding the wall breakline, review the current slope values for minimum, maximum, and mean.

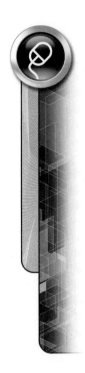

1. In the drawing, select the surface border or a triangle, press the right mouse button, and from the shortcut menu, select SURFACE PROPERTIES....

2. Click the **Statistics** tab and expand the **Extended** (statistics) section. Your report should look similar to Figure 4.27. The mean slope is around 6 percent, the lowest slope is zero, and the highest is just over 200 percent.

The mean and maximum slope will change greatly after the wall breakline is defined.

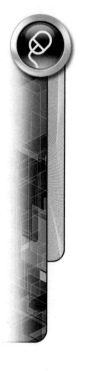

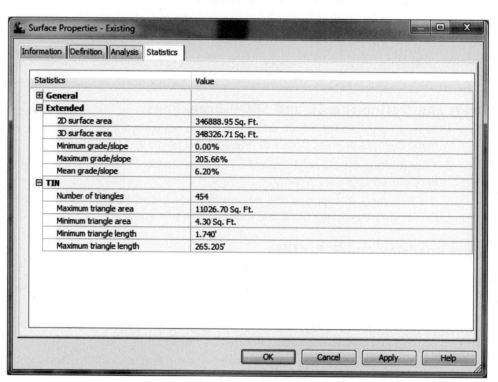

FIGURE 4.27

3. Click **OK** to exit the Surface Properties dialog box.

4. Make a new point group, **Headwall**, based on the raw description of **HDWL**. Use the **Raw Desc Matching** tab to identify the headwall points, and click **OK** to exit the dialog box.

5. After defining the Headwall point group, go into the Point Groups properties dialog box and make **Headwall** the top point group, followed by the No Show point group (see Figure 4.28).

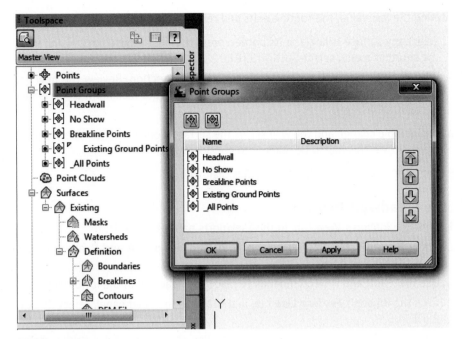

FIGURE 4.28

6. Use the 3DPOLY routine with the Transparent Commands toolbar's **Point Object** override and draw a 3D polyline between points 186, 187, 188, and 189.

7. In Prospector, in the Existing surface's Definition branch, select **Breaklines**, press the right mouse button, and from the shortcut menu, select ADD....

8. In the Add Breaklines dialog box, enter the description of **Headwall**, change the breakline type to **Wall**, and click **OK** to exit the dialog box.

9. Select the headwall 3D polyline, press ENTER, and select a point to the north-east of point 188 as the offset side.

The routine responds by prompting you for a method to set the offset side elevations.

10. Enter '**I**' for Individual heights and press ENTER to view each headwall vertex.

The survey crew measured a down distance from the wall's top to bottom. Because of this method, the offset elevation for each point is a difference in elevation.

11. At the Specify elevation prompt, for Delta enter '**D**' and press ENTER.

12. The prompt changes to Enter elevation difference for offset point at .... Use the elevation difference entries in Table 4.4 for the difference values.

TABLE 4.4

Point Number	Difference in Elevation
186	−2.5
187	−3.5
188	−3.5
189	−2.5

After entering the last value, the routine exits and rebuilds the surface.

13. Select any surface triangle or its border, press the right mouse button, and from the shortcut menu, select SURFACE PROPERTIES....

14. Click the **Statistics** tab and expand the **Extended** section. Review the new values. The average for slopes is now nearly 225 and the maximum slope is just over 350,000.

This is a considerable change and the wall skews the surface slope statistic. When reviewing the surface slopes, this has to be taken into consideration.

15. Click **OK** to exit the Surface Properties dialog box.

View the Headwall Profile

1. Click the Ribbon's **View** tab. In the Viewports panel, click NAMED.

2. In the **Named Viewports** tab, select the named viewport **A and P**, and click **OK** to exit the dialog box.

3. Use the ZOOM and PAN commands to place the headwall in the lower viewport.

4. Click the Ribbon's **Feature Line** tab, and in the Launch Pad panel, select QUICK PROFILE.

5. Select the quick profile polyline drawn earlier, the Create Quick Profile dialog box displays, click **OK** to close it, and in the upper viewport select the profile's insertion point.

6. Close the Panorama to view the quick profile.

7. If necessary, move the quick profile view polyline to view the newly defined headwall.

8. Make the bottom viewport the current viewport.

9. Click the Ribbon's **View** tab. In the Viewports panel, click Set VIEWPORT, and from the menu, select SINGLE.

10. Click the Ribbon's **Home** tab.

11. In the Layers panel, use the **Layer Properties Manager** to set layer **0** (zero) as the current layer and freeze the **Existing-Breakline, RDS, LOT, C-TOPO-FEAT,** and **CL** layers to focus on the surface and its triangulation.

12. Use the ZOOM EXTENTS command to view all of the surface and its triangles.

13. In Prospector, select the **Point Groups** heading, press the right mouse button, and from the shortcut menu, select PROPERTIES....

14. Select **No Show**, move it to the top of the list, and click **OK** to exit and hide the points.

15. At Civil 3D's top left, Quick Access Toolbar, click the **Save** icon to save the drawing.

EXERCISE 4-3

After completing this exercise, you will:

• Be familiar with the Edit Surface menu.
• Know Surface Properties' Definition tab settings and their effects.
• Be able to delete and control surface edits.
• Be able to review surface edit effects.
• Be familiar with Natural Neighbor Interpolation (NNI) smoothing.

This unit's exercise continues with the previous units drawing. If you did not complete the Unit 2 exercise, or starting with this exercise browse to the Chapter 04 folder of the CD that accompanies this textbook and open the Chapter 04 - Unit 3 drawing.

Exercise Setup

1. If not open, open the previous exercise's drawing or browse to the Chapter 04 folder of the CD that accompanies this textbook and open the *Chapter 04 – Unit 3* drawing.

Surface Elevations

A review of simple surface edits can be done using a surface's tool tip. The Surface Properties' Information tab has the tool tip toggle.

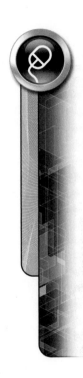

1. In the drawing, select the **Existing** surface border, press the right mouse button, and from the shortcut menu, select SURFACE PROPERTIES....

2. Click the **Information** tab and at the dialog box's lower left, notice the Show Tool-tips toggle; make sure it is toggled **ON**.

3. Click **OK** to close the dialog box and move the cursor to different surface locations, letting the cursor sit for a moment. The tooltip shows the elevation at the cursor's intersection.

Reviewing Slopes and Breakline Data

1. In the drawing, select **Existing** surface's border, press the right mouse button, and from the shortcut menu, select SURFACE PROPERTIES....

2. Click the **Statistics** tab and expand the **General** section to review its values.

Note the minimum, maximum, and the mean elevation. The mean elevation is about midway between the lowest and highest elevations.

3. Expand the **Extended** section to review its values.

The maximum grade (350,000+) is the result of the wall breakline. This single surface feature changes the average grade from around 6 percent to more than 220 percent.

4. Click the **Definition** tab.

5. In Definition Options, expand **Data Operations**, change Use Breaklines to **No**, click **OK** to exit the dialog box, and if necessary, click **Rebuild the Surface** to rebuild the surface with no breakline data.

6. Use the ZOOM and PAN commands to view the roadway's triangulation.

The surface returns to having only point data.

7. Select a surface triangle or the border, press the right mouse button, and from the shortcut menu, select SURFACE PROPERTIES....

8. In the **Definition** tab, expand **Data Operations**, toggle **Use Breaklines** to **Yes**, click **OK** to exit the dialog box, and if necessary, click **Rebuild the Surface**.

9. This restores the breakline data to the surface.

Temporarily Removing the Wall Breakline

To remove the impact of a single operation, move your attention to the Definition panel's lower half. It is here that one can toggle on or off the effect of one or more operations. By toggling off the wall breakline, it is temporarily removed from the surface. With it

removed, the Extended surface statistics return to their values before adding the wall breakline. After reviewing these sans wall statistics, you reenter the surface properties dialog box and toggle it on again.

1. Select a surface triangle or the border, press the right mouse button, and from the shortcut menu, select SURFACE PROPERTIES....

2. In the Operation Type, uncheck the **Wall breakline** toggle.

3. Click **OK** to exit the dialog box and click **Rebuild the Surface** to rebuild the surface with the new data mix.

4. If the Events vista display contains warnings about duplicate points, click the Panorama's **X** to close it and view the new triangulation.

The headwall triangles change.

5. In the drawing, select the **Existing** surface border, press the right mouse button, and from the shortcut menu, select SURFACE PROPERTIES....

6. Click the **Statistics** tab and expand the **Extended** section. Review the maximum and mean slope values.

The mean slopes return to their pre-wall values (they average around 6 percent and have a maximum of just over 200 percent).

7. Click the **Definition** tab. In Operation Type, toggle **ON** the **Wall breakline**, click **OK** to close the dialog box, and then click **Rebuild the Surface** to rebuild the surface.

Reviewing Build Settings

1. In the drawing, select a surface triangle or the border, press the right mouse button, and from the shortcut menu, select SURFACE PROPERTIES....

2. Click the **Statistics** tab and expand the **TIN** section. The minimum and maximum triangle lengths represent some very short and long surface triangles.

The question is, does a triangle leg of 260 feet represent a valid surface triangle? While building a surface, a surface tries to eliminate some extraneous peripheral triangles. However, if a maximum distance is not specified, a surface may have triangle legs around its periphery without supporting data.

Setting a Maximum triangle to control peripheral triangles may not completely eliminate the problem. After using a few values and viewing the results, you may have to edit the surface with delete lines or use a boundary to control the peripheral triangulation.

3. Click the **Definition** tab and in Definition Options, expand the **Build** section. In the Build section, change Use Maximum Triangle Length to **YES**, set the length to **100**, click **OK** to close the dialog box, and then click REBUILD THE SURFACE to rebuild the surface.

The surface border and the peripheral triangles change, reflecting the new settings. While triangles are moved in the north, some in the northwest are not removed. It would seem that deleting a triangle and/or defining a boundary is a better solution.

4. In the drawing, select the **Existing** surface border, press the right mouse button, and from the shortcut menu, select SURFACE PROPERTIES....

5. In the **Definition** tab Definition Options expand the **Build** section.

6. Toggle the Use Maximum Triangle Length to **NO**, click **OK** to close the dialog box, and then click **Rebuild the Surface** to rebuild the surface.

Deleting Triangles

1. In the Ribbon, click the View tab. In the Views panel's left side, click and restore the named view **Central,** and review the surface's eastern triangulation.

2. From the Existing surface's Definition branch, select **Edits**, press the right mouse button, and from the shortcut menu, select DELETE LINE (see Figure 4.29).

3. In the drawing, select the two triangle legs shown in Figure 4.27 and press ENTER twice to delete the lines and exit the command.

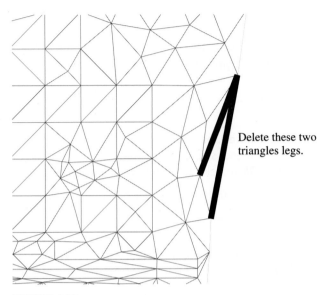

Delete these two triangles legs.

FIGURE 4.29

They are removed and surface displays a new surface border.

Prospector updates and indicates the edits by placing an icon to the left of Edits. Select the Edits heading to display a surface listed by the edits preview (see Figure 4.30). The edit also appears in Surface Properties' Definition panel's Operation Type.

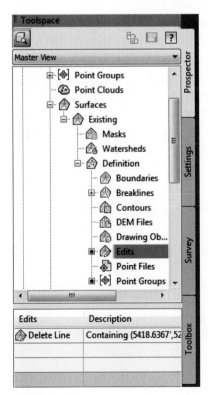

FIGURE 4.30

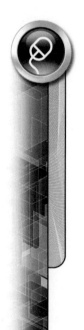

4. From the Definition branch, select **Edits** to display an edit list in preview with the delete line entry.

5. In the drawing, select the surface border, press the right mouse button, and from the shortcut menu, select SURFACE PROPERTIES....

6. If necessary, click the **Definition** tab.

Deleting multiple lines is an Operation Type entry at the panel's bottom. There is an option to toggle on or off the deletion's effect or remove it from the surface.

7. From the Operation Type list, select **DELETE LINE**, press the right mouse button, and review the shortcut menu commands. Remove From Definition... is one of the commands.

8. Press ESC to close the shortcut menu.

9. Click **OK** to exit the dialog box.

10. In Prospector's preview, select **Delete Lines** (see Figure 4.30), press the right mouse button, from the shortcut menu, select DELETE..., and in the Remove From Definition dialog box, click **OK** to delete the edit.

The Remove From Definition dialog box gives you the option of not deleting an edit.

11. If a Panorama is displayed, close the Events vista by clicking the **X** and view the new triangulation.

12. In the Existing surface's Definition branch, select **Edits**, press the right mouse button, from the shortcut menu, select DELETE LINE, and delete the two triangles identified in Figure 4.29.

Swap Edge

The Swap Edge routine changes a diagonal to the second solution. Use this routine to "tweak" a surface's triangulation in areas that do not need additional data or breaklines. Like Delete Lines, preview displays only the swapped edges' coordinates.

1. In the Existing surface's Definition branch, select **Edits**, press the right mouse button, and from the shortcut menu, select SWAP EDGE.

2. Select two or three diagonals to change their position and press ENTER to exit the routine.

3. In Preview, review the Swap Edge listings.

4. In Preview, select the **Swap Edge** entries, press the right mouse button, from the shortcut menu, select DELETE..., and in the Remove From Definition dialog box click **OK** to delete the edits and rebuild the surface.

5. If a Panorama is displayed, close the Events vista by clicking the **X** and view the new triangulation.

Surface Point Editing — Delete Point

If you want to delete a surface point, the surface point is removed, but the cogo point remains in the drawing and its assigned point group.

1. From the Ribbon's View tab, in the Views panel, select and restore the named view **Northeast**.

2. In Prospector, click the **Point Groups** heading, press the right mouse button, from the shortcut menu, select PROPERTIES..., move **Breakline Points** to the top position and **No Show** to the second position to view the northeastern swale points, and click **OK** to exit the dialog box.

3. Use the ZOOM command to better view points 107, 111, and 112.

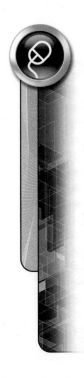

4. In the Existing surface's Definition branch, select the **Edits** heading, press the right mouse button, and from the shortcut menu, select DELETE POINT.

5. When prompted for a point, click near point number 110 and press ENTER twice to exit the routine.

After deleting the surface point, point 110 remains in the drawing. Preview lists only the deleted point's coordinates.

6. In the Existing surface's Definition branch, select the **Edits** heading to preview the edits list.

7. From the Edits list, select the **Delete Point** entry, press the right mouse button, from the shortcut menu, select DELETE..., and in the Remove From Definition dialog box click **OK** to restore the deleted point to the surface.

8. If a Panorama is displayed, close the Events vista by clicking the **X** and view the new triangulation.

9. In Prospector, select the **Point Groups** heading, press the right mouse button, from the shortcut menu, select PROPERTIES..., move **No Show** to the top position, and click **OK** to exit the dialog box.

10. Use the ZOOM EXTENTS command to view the entire site.

Boundaries

The surface under review has several points in the west and northwest that seem to be peripheral to the site. These points could be deleted from the surface. However, an outer boundary limits the data to the points inside its boundary.

1. In Ribbon's Home tab, Layers panel, use Layer Properties Manager and create a new layer: **Existing-Boundary**. Make it the current layer, assign it a color, and click **X** to create the layer and exit the panel.

2. Start the POLYLINE routine and draw a closed boundary similar to the one in Figure 4.31.

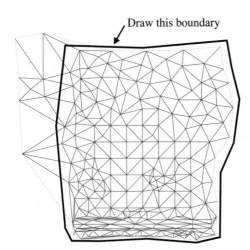

Draw this boundary

FIGURE 4.31

3. In the Existing surface's Definition branch, select **Boundaries**, press the right mouse button, and from the shortcut menu, select ADD....

4. In the Add Boundaries dialog box, name the boundary **Main Site**, set its type to **Outer**, toggle **OFF** Non-destructive breakline, click **OK**, and select the boundary polyline.

5. Open the **Layer Properties Manager**, make the current layer **0** (zero), freeze the layers **Existing-Boundary** and **V-NODE**, and click **X** to exit the panel.

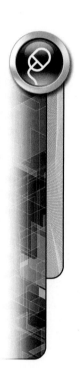

6. At Civil 3D's top left, Quick Access Toolbar, click the **Save** icon to save the drawing.

The surface contains all the needed edits and changes and is ready for analysis. However, it is difficult to decide whether to do any surface smoothing. The decision about smoothing happens after viewing the initial contours.

The next unit reviews the distribution of surface slopes and elevations.

EXERCISE 4-4

After completing this exercise, you will:

- Define surface slopes and elevations analysis styles.
- Define surface styles using Equal Interval, Quantile, and Standard Deviation parameters.
- Define a watershed surface style.
- Assign and process style settings in the Surface Properties' Analysis tab.
- Create water drop paths across a surface.

This unit's exercise continues with the Unit 3 exercise drawing. If you did not complete the Unit 3 exercise, or starting with this exercise browse to the Chapter 04 folder of the CD that accompanies this textbook and open the Chapter 04 - Unit 4 drawing.

Exercise Setup

1. If not open, open the previous exercise's drawing or browse to the Chapter 04 folder of the CD that accompanies this textbook and open the *Chapter 04 – Unit 4* drawing.
2. Click the **Settings** tab.
3. In Settings, expand the **Surface** branch until you are viewing the Surface Styles list.

Slope Surface Styles

The first slope style ranges by Equal Interval. This style divides the overall range by the number of ranges. In the current exercise, the range of slopes is just over 350,000. If a style uses 8 ranges, each range will represent just under 44,000 percent (0–44,000, 44,000–88,000, etc.)

1. From the list of styles select **Slope Banding (2D)**, press the right mouse button, and from the shortcut menu, select COPY....

The Surface style dialog box opens.

2. Click the **Information** tab, name the style **Slope and Arrows – Equal Interval**, for the description enter **Slope triangles and arrows by equal interval**, and click **Apply**.
3. Click the **Analysis** tab.
4. Expand the **Slopes** section and adjust your settings to match those in Figure 4.32.
5. Expand the **Slope Arrows** section and adjust your settings to match those in Figure 4.33.
6. Click the **Display** tab.
7. At the display list's bottom, turn **ON** the following components: **Border**, **Slopes**, and **Slope Arrows** (see Figure 4.34).
8. Click **OK** to exit the dialog box.

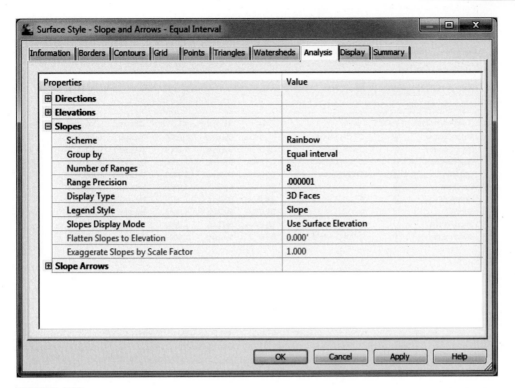

FIGURE 4.32

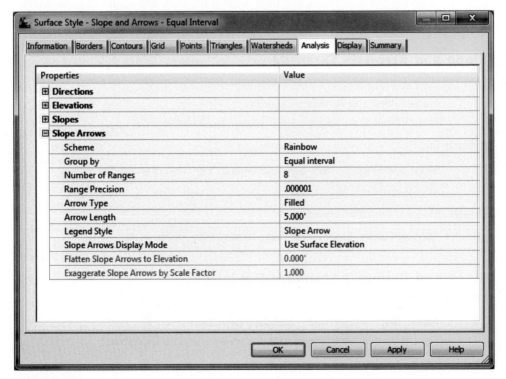

FIGURE 4.33

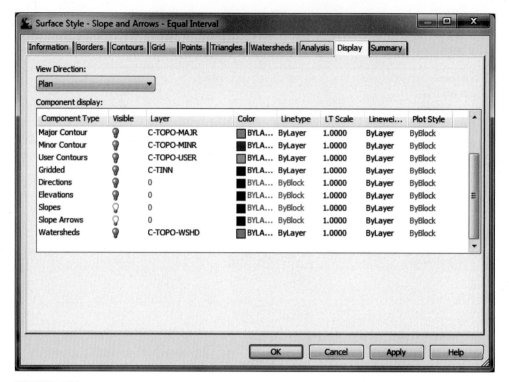

FIGURE 4.34

9. Create the remaining two styles, **Slope and Arrows – Quantile** and **Slope and Arrows – Standard Deviation**, by selecting **Slope and Arrows – Equal Interval** from the list of styles, pressing the right mouse button, and selecting COPY... from the shortcut menu, and then rename the style and adjust its parameters.

Create the new styles using the following values:

Slope and Arrows – Quantile
Information tab:

> Name: Slope and Arrows – Quantile
> Description: Slope triangles and arrows by Quantile.

Analysis tab:

> Slope and Slope Arrows: Same settings as Figure 4.32 and 4.33
> Group by: Quantile.

Display tab:

> Turn on Border, Slopes, and Slope Arrows (see Figure 4.34).

Slope and Arrows – Standard Deviation:
Information tab:

> Name: Slope and Arrows – Standard Deviation
> Description: Slope triangles and arrows by Standard Deviation.

Analysis tab:

> Slope and Slope Arrows: Same settings as Figure 4.32 and 4.33
> Group by: Standard Deviation.

Display tab:

> Turn on Border, Slopes, and Slope Arrows (see Figure 4.34).

Elevation Surface Styles

1. From the list of styles, select **Elevation Banding (2D)**, press the right mouse button, and from the shortcut menu, select COPY....

The Surface Style dialog box opens.

2. Click the ***Information*** tab, for the style's name enter **Elevations – Equal Interval**, and for the description enter **Elevations by Equal Interval**.

3. Click the ***Analysis*** tab.

4. Expand the **Elevations** section and adjust your settings to match those in Figure 4.35.

5. Click the ***Display*** tab, click off all of the components, and click **ON Elevations** (see Figure 4.36).

6. Click the View Direction drop list and from the list select Model.

7. Click the ***Display*** tab, click off all of the components, and click **ON Elevations.**

8. Click the ***OK*** button to exit the dialog box.

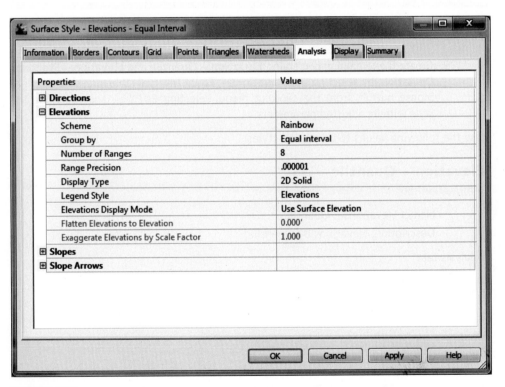

FIGURE 4.35

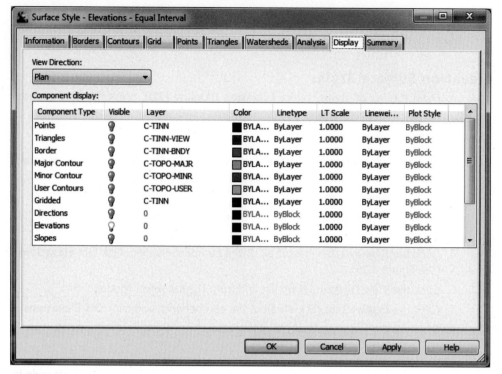

FIGURE 4.36

9. Create the remaining two styles, **Elevation – Quantile** and **Elevation – Standard Deviation**, by selecting the **Elevations – Equal Interval** style, pressing the right mouse button, from the shortcut menu, selecting COPY..., and renaming the style with its adjusted parameters.

Use the following values to create the styles:

Elevation – Quantile

Information tab:

> Name: Elevations – Quantile
> Description: Elevations by Quantile.

Analysis tab:

> Set the same settings as Figure 4.35
> Group by: Quantile.

Display tab:

> Match the settings found in Figure 4.36.

Elevation – Standard Deviation

Information tab:

> Name: Elevations – Standard Deviation
> Description: Elevations by Standard Deviation.

Analysis tab:

> Set the same settings as Figure 4.35
> Group: Standard Deviation.

Display tab:

Match the settings found in Figure 4.36.

Analyzing Surface Slopes

There are two steps to using an analysis style: assign the style and then run the analysis.

1. Click the *Prospector* tab.

2. In Prospector, expand **Surfaces** until you are viewing the surface list, from the list select **Existing**, press the right mouse button, and from the shortcut menu, select SURFACE PROPERTIES....

3. Click the *Information* tab and change the Surface Style to **Slopes and Arrows – Equal Interval**.

4. Click the *Analysis* tab, at the top click the **Analysis type** drop-list arrow, from the list select **Slopes**, review the range number (it should be **8**), and click the **Run Analysis** button (the down arrow to the right of Range number) to produce the slope range details.

5. Change the Analysis type to **Slope Arrows**, review the range number (it should be **8**), and click the **Run Analysis** button (the down arrow to the right of Range number) to produce the arrow range details.

6. Click *OK* to view the surface with colorized triangles and arrows.

All of the triangles are red, except for those around the wall. This type of analysis style is not much use with skewed data.

7. Repeat Steps 2–6 and assign and process the **Slope and Arrows – Standard Deviation** for both Slopes and Slope Arrows. Before viewing the new triangles and arrows, review the range details to see if they better represent the slopes on the surface.

As with the Slopes and Arrows – Equal Interval style, the Slopes and Arrows – Standard Deviation style shows that all but a few of the triangles are red (the first range group). Again, this type of style does not work well with skewed data.

8. Repeat Steps 2–6 and assign and process the **Slopes and Arrows – Quantile** for both Slopes and Slope Arrows. Before viewing the new triangles and arrows, review the range details to see if they better represent the slopes on the surface.

This surface style handles the skewed data better than the Equal Interval or Standard Deviation.

The range details for this style calculate range breaks that fall between integer slope numbers (i.e., 2, 4, etc.). You can edit the range breaks after processing the analysis and adjust each range to a specific minimum and maximum value. The style recalculates the range membership using the new range values.

Modifying Slope Range Values

1. In the drawing, click the surface border or triangle, press the right mouse button, and from the shortcut menu, select SURFACE PROPERTIES....

2. If necessary, click the *Analysis* tab.

3. Set the analysis to **Slopes** and in Range Details change the range values to those listed in Table 4.5. Start editing the ranges at the eighth range and work up to the first.

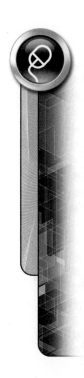

TABLE 4.5

Range ID:	Minimum Slope:	Maximum Slope:
1	0.0	2.0
2	2.0	4.0
3	4.0	6.0
4	6.0	8.0
5	8.0	10.0
6	10.0	100.0
7	100.00	210.00
8	210.00	351,000

4. After resetting the slope ranges, click the **Apply** button to recalculate range membership.

5. Change the analysis to **Slope Arrows** and in the Range Details change the range values to those set in Step 3. Start editing the ranges at the eighth range and work up to the first.

6. After resetting slope arrows for the ranges, click the **Apply** button to recalculate the range membership.

7. Click **OK** to exit the dialog box and display the range grouping.

Create a Slope Range Legend

1. On the Ribbon click the Annotate tab. In the Labels & Tables panel, click the Add table drop-list arrow and from the shortcut menu, select, ADD SURFACE LEGEND TABLE. In the command line for the legend type enter 'S' for slopes, press ENTER, for a dynamic table type 'D', press ENTER, and select a point to the surface's right to create a legend.

2. Use the Zoom and Pan commands to review the table.

3. Use the ERASE command to erase the slope legend from the drawing.

4. Use the **ZOOM EXTENTS** command to view the entire site.

Analyzing Surface Elevations

1. In Prospector, Surfaces branch, select the **Existing** surface, press the right mouse button, and from the shortcut menu, select SURFACE PROPERTIES....

2. Click the **Information** tab and change the Surface Style to **Elevations – Equal Interval**.

3. Click the **Analysis** tab, change the Analysis type to **Elevations**, check the range number (it should be **8**), and click the **Run Analysis** button (the down arrow to the right of Range number) to produce elevation range details (see Figure 4.37).

4. Click **OK** to view the surface with colorized triangles.

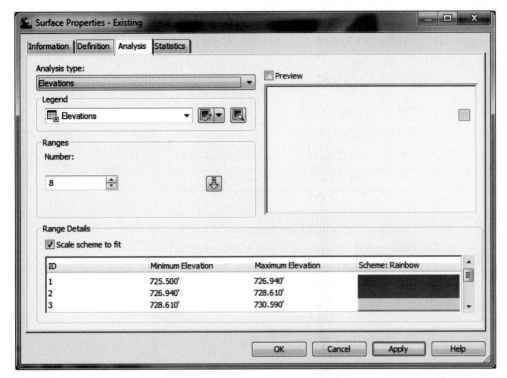

FIGURE 4.37

The Elevations – Equal Interval surface style correctly displays the surface elevations.

    5. Repeat Steps 1–4 assigning and processing Elevations – Quantile.

Before viewing the display, review the range details to see if they better represent the surface elevations.

The surface elevation map is slightly different from the previous one, but it still displays a good surface elevation map.

    6. Repeat Steps 1–4 assigning and processing Elevations – Standard Deviation.

    7. At Civil 3D's upper left, Quick Access Toolbar, click the **Save** icon to save the drawing.

Before viewing the display, review the range details to see if they better represent the surface elevations.

The surface elevation map is slightly different from the previous one, but it still displays a good surface elevation map.

All three methods correctly represent the elevation information, but in slightly different range values. It would be better if the ranges were at specific whole number elevations (i.e., 725, 728, etc.).

Modifying Elevation Range Values

    1. In Existing's Surface Properties dialog box reassign and reprocess the **Elevation – Equal Interval** surface style.

    2. Edit the minimum and/or maximum range values from the last to first elevation range using Table 4.6 as a guide.

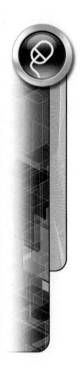

TABLE 4.6

Range ID:	Minimum Elevation:	Maximum Elevation:
1	725.5	728.0
2	728.0	730.0
3	730.0	732.0
4	732.0	734.0
5	734.0	736.0
6	736.0	739.5

3. Click **Apply** to recalculate the ranges.

4. Click **OK** to exit and display a new elevation map.

5. At Civil 3D's upper left, Quick Access Toolbar, click the **Save** icon to save the drawing.

All of these styles emphasize and create a 2D elevation map.

Create a 3D Elevation Surface Style

This surface style creates a 3D elevation view with user-defined elevation ranges.

1. Click the **Setting** tab.

2. If necessary, expand the **Surface** branch until you view the Surface Styles list.

3. From the list select **Elevations – Equal Interval**, press the right mouse button, and from the shortcut menu, select COPY....

4. Click the **Information** tab and change the style name to **Elevations – Equal Interval – 3D**.

5. Click the **Analysis** tab and expand the **Elevations** section.

6. In the **Analysis** tab's Elevations section, change the following values (see Figure 4.38):

Scheme: Rainbow

Number of Ranges: 6

Display Type: 3D Faces

Elevations Display Mode: Exaggerate Elevation

Exaggerate Elevation by Scale Factor: 3

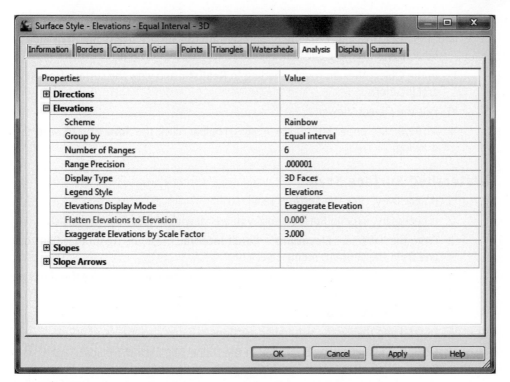

FIGURE 4.38

7. Click the *Display* tab, at the top left change the View Direction: to **Model**, toggle **OFF** all of the components, and toggle **ON** only **Elevations**.

8. Click *OK* to exit the dialog box.

9. In the drawing, select anything representing the surface, press the right mouse button, and from the shortcut menu, select SURFACE PROPERTIES....

10. Click the *Information* tab and change the Surface style to **Elevations – Equal Interval – 3D**.

11. Click the *Analysis* tab, set the Analysis type to **Elevations**, make sure the range number is **5**, click the **Run Analysis** button (the down arrow to the right of the Range number), and in the Range Details area review the elevation range values.

12. Edit the range minimums and maximums to the values in Table 4.7. Start the editing at the highest range and work down to the lowest range.

TABLE 4.7

Range ID:	Minimum Elevation:	Maximum Elevation:
1	725.5	728.0
2	728.0	730.0
3	730.0	732.0
4	732.0	734.0
5	734.0	736.0
6	736.0	739.5

13. Click **OK** to exit the dialog box.

14. In the drawing select anything representing the surface, press the right mouse button, and from the shortcut menu, select OBJECT VIEWER....

15. In the Object Viewer, toggle shading to **Realistic** and view the surface from several locations.

16. Exit the Object Viewer.

17. At Civil 3D's upper left, Quick Access Toolbar, click the **Save** icon to save the drawing.

Watershed Surface Style

1. If necessary, click the **Settings** tab.

2. If necessary, expand **Surface** until you are viewing the Surface Styles list.

3. From the list select **Elevations – Equal Interval**, press the right mouse button, and from the shortcut menu, select COPY....

4. Click the **Information** tab, rename the style **Watersheds**, and for the description enter **Watershed**.

5. Click the **Watersheds** tab.

6. Expand and review the watershed sections.

7. Expand the **Surface** section, click in the Value cell, click the ellipsis, and in the Surface Watershed Label Styles dialog box, click the drop-list arrow, and select **Watershed**. Click **OK** to exit the dialog box and return to the Surface Style dialog box.

8. Click the **Display** tab, scroll to the list's bottom and only toggle **ON Watersheds** and turn off the remaining components.

9. Click **OK** to create the style and exit the dialog box.

Watershed Analysis

1. In the drawing select anything representing the surface, press the right mouse button, and from the shortcut menu, select SURFACE PROPERTIES....

2. Click the **Information** tab and set the Surface Style to **Watersheds**.

3. Click the **Analysis** tab and adjust your settings to those in Figure 4.39.

4. Click the **Run Analysis** button (the down arrow to the right of Minimum Average Depth number).

5. Click **OK** to exit the dialog box and view the watershed analysis.

6. At Civil 3D's upper left, Quick Access Toolbar, click the **Save** icon to save the drawing.

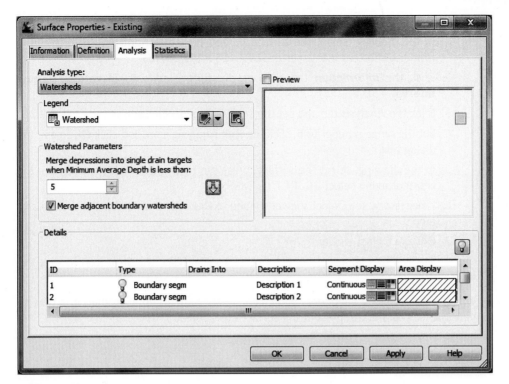

FIGURE 4.39

7. Click the **Prospector** tab.

8. In Prospector, Surfaces branch, expand **Existing** until you are viewing Masks and Watersheds.

9. Click the **Watersheds** heading and preview the lists of the Existing surface's watersheds.

10. From Preview, select a watershed, press the right mouse button, and from the shortcut menu, select ZOOM TO.

11. Select a couple more watersheds and use the ZOOM TO command to view their drawing location.

Water Drop Analysis

Water Drop draws water's path across a surface. This path may be a critical review of existing and design conditions.

1. Use the ZOOM EXTENTS command to view the entire site.

2. On the Ribbon, click the Analyze tab. In the Ground Data panel, click **Water Drop**.

3. In the Water Drop dialog box, click **OK** to accept the values and exit the dialog box.

4. Select points in different watersheds and view their water trails over the surface. When done, press ENTER to exit the command.

5. Open **Layer Properties Manager**, freeze the layer **C-TOPO-WDRP**, and click **X** to exit the panel.

6. At Civil 3D's upper left, Quick Access Toolbar, click the **Save** icon to save the drawing.

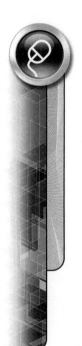

Surface Elevation – 3D

1. In the drawing, select the surface border, press the right mouse button, and from the shortcut menu, select **SURFACE PROPERTIES**....

2. Click the **Information** tab and set the surface style to **Elevation – Equal Interval – 3D**.

3. Click the **Analysis** tab and set the analysis type to **Elevations**.

4. Set the range number to **6**, click the **Run Analysis** icon, and click **OK** to exit the dialog box.

5. In the drawing, select the surface, press the right mouse button, and from the shortcut menu select OBJECT VIEWER.

6. If necessary, set Object Viewer's mode to **Realistic** and review the surface as a model.

7. Exit the Object viewer.

8. At Civil 3D's upper left, Quick Access Toolbar, click the **Save** icon to save the drawing.

If the current surface components and characteristics are acceptable, then it is time to create the contours that represent the surface elevations.

EXERCISE 4-5

After completing this exercise, you will:

- Be able to modify a contour style.
- Be able to create contour labels.
- Be able to use Natural Neighbor smoothing on a surface.
- Be able to create slope labels.
- Be able to create spot elevation labels.

This exercise continues with the previous unit's exercise drawing. If you did not complete the Unit 4 exercise, or starting with this exercise browse to the Chapter 04 folder of the CD that accompanies this textbook and open the Chapter 04 - Unit 5 drawing.

Exercise Setup

1. If not open, open the previous exercise's drawing or browse to the Chapter 04 folder of the CD that accompanies this textbook and open the *Chapter 04 – Unit 5* drawing.

Contour Surface Style

1. Click the **Settings** tab.

2. In Settings, expand the **Surface** branch until you view the Surface Styles list.

3. Select the style **Contours 1 ' and 5 ' (Background)**, press the right mouse button, and from the shortcut menu, select EDIT....

4. Click the **Contours** tab and review the settings for section 3D Geometry, Contour Display Mode.

5. Expand the **Contour Intervals** section and review its values.

6. Expand the **Contour Depressions** section and toggle Display Depression Contours to **True**.

7. Click **OK** to exit the dialog box.

8. In Settings, expand the **Surface** branch until you view a list of Contour Label styles.

9. Select the **Existing Major Labels** style, press the right mouse button, and select EDIT... to display the Label Style Composer.

10. Click the *General* tab.

The *General* tab sets the label's visibility, layer, insertion orientation, and plan readability.

11. Click the *Dragged State* tab to review its settings.

The Dragged State settings control how a label reacts if it is dragged from its original location.

12. Click the *Layout* tab to review its settings.

The Layout settings define and format the label. The panel's General section defines how the label attaches to a contour, the Text section defines the text's format and its location relative to the contour, and the Border section, if used, defines a box surrounding the label.

13. In the **Text** section, click in the **Contents** value cell, then click the cell's ellipsis to display the Text Component Editor – Contents dialog box (see Figure 4.40).

To change a modifier's value, on the dialog box's right side highlight the format string, on the left side click once in a Value cell, and from the list select the new value. To transfer the new value from the left to the highlighted format on the right, click the arrow at the top center of the dialog box.

14. Click *Cancel* until you have exited the dialog boxes and returned to the command line.

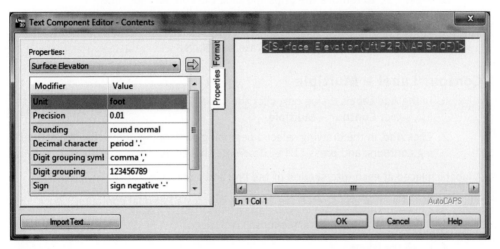

FIGURE 4.40

Creating and Labeling Contours

Create surface contours by assigning a contour surface style.

1. Click the *Prospector* tab.

2. In Prospector, expand the **Surfaces** branch until you are viewing the surface list, from the list select **Existing**, press the right mouse button, and from the short-cut menu, select SURFACE PROPERTIES....

3. Click the *Information* tab, change the Surface Style to **Contours 1 ' and 5 ' (Background)**, and click *OK* to exit the dialog box.

Assigning this style makes the surface display elevations as contours.

4. In the Ribbon, select the **Annotate** tab. In the Labels & Tables panel's left, click the Add Labels drop-list arrow, select Surface, and in its flyout, select ADD SURFACE LABELS to display the Add Labels dialog box.

5. Click the **Label Type's** drop-list arrow and from the list, select **Contour - Single**.

The dialog box changes to show entries for Major, Minor, and User contour label styles. These styles are from the Edit Feature Settings values.

6. Click **Add** and in the drawing select a few contours where there is to be a label. When you are done, press ENTER to exit the routine.

Each contour type (major or minor) receives the correct label type.

7. Use the ZOOM command to better view a label.

8. Click a contour label to view its grips.

A label has a text grip and two line grips.

9. Select one of the grips and drag the line over another contour until you are viewing a new contour label.

A single contour label can become a multiple contour label.

10. Press ESC to remove the grips.

11. Select a label, right mouse click, and from the shortcut menu, select PROPERTIES....

12. In the Properties palette, Labels section, if necessary, set Display Contour Label Line to **false**, and click **X** to close the palette.

13. Press ESC to remove the grips and hide the contour label line.

Selecting a label displays the hidden label line.

14. Use the ERASE command and erase the labels.

Contour Label – Multiple

1. In the Add Labels dialog box, click the **Label Type** drop-list arrow, and from the list, select **Contour – Multiple**.

2. Click **Add**, in the drawing select a beginning and ending point of a line intersecting contours, and press ENTER to create the labels.

A label is placed at each intersection of the line and a contour.

3. Select a label and stretch the line so it crosses additional contours.

New labels appear at the line intersections.

4. Erase the labels.

Contour Label – Multiple at Interval

1. In the Add Labels dialog box, click the **Label Type** drop-list arrow, and from the list, select CONTOUR – MULTIPLE AT INTERVAL.

2. Click **Add**; in the drawing select a beginning and ending point of a line intersecting contours, in the command line for the interval enter **300**, and press ENTER to create the labels.

3. Use ZOOM command to view the new labels.

4. Activate the grips of some labels and relocate them in the drawing.

5. Close the **Add Labels** dialog box.

6. Use the UNDO command and remove the labels from the drawing.

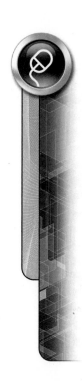

7. At Civil 3D's top left, Quick Access Toolbar, click the **SAVE** icon to save the drawing.

Surface Smoothing

The surface has several chevron contours. The reason for these contours is the lack of data. The problem with contour smoothing or splining is the potential for creating crossing contours. Edits' Surface Smoothing creates interpolated data, which results in smoother contours.

1. In the drawing, select a contour, press the right mouse button, from the shortcut menu select SURFACE PROPERTIES..., and click the **Information** tab.

2. In the **Information** tab, set the current surface style to **Border Triangles Points** and click **OK** to exit the dialog box.

3. On the Ribbon, click the View tab. In the Views panel's left, select the named view **Central** to set the current view.

4. If necessary, click the **Prospector** tab.

5. In Prospector, expand the **Existing** surface until you are viewing its Definition list.

6. Select **Edits**, press the right mouse button, and from the shortcut menu, select SMOOTH SURFACE....

7. Set the smoothing method to **Natural Neighbor Interpolation**, Output Locations to **Grid based**, and set an X and Y spacing of **5**.

8. Click in the **Select Output Region's** value cell; at the cell's right side click its ellipsis, in the command line for the rectangle option enter '**E**', and select a region covering the surface's central area. After selecting a region, you return to the Smooth Surface dialog box.

9. Click **OK** to exit the Smooth Surface dialog box.

The command's effect creates new grid triangulation in the selected region. The interpolation smoothes the surface contours (see Figure 4.41).

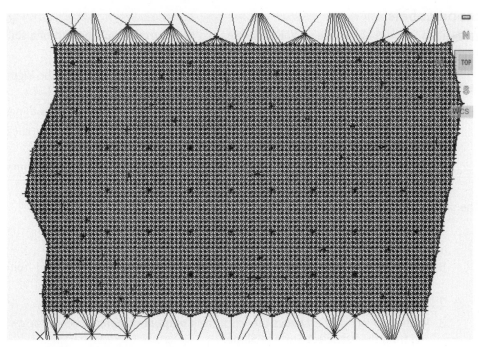

FIGURE 4.41

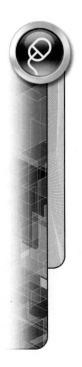

10. In the drawing, select something representing the surface, press the right mouse button, and from the shortcut menu, select SURFACE PROPERTIES....

11. In the **Information** tab, set the current surface style to **Contours 1 ' and 5 ' (Background)** and click **OK** to exit the dialog box.

12. At Civil 3D's top left, Quick Access Toolbar, click the **Save** icon to save the drawing.

The contours are considerably smoother than the original ones (see Figure 4.42).

FIGURE 4.42

Slope Labels

Add Labels includes Slope labels. While triangles and slope arrows could be part of a surface's documentation, there is a need to annotate critical surface slopes.

1. In the drawing, select something representing the surface, press the right mouse button, and from the shortcut menu, select SURFACE PROPERTIES....

2. Click the **Definition** tab and in **Operations Type** toggle **OFF** surface smoothing.

3. Click **OK** to exit the dialog box and in the Rebuild Surface dialog box click **Rebuild the Surface**.

4. If a Panorama displays a warning about duplicate points, close the **Panorama Event Viewer**.

The surface returns to the previous state.

5. Click the **Settings** tab and expand the **Surface** branch until you are viewing the list of Slope label styles.

6. From the list of styles select **Percent**, press the right mouse button, and from the shortcut menu, select EDIT....

7. If necessary, click the **Layout** tab.

8. At the top left of the **Layout** tab, click the **Component name** drop-list arrow, and from the list select **Surface Slope**.

9. In the Text section, click in the **Contents** value cell (an ellipsis appears) and click the ellipsis to display the Text Component Editor. Review the format values for the component.

The dialog box's Properties portion contains the component's format (see Figure 4.43).

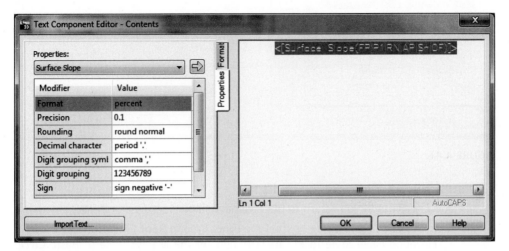

FIGURE 4.43

10. Click **OK** until you return to the command line.
11. In the Ribbon, select the **Annotate** tab. At the Labels & Tables panel's left, click the Add Labels drop-list arrow, select Surface, and in its flyout select ADD SURFACE LABELS to display the Add Labels dialog box.
12. If necessary, with Feature set to **Surface**, change the Label Type to **Slope**, and the Slope Label Style to **Percent**.
13. Click **Add**.
14. If prompted to select a surface, select the surface border or contour identifying the surface.
15. In the command line make sure the slope method is one-point, press ENTER, and in the drawing select a few points, labeling their slope.

The labels annotate the percentage and direction of the down slope by orientating the text and arrow toward the down slope direction.

16. UNDO until removing the labels.

Spot Elevation Labels

1. If necessary, click the **Settings** tab, and expand the **Surface** branch until you are viewing the Spot Elevation label styles list.
2. Select **EL:100.00**, press the right mouse button, and from the shortcut menu, select **EDIT**....
3. In the **Layout** tab's Text section, click in the **Contents** value cell displaying an ellipsis, and click the ellipsis displaying the Text Component Editor.
4. Highlight the component format on the right to review its settings on the left (see Figure 4.44).

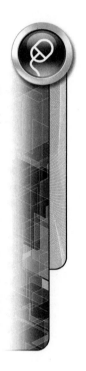

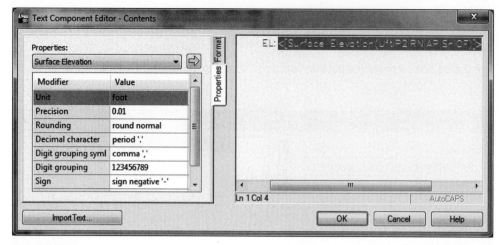

FIGURE 4.44

5. Click **OK** until you return to the command line.

6. In the Add Labels dialog box, set the Label Type to **Spot Elevation**, set the Spot Elevation Label Style to **EL: 100.00**, and set the Marker Style to **Basic Circle with Cross**.

7. Click **Add**.

8. If prompted for a surface, select the border or a contour representing the surface.

9. In the drawing, select four or five points to create spot elevation labels.

10. When you are done, press ENTER.

11. In the drawing, select one of the spot elevation labels to display its grips.

The square grip relocates the label.

12. In the drawing, select a square grip and relocate the label.

The diamond grip relocates the elevation point and updates the label's elevation.

13. In the drawing, select a diamond grip and relocate the label's location.

The circular grip of a dragged label resets it to its original location.

14. Click a dragged label's circular grip to reset it to its original label layout.

15. Erase the labeling from the drawing.

16. At Civil 3D's top left, Quick Access Toolbar, click the **Save** icon to save the drawing.

Creating an Expression

An expression uses an object property. The following exercise section creates a spot elevation label using the surface's elevation and adding 0.5 feet.

1. If necessary, click the **Settings** tab.

2. If necessary, in Settings, expand the **Surface** branch until you are viewing the Label Styles, Spot Elevations styles list.

3. From the list, select **Expressions**, press the right mouse button, and from the shortcut menu, select NEW....

4. For the Expression Name, enter **TFOC** and for its description enter **Top of Face-of-Curb** (see Figure 4.45).

5. Click the ***Insert Property*** button, (first large button on the right) to display the expression properties list and select **Surface Elevation** from the list.

This adds surface elevation to the expression.

6. On the calculator click the plus sign to add it to the expression.
7. Click the **0** (zero), a **point (.)**, and **5** to create a 0.5 entry after the plus sign.
8. Click **OK** to exit the dialog box.

The label type's Expression entry has an icon showing that it has content.

9. Click the **Expression** heading for Spot Elevation to list TFOC in preview.

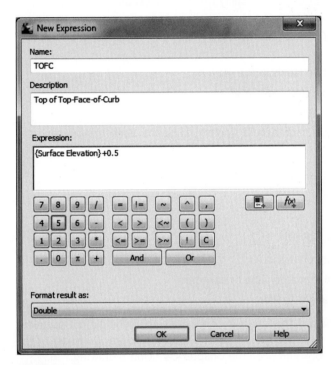

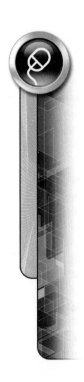

FIGURE 4.45

Adding an Expression to a Label

The expression is now a spot label surface property available as a label component (see Figure 4.38).

1. From the list of Spot Elevation label styles, select **EL:100.00**, press the right mouse button, and from the shortcut menu, select COPY....
2. Click the ***Information*** tab and change the style name to **Gutter and TFOC**. For the description enter **Gutter and Top of Face-of-Curb**.
3. Select the ***Layout*** tab.
4. In the Text section, click in the **Contents** value cell and then click its ellipsis to display the Text Component Editor.
5. On the right side of the dialog box click the text **EL:** and change it to **Gutter:**.
6. Place the cursor after the Gutter's format string (after the ›, greater than sign) and press ENTER to make a new line.
7. For text enter **F-TOC:**, leaving a space after the colon.

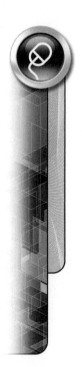

8. On the left side, click the **Properties** drop-list arrow and from the list select **TFOC**.

9. On the left side, change the modifier Precision by clicking in the **Value** cell, and selecting **0.01** from the list.

10. At the top center click the arrow to create a format string for TFOC.

11. Click **OK** until you exit and return to the command line.

12. At Civil 3D's top left, Quick Access Toolbar, click the **Save** icon to save the drawing.

13. If necessary, in the Ribbon, select the **Annotate** tab. in the Labels & Tables panel's left, click the Add Labels drop-list arrow, select Surface, and in its flyout select ADD SURFACE LABELS to display the Add Labels dialog box.

14. In Add Labels, change Label type to **Spot Elevation**, change Spot Elevation label style to **Gutter and TFOC**, and, if necessary, change the Marker style to **Basic Circle with Cross**.

15. Click **Add,** and place two labels in the drawing.

16. In the Add Label dialog box, click **Close**.

17. Erase the labels from the drawing.

18. At Civil 3D's top left, Quick Access Toolbar, click the **Save** icon to save the drawing.

EXERCISE 4-6

After completing this exercise, you will:

- Be able to place points on a grid.
- Be able to place points on a polyline or contour.
- Be able to export a LandXML file.

This exercise continues with the previous unit's exercise drawing. If you did not complete the Unit 5 exercise, or starting with this exercise browse to the Chapter 04 folder of the CD that accompanies this textbook and open the Chapter 04 - Unit 6 drawing.

Exercise Setup

1. If not open, open the previous exercise's drawing or browse to the Chapter 04 folder of the CD that accompanies this textbook and open the *Chapter 04 – Unit 6* drawing.

Point Creation and Point Identity Settings

1. If necessary, click the **Prospector** tab.

2. In Prospector, select the Points heading, right mouse click, and from the shortcut menu select CREATE....

3. On the Create Points toolbar's right side, click the **Expand the Create Points dialog** chevron.

4. Expand the **Points Creation** section and set the following values:

Prompt For Descriptions: **Automatic**

Default Description: **EXISTING**

5. Expand the **Point Identity** section, set the Next Point Number to **400**, and press ENTER.

6. On the Create Points toolbar's right side, click the **Collapse the Create Points dialog** chevron.

7. In Layer Properties Manager, thaw the **V-NODE** layer, and click **X** to exit the palette.

8. If necessary, click the *Prospector* tab.

9. Select the **Point Groups** heading, press the right mouse button, and from the shortcut menu, select PROPERTIES....

10. Move **_All Points** to the list's top, and click *OK* to exit the dialog box.

If Civil 3D issues a warning about duplicate points in the surface, close the Panorama's Event Viewer vista.

Points on Grid

Points on Grid creates points on a user-defined grid and assigns surface elevations as their elevations.

1. In the Create Points toolbar, from the Surface's icon stack, click the drop-list arrow and from the list select **ON GRID** (see Figure 4.46).

2. The routine prompts you to select a surface object; select a surface component to identify it.

3. Next, the routine prompts you for a grid base point. Select a point at the surface's lower left.

4. The command line prompts you for a rotation angle; press ENTER to accept zero.

5. The command line prompts you for grid spacings; for grid X and Y spacing enter **30**.

6. Finally, in the drawing select a point and upper-right grid limit within the surface's interior.

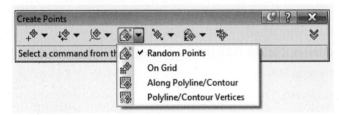

FIGURE 4.46

A box is displayed to represent the grid area. Within the grid area is another box showing the grid spacing (lower-left corner).

7. The command line asks if you want to change the grid definition. Answer **NO** by pressing ENTER to continue to the next prompt.

The routine creates points using the automatic description (EXISTING).

8. The command line prompts you to select another surface object. Press ENTER to exit the routine and return to the command line.

9. In Prospector, click the **Points** heading to preview the new points.

The new points in preview start with point number **400**.

10. In Prospector, select the **Point Groups** heading, press the right mouse button, and from the shortcut menu, select NEW....

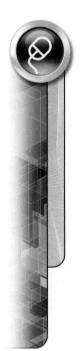

11. In the Point Group Properties dialog box, *Information* tab, for the point group name enter **Points From Existing**.

12. Click the *Include* tab, toggle **ON With Numbers Matching**, and enter in the point numbers from **400–800** (or whatever ending point number is appropriate for the points in your drawing).

13. Click **OK** to exit the Point Group Properties dialog box.

Points on a Polyline/Contour

1. On the Ribbon, select the View tab. At the Views panel's left, select the named view **Central** to restore the view.

2. Use the ZOOM WINDOW command and zoom in on the surface.

3. On the Create Points toolbar's right side, click the **Expand the Create Points dialog** chevron.

4. In the **Points Creation** section, set the Default Description value to **CONTOUR**.

5. On the Create Points toolbar's right side, click the **Collapse the Create Points dialog** chevron.

6. To the right of the Surface icon stack, click the drop-list arrow and select ALONG POLYLINE/CONTOUR from the list.

7. In the drawing, select something representing the surface, press ENTER to accept the default distance (**10**), and select a contour.

8. Press ENTER to exit the routine.

9. In Prospector, select the **Point Groups** heading, press the right mouse button, and from the shortcut menu, select NEW....

10. In the *Information* tab, assign the point group the name **From Contour**.

11. Click the *Include* tab, toggle **ON With Raw Description Matching**, enter **CONTOUR**, and click **OK** to create the point group From Contour.

12. In Prospector, select the heading **Point Groups**, press the right mouse button, and from the shortcut menu, select PROPERTIES....

13. Select **No Show** and move it to the list's second position.

14. Click **OK** to exit the dialog box.

Only the contour points are displayed.

15. You may need to expand the Point Groups heading to select any out of date groups, right click and select Update.

16. Select the heading **Point Groups**, press the right mouse button, and from the shortcut menu, select PROPERTIES....

17. Select the point group **No Show** and move it to the list's top.

18. Click **OK** to exit the Point Group Properties dialog box.

No points display.

19. Click the Create Points toolbar's red **X** to close it.

20. In Prospector, Point Groups select the **From Contour** point group, press the right mouse button, and from the shortcut menu, select DELETE POINTS.... Click **OK** to answer yes to the Are you sure dialog box.

21. In Prospector, select the point group **From Contour**, press the right mouse button, and from the shortcut menu, select DELETE.... Click **Yes** to the Are you sure dialog box.

22. In Prospector, select the point group **Points From Existing**, press the right mouse button, and from the shortcut menu, select DELETE POINTS.... Click **OK** to answer Yes to the Are you sure dialog box.

23. In Prospector, select the point group **Points From Existing**, press the right mouse button, and from the shortcut menu, select DELETE.... Click **Yes** to the Are you sure dialog box.

24. At Civil 3D's top left, Quick Access Toolbar, click the **Save** icon and save the drawing.

LandXML Export Settings

1. Click the **Settings** tab.

2. At the top select the drawing name, press the right mouse button, and from the shortcut menu, select EDIT LANDXML SETTINGS....

3. Click the **Export** tab, expand the **Data Settings** section, and change the Imperial Units value to **survey foot**.

4. Expand the **Surface Export Settings** section, set Surface Data to **points and faces**, and Watersheds to **off**.

5. Click **OK** to exit the dialog box.

Exporting a Surface and Point Groups

1. Click the **Prospector** tab.

2. In Prospector, from the Surfaces branch, select the surface **Existing**, press the right mouse button, and from the shortcut menu, select EXPORT LANDXML....

3. In the Export to LandXML dialog box, toggle **ON** the surface **Existing** and point group **Breakline Points** to include them in the LandXML file.

4. Click **OK** to continue the exporting process.

5. In the Export LandXML dialog box, browse to the folder **Civil 3D Projects**, name the file, and click **Save** to create the file.

This completes the Civil 3D surfaces chapter. Chapter 10 focuses on designing a second surface and calculating its earthwork volumes (existing ground and a design surface). This chapter focused on surface design tools and the calculating earthwork volumes process.

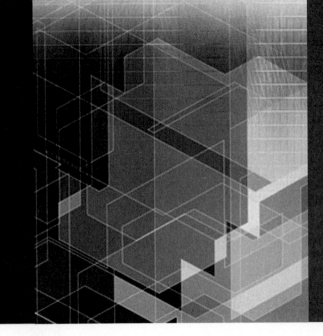

Alignments

After completing this exercise, you will:

- Be familiar with alignment settings.
- Be familiar with alignment Edit Features Settings.
- Be familiar with alignment object styles.
- Be familiar with the Major & Minor label set.
- Be familiar with design check expressions and sets.

Drawing Setup

This exercise starts with the new drawing, using the *Chapter 05 – Unit 1.dwt* file, found in the Chapter 05 folder of the CD that accompanies this textbook.

1. If not in Civil 3D, double-click its desktop icon to start the application.

2. Close the drawing and do not save it.

3. At Civil 3D's top left, click Civil 3D's drop-list arrow and from the Application Menu, select NEW. In the Select template dialog box, browse to the Chapter 05 folder of the CD that accompanies this textbook and open the *Chapter 05 – Unit 1.dwt* file. Click **Open** to use the template file.

4. At Civil 3D's top left, click Civil 3D's drop-list arrow and from the Application Menu, highlight Save As, from the flyout menu select AutoCAD Drawing, browse to the Civil 3D Projects folder, for the drawing name enter **Alignments**, and click **Save** to save the file.

Edit Drawing Settings

Settings' Edit Drawing Settings values affect the alignment object and its labeling. The settings include layer-naming conventions and abbreviations for alignment geometry points.

1. Click the **Settings** tab.
2. In Settings, at its top, select the drawing name, press the right mouse button, and from the shortcut menu, select EDIT DRAWING SETTINGS....
3. Click the **Object Layers** tab.
4. In Object Layers, set the Modifier for Alignment, Alignment-Labeling, and Alignment Table to **Suffix**, and for the modifier value enter a dash followed by an asterisk (**-***) (see Figure 5.1).
5. Click the **Abbreviations** tab and review its settings for alignment points.
6. Click **OK** to save the changes and close Edit Drawing Settings.

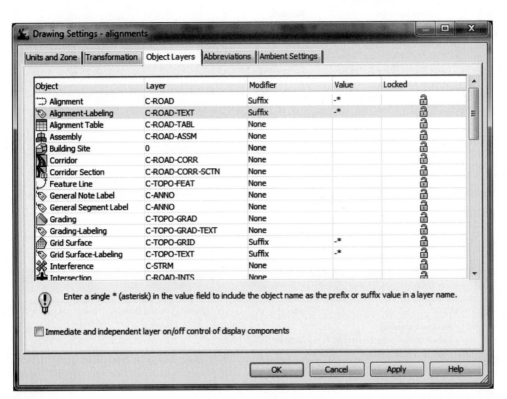

FIGURE 5.1

Edit Feature Settings — Alignment

Alignment's Edit Feature Settings assign several initial values and styles for creating and labeling alignments (see Figures 5.2, 5.3, and 5.4).

1. In Settings, select the Alignment heading, press the right mouse button, and from the shortcut menu, select EDIT FEATURE SETTINGS....
2. Expand the Default Styles, Default Name Format, Station Indexing, Criteria-Based Design Options, and Dynamic Alignment Highlight Options sections to review their values.
3. In the Criteria-Based section change the speed to **25** and click **Apply**.
4. Click **OK** to close the dialog box.

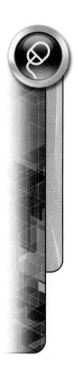

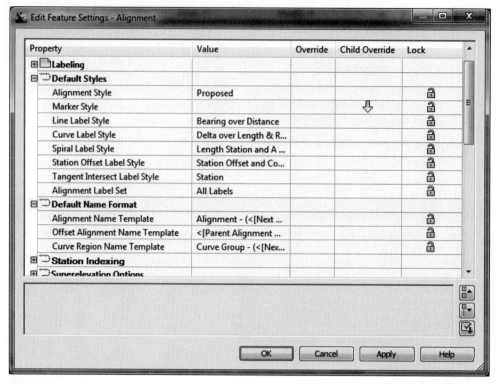

FIGURE 5.2

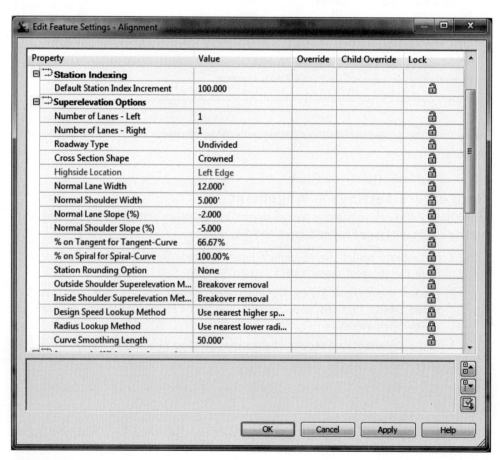

FIGURE 5.3

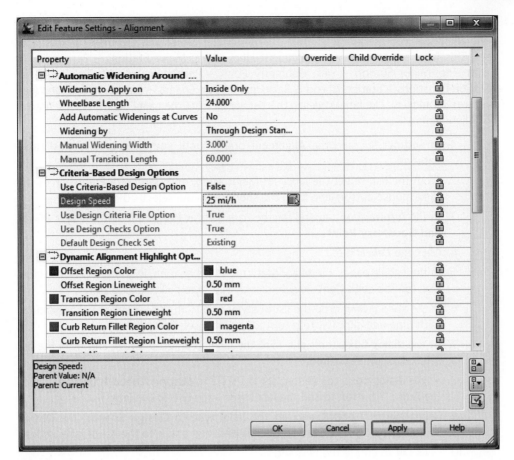

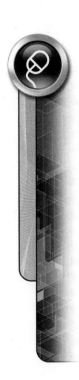

FIGURE 5.4

Alignment Object Styles

This and the following exercises use two alignment object styles: Layout and Existing. The Layout style applies different colors to each alignment segment type, and the Existing style assigns the same color to all alignment segments.

1. In Settings, expand the Alignment branch until you are viewing the Alignment Styles list.

2. From the list, select the style **Layout**, press the right mouse button, and from the shortcut menu, select EDIT....

3. Click the **Design** tab. Toggle **ON** and change the Grip edit behavior value to **5.0**.

4. Click the **Markers** tab to review its settings and change the arrowhead size to **0.2**.

5. Click the **Display** tab to review the components settings for an existing alignment. Each component may have a different color assignment.

6. Click **OK** to exit the style.

7. From the list, select the style **Existing**, press the right mouse button, and from the shortcut menu, select EDIT....

8. Click the **Design** tab to review its settings.

9. Click the **Markers** tab to review its settings.

10. Click the **Display** tab to review the components settings for a proposed alignment. Each component is of the same color.

11. Click **OK** to exit the Existing style.

Alignment Label Set and Label Styles

The alignment you are going to draft uses the Major Minor and Geometry Points label set. This label set contains five alignment label styles: Major Stations – Perpendicular with Line; Minor Station – Tick; Geometry Point – Perpendicular with Tick and Line; Station Equations – Station Ahead & Back; and Design Speeds – Station over Speed (see Figures 5.5 and 5.6). The label styles in this set are all from the Label Styles' Station branch of Alignments.

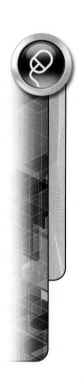

1. In Settings, expand the Alignment branch until you view Label Styles' Label Sets list.

2. From the list, select the label set Major Minor and Geometry Points, press the right mouse button, and from the shortcut menu, select COPY....

3. In the *Information* tab and for the name, enter **Perpendicular – Major Minor and Geometry Points**.

4. Click the *Labels* tab.

5. In the Labels panel, the Major Stations' Styles column, in the cell click the label icon (on the cell's right side) to display the style picker dialog box.

6. In the style picker, drop the list of styles and from the list select **Perpendicular with Line** and click OK to return to the label set dialog box.

7. At the dialog box's top right, click the label Type drop arrow, and from the list select **Station Equations**.

8. At the dialog box's top center, set the style to **Station Ahead & Back**, and click the Add>> button to add the label type and style to the label list.

9. Repeat Steps 7 and 8, and set the label type to **Design Speeds**, the Design Speeds Label Style to **Station over Speed**, and add it to the list of label styles.

10. Click **OK** to create the new set.

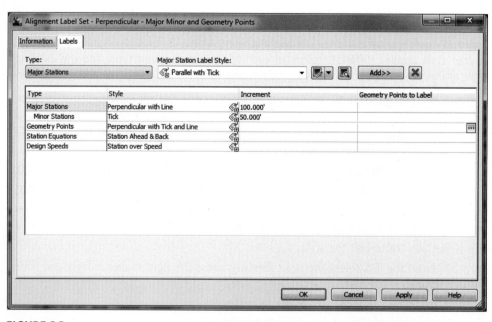

FIGURE 5.5

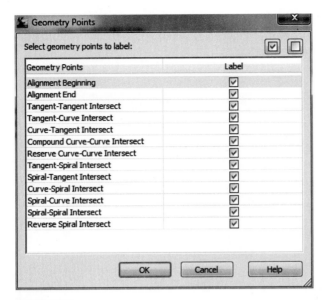

FIGURE 5.6

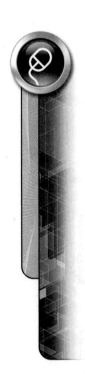

Review Label SetStyles

1. In Settings, Alignment Label Styles, expand the Station branch until you are viewing Major Station's styles list.

2. From the styles list, select **Perpendicular with Line**, press the right mouse button, and from the shortcut menu, select EDIT....

3. In the Label Style Composer dialog box, click Layout to review its contents. The text anchors on the Line component.

4. At the panel's top right, click the Component name drop-list arrow and select the Line component to review its anchor point. It is the feature (station).

5. Click **OK** to exit the dialog box.

Design Checks

Design Checks set minimum and maximum values as expressions in a check set.

1. In the Settings, Alignment branch, expand Design Checks until you are viewing the Design Check Sets list.

2. Under Design Checks select Curve, press the right mouse button, and from the shortcut menu, select NEW....

3. In the New Design Check calculator for the check name, enter **Minimum Radius**.

4. In the New Design Check calculator, in the middle right, click the icon Insert property (first button on the right), and from the list, select **Radius**. Next, click the equal to or greater than icon ($> =$) and then enter **75.00** for the minimum radius. Then review your expression and compare it to Figure 5.7. When you are done entering the values, click **OK** to create the check.

5. Under Design Checks select Line, press the right mouse button, and from the shortcut menu, select NEW....

6. In the New Design Check calculator, for the check name enter **Minimum Length**.

7. In the New Design Check calculator, in the middle right, click the icon Insert property, and from the list select **Length**. Next, click the equal to or greater

than icon (> =) and then enter **100.00** for the minimum length. Then review your expression and compare it to Figure 5.7. When you are done entering the values, click **OK** to create the check.

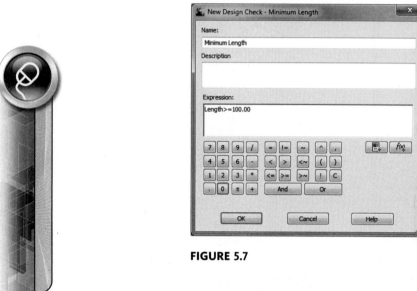
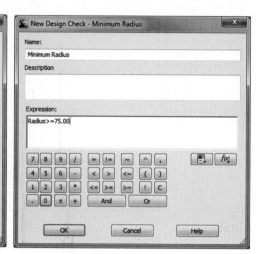

FIGURE 5.7

Create a Design Check Set

1. In the Settings, Alignment branch, select Design Check Sets, right mouse click, and from the shortcut menu, select NEW....

2. In the Alignment Design Check Set dialog box, click the **Information** tab and for the name enter **Dupage Design**.

3. Click the **Design Checks** tab.

4. In the upper left, click the Type: drop-list arrow and set the type to **Line**. Line Checks should list **Minimum Length**. Click **Add >>** to add it to the check list.

5. In the upper left, click the Type: drop-list arrow and set the type to **Curve**. Curve Checks should list **Minimum Radius**. Click **Add >>** to add it to the check list.

6. Click **OK** to exit the dialog box.

7. At Civil 3D's top left, Quick Access Toolbar, click the **SAVE** icon to save the drawing.

This completes the exercise setup and styles review for this chapter. Next, you learn to define and draft alignments.

EXERCISE 5-2

After completing this exercise, you will:

- Be able to create an alignment by importing a LandXML file.
- Be able to create an alignment from a polyline.
- Be able to create an alignment using the Alignment Layout Tools toolbar.
- Be able to use transparent commands while laying out a centerline.
- Be able to create an alignment from cogo points.

Exercise Setup

This exercise continues with the previous exercise's drawing, Alignments. If you did not complete the previous exercise, browse to the Chapter 05 folder of the CD that accompanies this textbook and open the *Chapter 05 – Unit 2.dwg*.

1. If not open, open the previous exercise's Alignments drawing, or browse to the Chapter 05 folder of the CD that accompanies this textbook, and open the *Chapter 05 – Unit 2* drawing.

2. In Layer Properties Manager, or by using the Layer Freeze icon, freeze the contour and boundary layers **3EXCONT**, **3EXCONT5**, and **Boundary**, and click **X** to exit the palette.

Import an Alignment

The LandXML file contains alignment definitions and can be imported to create alignments. The LandXML files for this section are in the Chapter 05 folder of the CD that accompanies this textbook.

1. On the Ribbon, click the ***Insert*** tab. On the Import panel, select LANDXML.

2. In Import LandXML, browse to the Chapter 05 folder of the CD that accompanies this textbook, and from the files select **Existing Road.xml**, and click OPEN.

3. In the Import LandXML dialog box, click **OK** to import the Senge Drive alignment.

Create Alignment from Objects

Create Alignment from Objects converts a polyline into an alignment.

1. On the Ribbon, click the ***View*** tab. At the left side of the Views panel, select and restore the named view **Existing Road**.

2. Click Ribbon's ***Home*** tab, Create Design Panel, click the drop-list arrow for Alignment, from the shortcut menu select CREATE ALIGNMENT FROM OBJECTS, in the drawing near its southern end, select the red polyline (C-ROAD-CTLN), and press ENTER to continue.

3. The command line prompts to accept or reverse the alignments direction. Press ENTER to accept the direction and to continue.

4. The Create Alignment from Objects dialog box opens. Using Figure 5.8 as a guide, enter the following values: for Name, leaving the counter, enter **OMalley Phase 2**; for the type set it to **Centerline**, for Site, click the Create New icon on the right, for the name enter **Site 1**, click *OK*; for the Alignment Style assign **Existing**; for the Alignment Label Set assign **Perpendicular – Major Minor and Geometry Points**; toggle **OFF** Add curves between tangents, and toggle **ON** Erase existing entities.

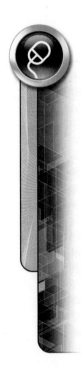

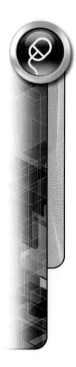

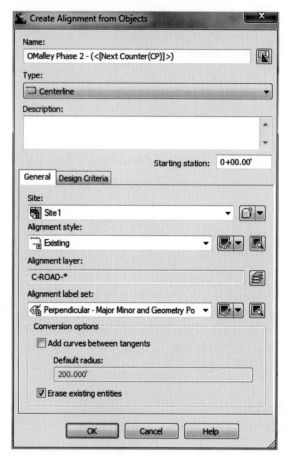

FIGURE 5.8

5. Click the **_Design Criteria_** tab.

6. Using Figure 5.9, set the following values. Set the Starting design speed to **25**, toggle **ON** Use criteria-based design, toggle **OFF** Use design criteria file, toggle **ON** Use design check set, click the check set drop-list arrow, and from the list select **Dupage Design**.

7. Click **_OK_** to exit and define the alignment.

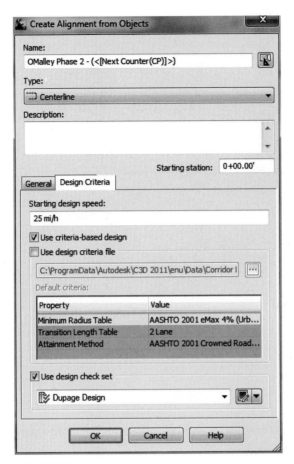

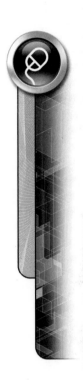

FIGURE 5.9

The alignment violates the 100 foot minimum tangent rule.

8. Place the cursor over the violation icon and a tooltip lists the violation.

The stationing should start at the alignment's northern end and increase toward the south.

9. If the stationing increases from the south to north, select the Alignment. From the Ribbon's Alignment: OMalley Phase 2 – (1) tab, select the Modify panel's title to unfold the panel and select Reverse Direction, and in the warning dialog box click **OK** to accept the change and reverse its direction.

10. Use the ZOOM EXTENTS command to view the entire site.

11. At Civil 3D's top left, Quick Access Toolbar, click the **Save** icon to save the drawing.

Offsetting Preliminary Centerline Tangents

The engineer wants the alignment to start at two points at the site's southwestern corner (point numbers 1 and 2). The centerline should be about 185 feet in from the Phase I and II boundary north of the points. Lines on the P-CL layer represent lines to be offset to create centerline line segments (see Figure 5.10).

1. On the Ribbon, click the **View** tab. In the Views panel's left, select and restore the named view **Proposed Starting Point**.

2. Click Ribbon's **Home** tab. In the Layers panel, click ISOLATE, in the upper right of the view, select the entity on the **P-CL** layer (objects in beige) and press ENTER.

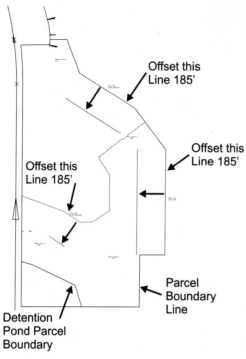

Offset this Line 185'

Offset this Line 185'

Offset this Line 185'

Parcel Boundary Line

Detention Pond Parcel Boundary

FIGURE 5.10

3. Use the OFFSET command, set the distance to **185**, offset the line in the upper right, southwest, and exit the offset command.

4. Use the ZOOM and PAN commands to view the remaining two lines.

5. Use OFFSET command and offset the eastern line, west; the northeastern line, southwest; and then exit the command.

6. Click the Ribbon's **Home** tab. In the Layers panel's right side, click unISOLATE.

Define the Site Boundary

Next, you define the boundary as a parcel (site).

1. In Ribbon's Home tab, Create Design panel's left side, click the drop-list arrow of the **Parcel** icon. From the shortcut menu select CREATE PARCEL FROM OBJECTS, select the parcel's outer boundary (see Figure 5.3), press ENTER, and click **OK** to accept the defaults in the Create Parcels – From Objects dialog box.

Define Detention Pond Parcel

1. In Ribbon's Create Design panel's left side, click the drop-list arrow of the **Parcel** icon. From the shortcut menu select CREATE PARCEL FROM OBJECTS, select the Detention Pond boundary located in the site's southwestern corner (see Figure 5.3), press ENTER, and click **OK** to accept the defaults in the Create Parcels – From Objects dialog box.

2. At Civil 3D's upper left, Quick Access Toolbar, click the **Save** icon to save the drawing.

Drafting the Rosewood Alignment

The Rosewood alignment starts at the site's lower-left (southwest) side (point number 1) and winds its way north to its connection with the OMalley centerline. Begin drafting the alignment with points 1 and 2 using transparent command point filters. In the Alignment Layout Tools toolbar other commands convert the previously offset lines to fixed tangent segments. Finally, free curves connect the converted tangent segments creating the finished alignment.

1. Make sure the Transparent Commands toolbar is visible.

2. Click the Ribbon's *View* tab. At the Views panels left, select and restore the named view **Proposed Starting Point**.

3. Click the Ribbon's Home tab. In the Create Design panel, click the *Alignment* icon, and from the shortcut menu select ALIGNMENT CREATION TOOLS.

4. In the Create Alignment – Layout dialog box, for the alignment Name and leaving the counter, enter **Rosewood**, set the alignment type to **Centerline**, create the alignment in **Site 1**, set the Alignment Style to **Layout**, and set the Alignment Label Set to **Perpendicular – Major Minor and Geometry Points**.

5. Click the *Design Criteria* tab and set the Starting design speed to **25**, toggle **ON** Use criteria-based design, toggle **OFF** Use design criteria file, toggle **ON** Use design check set, click the check set drop-list arrow, and from the list select **Dupage Design**.

6. Click *OK* to continue.

Tangent 1 – Fixed Tangent

The first segment is a fixed segment drawn with Fixed Line – Two Points and references points 1 and 2 with a transparent command point filter.

1. From the Alignment Layout Tools toolbar, click the *Fixed Line – (Two points)* icon (fifth in from the left).

2. Next, in the Transparent Commands toolbar click the point filter *Point Object* ('PO), and in the drawing select anywhere on point 1 and then anywhere on point 2.

3. Press ENTER twice to return to the command line.

A fixed alignment segment appears between the points with an arrow pointing from left to right, and it violates the minimum tangent length criteria (see Figure 5.11).

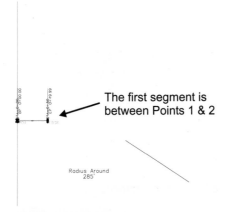

The first segment is between Points 1 & 2

FIGURE 5.11

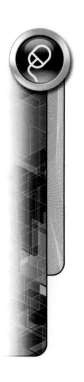

Tangent 2 – Converted Offset Line

The second tangent is the southern offset line converted to a fixed segment.

1. From the Alignment Layout Tools toolbar, click the **Convert AutoCAD line and arc** icon (to the right of the Spiral type icon stack), in the drawing select the offset line's western end, and press ENTER.

The converted segment's direction is the same and it is the first tangent.

Curve 1 – Free Curve

The first curve has two dependent segments, before and after, and is a free curve segment.

1. Use the ZOOM command to better view both entities' ends.
2. From the Alignment Layout Tools toolbar, click the Curve drop-list icon (sixth icon in from the left), and from the Curve icon stack, select FREE CURVE FILLET (BETWEEN TWO ENTITIES, RADIUS).
3. The command line prompts you to select the first entity. Select Tangent 1's eastern end, and when you are prompted for the next entity, select Tangent 2's western end.
4. The command line prompts you if the curve is Less than 180 degrees. Press ENTER to accept the value and continue.
5. The command line prompts you for the last parameter, radius. For the radius type **285** and press ENTER twice to create the curve and exit to the command line.

The Curve 1 is tangent at Tangent 1's end and slightly down from Tangent 2's western end (see Figure 5.12).

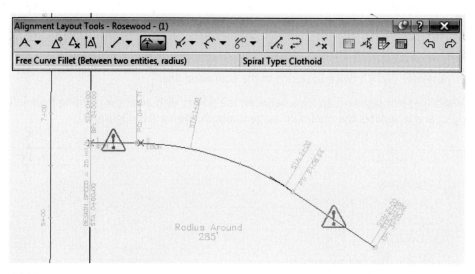

FIGURE 5.12

Tangent 3 – Converted Offset Line

The third tangent is the eastern offset line converted to a fixed segment.

1. Use the ZOOM and PAN commands to view Tangent 2 and the eastern offset line.

2. From the Alignment Layout Tools toolbar, click the **Convert AutoCAD line and arc** icon (to the right of the Spiral type icon stack), in the drawing select the line's southern end, and press Enter.

The converted segment's direction should be the same as the alignment.

3. If necessary, on the Ribbon, click the Home tab. Click Layer Control's drop-list arrow, and then scroll to and turn off the layer P-CL.

Curve 2 – Free Curve

The second curve has two dependent segments, before and after, and is a free curve segment.

1. From the Alignment Layout Tools toolbar, click the Curve drop-list icon (sixth icon in from the left), and from the Curve icon stack, select FREE CURVE FILLET (BETWEEN TWO ENTITIES, RADIUS).

2. The command line prompts you to select the first entity. Select Tangent 2's eastern end, and when prompted for the next entity, select Tangent 3's southern end.

3. The command line prompts you if the curve is Less than 180 degrees. Press ENTER to accept the value and continue.

4. The command line prompts you for the last parameter, radius. For the radius type **310** and press ENTER twice to create the curve and exit to the command line.

The Curve 2 is tangent at Tangent 2's end, up from Tangent 3's southern end (see Figure 5.13).

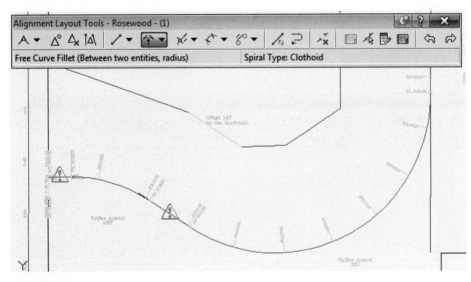

FIGURE 5.13

Tangent 4 – Converted Offset Line

The fourth tangent is the northern offset line converted to a fixed segment.

1. From the Layers panel, click the Previous icon to turn on the **P-CL** layer.

2. Use the ZOOM and PAN commands to view Tangent 3 and the northern offset line.

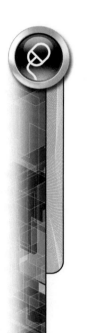

---

3. From the Alignment Layout Tools toolbar, click the **Convert AutoCAD line and arc** icon (to the right of the Spiral type icon stack), in the drawing select the line's western end, and press ENTER.

The converted segment's direction is in the opposite direction of the third tangent.

4. Click the Layer Control's drop-list arrow, and then scroll to and turn off the **P-CL** layer.

5. From the Alignment Layout Tools toolbar, click the **Reverse Sub-entity Direction** icon (the next icon to the right), in the drawing select the line's southern end, and press ENTER.

The converted segment's direction changes to be the same as the alignment.

Curve 3 – Free Curve

The third curve has two dependent segments, before and after, and is a free curve segment.

1. From the Alignment Layout Tools toolbar, click the Curve drop-list icon (sixth icon in from the left), and from the Curve icon stack, select FREE CURVE FILLET (BETWEEN TWO ENTITIES, RADIUS).

2. The command line prompts you to select the first entity. Select Tangent 3's northern end, and when prompted for the next entity, select Tangent 4's southern end.

3. The command line prompts you if the curve is Less than 180 degrees. Press ENTER to accept the value and continue.

4. The command line prompts you for the last parameter, radius. For the radius type **310** and press ENTER twice to create the curve and exit to the command line.

The Curve 3 is down from Tangent 3's northern end and slightly up from Tangent 4's southern end (see Figure 5.14).

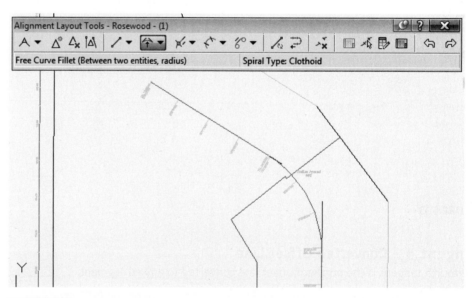

FIGURE 5.14

Tangent 5 – Converted Line

The fifth tangent is a line converted to a fixed segment.

1. Use the ZOOM and PAN commands to view Tangent 4 and the line just south of the OMalley alignment.

2. From the Alignment Layout Tools toolbar, click the **Convert AutoCAD line and arc** icon (to the right of the Spiral type icon stack), in the drawing select the line's southern end, and press ENTER.

The converted segment's direction is in the opposite direction of the third tangent and violates Minimum Length.

3. From the Alignment Layout Tools toolbar, click the **Reverse Sub-entity Direction** icon (the next icon to the right), in the drawing select the line's southern end, and press ENTER.

The converted segment's direction changes to be the same as the alignment.

Curve 4 – Free Curve

The third curve has two dependent segments, before and after, and is a free curve segment.

1. From the Alignment Layout Tools toolbar, click the Curve drop-list icon (sixth icon in from the left), and from the Curve icon stack, select FREE CURVE FILLET (BETWEEN TWO ENTITIES, RADIUS).

2. The command line prompts you to select the first entity. Select Tangent 4's northern end, and when prompted for the next entity, select Tangent 5's southern end.

3. The command line prompts you if the curve is Less than 180 degrees. Press ENTER to accept the value and continue.

4. The command line prompts you for the last parameter, radius. For the radius enter **335** and press ENTER twice to create the curve and exit to the command line.

The Curve 4 is tangent down from Tangent 5's southern end and tangent to Tangent 4 (see Figure 5.15).

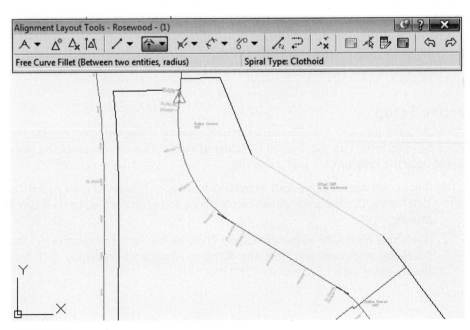

FIGURE 5.15

5. Click the red **X** on the Alignment Layout Tools toolbar to close it.

6. At Civil 3D's top left, Quick Access Toolbar, click the **Save** icon to save the drawing.

Alignment from Points

This exercise uses a different drawing. You must browse to the Chapter 05 folder of the CD that accompanies this textbook and open the *Alignment from Crown Points.dwg*.

1. Browse to the Chapter 05 folder of the CD that accompanies this textbook, and open the *Alignment from Crown Points* drawing.

2. Use the Zoom Extents command to view the entire roadway.

3. From the Home tab, the Create Design panel, click on the Alignments icon to display its command list.

4. From the command list, select Create Best Fit Alignment to display the Create Alignment dialog box.

5. At the dialog box's top, change the Input Type to COGO Points.

6. Set the Path 1 group to the point group Crown and click OK to create the alignment and to display the Best Fit Report.

7. Review the Best Fit Report and when done, dismiss it.

8. Use the Zoom and Pan commands to better view the alignment.

9. From the Menu Browser, select Save as… and save the drawing as Alignment from Points.

10. Exit the drawing.

This completes importing and drafting alignment segments. When using the Alignment Layout Tools toolbar, the resulting alignment is a combination of fixed, floating, and free segments. You can also convert objects or polylines into alignments, or import them from LandXML files or LDT projects. The next step is to review their segment values to see if they meet the minimum requirements.

EXERCISE 5-3

After completing this exercise, you will:

- Review an entire alignment's values.
- Review an alignment's sub-entity values.
- Select and view various alignments reports.

Exercise Setup

This exercise continues with the previous exercise's drawing. If you did not complete the previous exercise, browse to the Chapter 05 folder of the CD that accompanies this textbook and open the *Chapter 05 – Unit 3.dwg* file.

1. If not open, open the previous exercise's drawing, or browse to the Chapter 05 folder of the CD that accompanies this textbook and open the *Chapter 05 – Unit 3* drawing.

2. If you have closed the Alignment Layout Tools toolbar, in the drawing select the Rosewood alignment and from the Ribbon's Alignment: Rosewood – (1) tab, Modify panel, select GEOMETRY EDITOR.

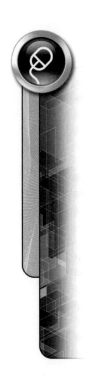

Alignment Entities Vista

The Alignment Grid View icon displays all critical alignment values in a single vista.

1. In the Alignment Layout Tools toolbar, click the **Alignment Grid View** icon (rightmost icon). Your Alignment Entities vista should look similar to Figure 5.16.

The Panorama is displayed with the Alignment Entities vista displaying all alignment values, including any criteria violations. Editable values are black.

2. Scroll horizontally through the Panorama to review the vista's values.
3. Click the Panorama's **X** to close it.

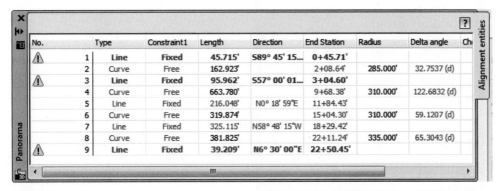

No.		Type	Constraint1	Length	Direction	End Station	Radius	Delta angle	Ch
⚠	1	**Line**	**Fixed**	**45.715'**	**S89° 45' 15...**	**0+45.71'**			
	2	Curve	Free	162.923'		2+08.64'	285.000'	32.7537 (d)	
⚠	3	**Line**	**Fixed**	**95.962'**	**S57° 00' 01...**	**3+04.60'**			
	4	Curve	Free	663.780'		9+68.38'	310.000'	122.6832 (d)	
	5	Line	Fixed	216.048'	N0° 18' 59"E	11+84.43'			
	6	Curve	Free	319.874'		15+04.30'	310.000'	59.1207 (d)	
	7	Line	Fixed	325.115'	N58° 48' 15"W	18+29.42'			
	8	Curve	Free	381.825'		22+11.24'	335.000'	65.3043 (d)	
⚠	9	**Line**	**Fixed**	**39.209'**	**N6° 30' 00"E**	**22+50.45'**			

FIGURE 5.16

Sub-Entity Editor

The Sub-entity Editor's Alignment Layout Parameters dialog box displays a selected segment's values.

1. On the Ribbon, click the **View** tab. At the Views panel's left, select and restore the named view **Proposed Starting Point**.
2. In the Alignment Layout Tools toolbar, select the **Sub-entity Editor** icon (the second icon in from the right-hand side) to display the Alignment Layout Parameters dialog box.
3. In the Alignment Layout Tools toolbar, click the **Pick Sub-entity** icon to the left of the Sub-entity Editor icon and select Curve 1. Your Editor should look similar to Figure 5.17.

The Alignment Layout Parameters dialog box displays the selected segment's values, including any criteria violations. Editable values are black. You may need to select the Show More and Expand All Categories button at the upper-right corner of the Alignment Layout Parameters dialog box to view more details.

4. Select Tangent 2 and review its values; notice its criteria violation.
5. Close the Sub-entity Editor by clicking its red **X**.
6. Close the Alignment Layout Tools toolbar by clicking its red **X**.

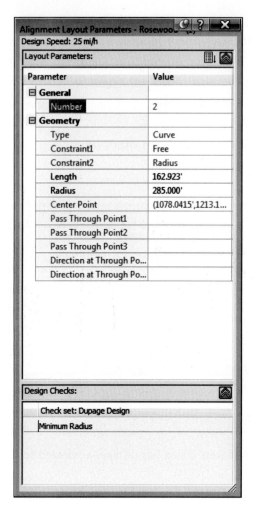

FIGURE 5.17

Toolbox Reports

Toolbox creates assorted reports for selected alignments.

1. If the **Toolbox** tab is not displayed in the Toolspace, in the Palettes panel, to the right of the Toolspace icon, click the **TOOLBOX** icon.

2. If necessary, click the **Toolbox** tab.

3. In Toolbox, expand Reports Manager and the Alignment section until you are viewing its reports.

4. From the list of reports select **Station_and_Curve**, press the right mouse button, and from the shortcut menu, select EXECUTE....

5. The Export to XML Report dialog box is displayed and lists all alignments. Click **OK** to create reports for each alignment.

You can toggle off the alignments that do not need a report.

Internet Explorer is displayed with a report for each alignment.

6. Scroll through the report until you can view Rosewood's values, and then review them.

7. Close Internet Explorer.

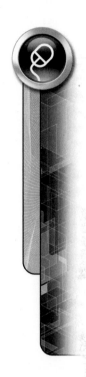

8. From the list of reports, select ALIGNMENT DESIGN CRITERIA VERIFICA-
TION, press the right mouse button, and from the shortcut menu, select
Execute....

9. The Create Reports – Alignment Design Criteria Verification Report dialog box
opens. Toggle OFF all alignments except for Rosewood – (1), and click CREATE
REPORT.

Internet Explorer displays a report marking each violation.

10. Scroll through the report to review Rosewood's values.

11. Close Internet Explorer.

12. In the Create Reports – Alignment Design Criteria Verification Report dialog
box, click **Done** to close the dialog box.

13. At Civil 3D's top left, Quick Access Toolbar, click the **Save** icon to save the
drawing.

Inquiry Tools

1. In the Ribbon, click the Analyze tab. In the Inquiry panel, select INQUIRY
TOOL.

2. In the Inquiry Tool palette, click the Select an inquiry type's drop-list arrow, expand the Alignment section, and from the list select **Alignment Station and
Offset at Point**.

3. In the Select Alignment dialog box, select the alignment Rosewood – (1), and
click **OK** to continue.

4. In the drawing, select a point near the Rosewood alignment.

The Inquiry Tool palette displays the selected point's coordinates, station, and offset.

5. Select a few more locations and note the values listed in the Inquiry Tool
palette.

6. In the Inquiry Tool palette, click the Select an inquiry type's drop-list arrow and
from the list select ALIGNMENT TWO STATIONS AND OFFSETS AT POINT.

7. In the Select First Alignment dialog box, select Rosewood – (1), and click **OK** to
continue.

8. In the Select Second Alignment dialog box, select Senge Drive and click **OK** to
continue.

9. In the drawing, select a point near the Rosewood alignment.

The Inquiry Tool palette shows the selected point's coordinates, station, and offset relative to the two alignments.

10. Select a few more locations and note their Inquiry Tool palette values.

11. Click the Inquiry Tool palette's **X** to close it.

12. At Civil 3D's top left, Quick Access Toolbar, click the **Save** icon to save the
drawing.

This completes an overview of alignment review methods. The Grid View and Sub-Entity
vista editors display all alignment segment values. Toolbox's reports create various alignment reports.

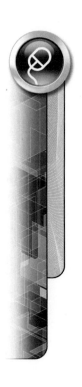

EXERCISE 5-4

After completing this exercise, you will:

- Graphically edit an alignment.
- Edit alignment values in the Grid View vista.
- Edit an alignment's values in the Sub-entities Editor dialog box.
- Review and edit an alignment from its Properties dialog box.
- Modify design speeds for an alignment.

Exercise Setup

This exercise continues with the previous exercise's drawing. If you did not complete the previous exercise, browse to the Chapter 05 folder of the CD that accompanies this text-book and open the *Chapter 05 – Unit 4.dwg* file.

1. If not open, open the previous exercise's drawing or browse to the Chapter 05 folder of the CD that accompanies this textbook and open the *Chapter 05 – Unit 4* drawing.

Graphical Editing — Tangent Segments

When you have selected any alignment segment, the entire alignment displays its editing grips. To edit the alignment, click a grip and move it to a new location. If you locate a grip that does not correctly resolve, a solution will not be displayed and you cannot pick that point.

There are four fixed tangents in the alignment. You can move all of the tangents by selecting their middle grip or by adjusting their bearing by selecting their end grip.

1. Use the ZOOM command to better view the eastern north/south tangent.
2. In the drawing, click the alignment to activate its grips.
3. Select the eastern tangent's middle grip (square) and move the alignment east and west.

As the tangent moves, the two free curves change to accommodate the tangent's motion.

4. Press ESC to release the tangent.
5. In the drawing, click the eastern tangent's southern extension grip and move the cursor east and west.

This grip changes the tangent's direction from its southern end while holding its northern point. As the grip relocates, the free curves adjust to maintain valid attachments.

6. Press ESC to release the tangent.

Curve Segments

When you select any free curve's grip and move its location, you adjust the curve's radius.

1. Use the ZOOM and PAN command and view Curve 2.
2. If necessary, click the alignment to activate its grips.
3. Activate the curve's middle grip and slowly move it toward the southeast.

Both tangents lengthen to accommodate the curve's changing radii.

4. Press ESC to release the curve.

Alignment Grid View

1. While the alignment is highlighted, from the Modify palette, select GEOMETRY EDITOR to display the Alignment Layout Tools toolbar.

2. On the toolbar's right, click the **Alignment Grid View** icon to display the Panorama with the Alignment Entities vista.

3. In the Alignment Entities vista, review each curve's radii. All radii are divisible by 5.

4. Click the **X** to close the Alignment Entities vista.

5. At Civil 3D's top left, Quick Access Toolbar, click the **Save** icon to save the drawing.

Sub-Entity Editor

The Sub-entity Editor's – Alignment Layout Parameters dialog box displays a selected segment's information.

1. In the Alignment Layout Tools toolbar, click the **Sub-entity Editor** icon to display the Alignment Layout Parameters dialog box.

2. In the Alignment Layout Tools toolbar, click the **Pick Sub-entity** icon. In the drawing, select the alignment Curve 2.

The curve's data appear in the editor. The current radius is **310** feet and is editable.

3. Click the Alignment Layout Parameters dialog box's red **X** to close it.

4. Click the Alignment Layout Tools toolbar's red **X** to close it and return to the command line.

Alignment Properties — Station Control

1. On the Ribbon, click the **View** tab. At the Views panel's left, select and restore the named view Station Equation.

2. In the drawing, select the alignment, from the Modify panel, select ALIGNMENT PROPERTIES icon.

3. Click the **Station Control** tab to review its contents.

4. In the dialog box, at the middle left click the **Add station equation** icon.

The dialog box disappears, and in the drawing a jig attaches to the alignment reporting stations.

5. Using an Intersection object snap, select the alignment's intersection with the Phase 1 line and return to the Alignment Properties dialog box.

6. In the dialog box, for station ahead enter **2500**, and set Increase/Decrease to Increasing.

7. Click **Apply** to modify the alignment's stationing.

The alignment's beginning station and length remain the same; however, the ending station changes to a higher value.

8. Review the new station values.

Alignment Properties — Design Speed

1. In the Alignment Properties dialog box, click the **Design Criteria** tab.

2. At the panel's top left, click the **Add design speed** icon twice.

3. For station 0+00 change the design speed to **10**.

4. In the new speed entry, for the start station enter **75** and for its speed enter **25**.

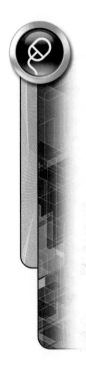

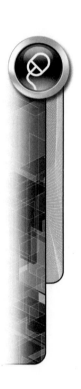

5. In the second new speed entry, for the station enter **215** and for its speed enter **30**.

Your Design Criteria panel should now look like Figure 5.18.

6. Click **OK**.

7. At Civil 3D's top left, Quick Access Toolbar, click the **Save** icon to save the drawing.

This ends the exercise on alignment editing. Alignment edits occur graphically, in a sub-entity or an overall editor, and in the Alignment Properties dialog box. The Alignment Properties dialog box is the only place where you can define station equations, design speeds, superelevation, and other important alignment values.

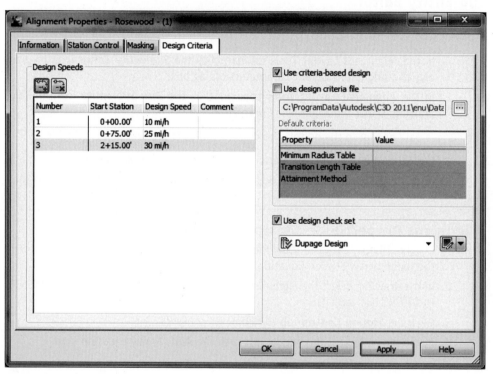

FIGURE 5.18

EXERCISE 5-5

After completing this exercise, you will:

- Be able to review alignment label sets.
- Be able to review alignment label styles.
- Be able to create new label styles.
- Be able to define a new label set.
- Be able to apply a label set to an alignment.
- Be able to add station and offset labels.
- Be able to add segment labels.

Exercise Setup

This exercise continues with the previous exercise's drawing. If you did not complete the previous exercise, browse to the Chapter 05 folder of the CD that accompanies this textbook and open the *Chapter 05 – Unit 5.dwg* file.

1. If not open, open the previous exercise's drawing or browse to the Chapter 05 folder of the CD that accompanies this textbook's and open the *Chapter 05 – Unit 5* drawing.

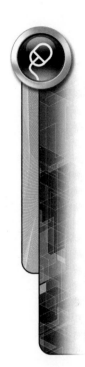

Label Sets — All Labels

When creating an alignment, the Create Alignment dialog box has an entry for assigning a Label Set.

1. Click the **Settings** tab.
2. In Settings, expand the Alignment branch until you view the Label Styles, Label Sets list.
3. From the list of label sets, select **All Labels**, press the right mouse button, and from the shortcut menu, select EDIT....
4. In the Alignment Label Sets dialog box, click the **Labels** tab and review the list of assigned label styles.
5. Click **Cancel** to exit the dialog box.

The label styles for a label set are from the Stations branch below Label Sets.

Major Station — Perpendicular Style

The label set uses a new style that places a major station's label perpendicular to an alignment. The label is orientated to the centerline direction and the text does not change when it is dragged away from its original position.

1. In Settings, expand the Alignment, Label Styles', Station, Major Station branches until you view its styles list.
2. From the list of Major Station styles, select **Perpendicular with Tick**, press the right mouse button, and from the shortcut menu, select COPY....
3. In the **Information** tab, change the Name to **Perpendicular**, and give the style a short description.
4. Click the **General** tab.
5. To make the label orientate to the centerline's path, click in the Plan Readability's Plan Readable value cell to display its drop-list arrow. Click the drop-list arrow, and change its value to **False**.
6. Click the **Layout** tab.
7. At the panel's top left, click the Component name drop-list arrow and select **Tick** from the list. Delete it by selecting the red **X** in the panel's upper-middle portion.

The current component name is now Station.

8. In the Text section, click in the Attachment's value cell to display its drop-list arrow. Click the drop-list arrow, and from the list select **Middle Center**.
9. In the Text section, click in the Y Offset value cell and change the offset value to **0** (zero).

Your Layout panel should look similar to Figure 5.19.

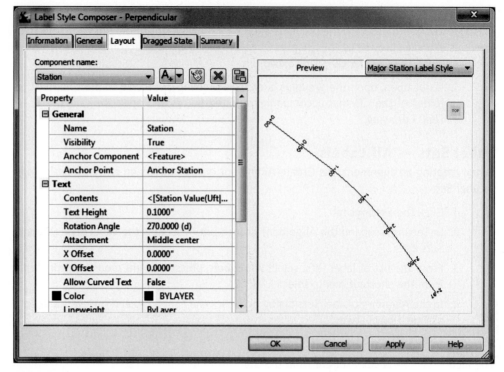

FIGURE 5.19

10. Select the ***Dragged State*** tab.

11. In the Leader section, click in the Visibility value cell to display its drop-list arrow. Click the drop-list arrow, and change it to **False**.

12. In the Dragged State Components section, click in the Display value cell to display its drop-list arrow. Click the drop-list arrow, and change the value to **As Composed**.

13. Click **OK** to create the style.

14. At Civil 3D's top left, Quick Access Toolbar, click the ***Save*** icon to save the drawing.

Perpendicular Label Set

The All Labels label set contains parallel label station styles. The new label set is a copy of All Labels except for the Major Station label.

1. If necessary, expand the Alignment, Label Styles, and Label Sets branches until you view the Label Sets list.

2. From the label sets list, select **All Labels**, press the right mouse button, and from the shortcut menu, select COPY....

3. Click the ***Information*** tab and for the name, enter **All Labels – Perpendicular**.

4. Click the ***Labels*** tab.

5. In the labels list under the Type heading, Major Stations type, at the right of the Style's cell, click the blue label icon to display the Pick Label Style dialog box.

6. In the dialog box, click the drop-list arrow, from the list select **Perpendicular**, and click **OK** until you exit the dialog boxes.

This changes the Major Stations label to Perpendicular and the Label Set is completed.

Changing Label Assignments

The Edit Alignment labels setting changes an alignment's labels.

1. In the drawing select the Rosewood – (1) alignment, press the right mouse button, and from the shortcut menu, select EDIT ALIGNMENT LABELS....

2. In the labels list under the Type heading, select Major Stations, and to the right of Add>>, click the red **X** to remove the label type and its style.

3. Repeat the previous step until all Types are removed.

4. At the bottom of the dialog box, click IMPORT LABEL SET....

5. In the Select Style Set dialog box, click the drop-list arrow. From the list select All Labels – Perpendicular, and click **OK** to return to the Alignment Labels dialog box (see Figure 5.20).

6. Click **OK** to exit the dialog box.

7. Use the ZOOM and PAN commands to better view the alignment labeling.

The major stationing labels should now be perpendicular to the alignment.

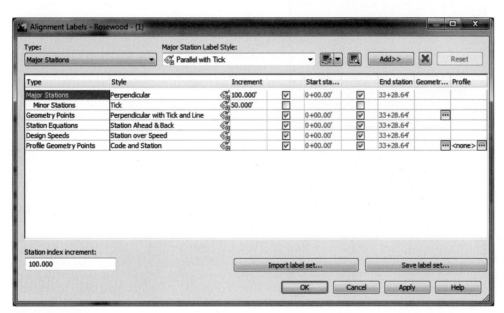
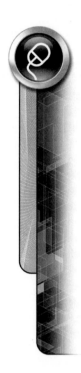

FIGURE 5.20

Add Labels – Station and Offset – Fixed Point

There are critical points along the centerline's path that need annotation: lot corners, existing trees, existing utilities, driveway intersections, and so on.

The file for this exercise is in the Chapter 05 folder of this textbook's CD.

1. Make sure the Transparent Commands toolbar is displayed.

2. On the Ribbon, click the **Insert** tab. On the Import panel, select LANDXML.

3. In the Import LandXML dialog box, browse to Station *Offset.xml*, select it, and click Open. The Import LandXML dialog box opens. In the Import LandXML dialog box, click **OK**.

The file is read and the points appear in the drawing.

4. On the Ribbon click the **View** tab. On the Views panel's left, select and restore the named view **Station Equation**.

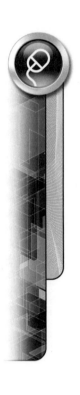

5. On the Ribbon, click the ***Annotate*** tab. On the Labels & Tables panel's left, click Add Labels' drop-list arrow, on the shortcut menu, click Alignment, and on the flyout select ADD ALIGNMENT LABELS.

6. If necessary, in Add Labels, change the Label type to **Station Offset – Fixed Point** and set the Station offset label style to **Station Offset and Coordinates**.

Your Add Labels dialog box should match Figure 5.21.

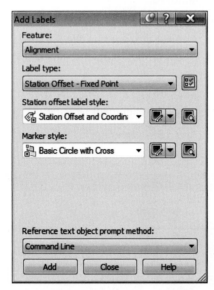

FIGURE 5.21

7. Click ***Add*** to start the labeling process.

8. The Add Labels command prompts you to select an alignment. In the drawing, select the Rosewood alignment.

The routine displays a station jig, and asks you to select a point.

You want a label that references the cogo points. Use the transparent command Point Object override. The transparent override affects only one selection. You need to reselect the Point Object icon for each point selection.

9. From the Transparent Commands toolbar, click the ***Point Object*** icon and in the drawing select one of the points.

10. Repeat the previous step, and annotate the remaining points.

11. When you finish annotating, press ENTER to end the command.

The labels may be on top of one another.

12. In the drawing, select a station/offset label to display its grip; select the grip (cyan diamond), and drag the label to a new position.

The label's Dragged State changes the label to stacked text with a leader.

13. Use the PAN command and pan the drawing to view more of the alignment.

Add Labels – Station and Offset

1. In the Add Labels dialog box, change the Label type to **Station Offset** and the Label style to **Station Offset and Coordinates**.

2. Click ***Add*** to start the labeling process.

3. The Add Label command prompts you to select an alignment. In the drawing, select the Rosewood alignment.

A station jig appears ready to select a station and offset.

4. In the drawing that is using the station jig, identify the station by selecting a point near the alignment. To identify the offset, select a second point to one side or the other of the alignment.

5. Repeat the previous step to create a few more station/offset labels. You can also type in values for the Station and Offset.

6. Press ENTER to exit the labeling, and in the Add Labels dialog box, click **Close** to exit it.

7. At Civil 3D's top left, Quick Access Toolbar, click the **Save** icon to save the drawing.

8. In the command line type DVIEW and press ENTER. In the command line, select object, enter **All**, and press ENTER twice. In the command line, enter **TWist** and press ENTER. Slowly move the cursor to rotate your view of the drawing.

All labels rotate to be plan-readable (horizontal).

9. Move the mouse to specify a rotation, pick a point in the drawing, and press ENTER to exit Dview.

10. In the Ribbon, select the **View** tab. On the Views panel, select NAMED VIEWS and create a New view named **Twisted**.

11. In the command line, type **Plan** and then press ENTER twice to restore the previous display.

Labels and Layouts

1. On the Status Bar, click the **Layout1** icon.

2. In the layout, to enter the viewport's model space by double-clicking in it.

3. From the Views panel, select and restore the named view **Twisted**. You may need to scroll down the list by clicking the down arrow.

4. On the Status Bar, set the Viewport Scale to **1″ = 60′**.

5. If necessary, in the command line, type REGENALL (REA) and press ENTER to resize the station offset labels.

The labels are resized appropriately for the current scale.

6. On the Status Bar, set the Viewport Scale to **1″ = 30′**.

7. If necessary, in the command line, type REGENALL (REA) and press ENTER to resize the station offset labels.

The labels are resized appropriately for the current scale.

8. From the Status Bar click the Model icon.

9. At Civil 3D's top left, Quick Access Toolbar, click the **Save** icon to save the drawing.

Alignment Segment Labels

There are times when you need to label an alignment's tangents and curves. You use the Add Labels dialog box to create these labels.

1. Click the **Annotate** tab. On the Labels & Tables panel's left, click Add Labels' drop-list arrow, on the shortcut menu, click Alignment, and on the flyout select ADD ALIGNMENT LABELS.

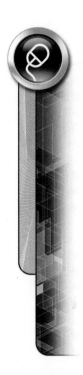

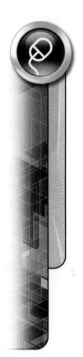

2. In Add Labels, Label type, click the drop-list arrow and from the list select **Multiple Segment**.

3. Click Table Tag Numbering to review its current numbers. After reviewing the numbers, click *OK* to return to the Add Labels dialog box.

4. Click *Add*. The command line prompts you to select an alignment; in the drawing, select the Rosewood alignment.

Segment labels appear along the alignment's tangents and curves.

5. In the Add Labels dialog box, click *Close*.

6. Use the ZOOM and PAN commands to view labels.

Alignment Table

Segment labels are the basis for an alignment's lines and curves table.

1. Use the ZOOM and PAN commands to view an empty area to the right of the site.

2. If necessary, click the *Annotate* tab. On the Labels & Tables panel, click Add Tables' drop-list arrow, on the shortcut menu, click Alignment, and on the flyout select ADD SEGMENT.

3. In the Alignment Table Creation dialog box, set the table to By alignment, click its drop-list arrow, and from the list select Rosewood – (1). All remaining settings stay the same. Click *OK* to close the dialog box.

4. As you exit, a table hangs on the cursor. In the drawing, select a point to locate the table.

5. At Civil 3D's top left, Quick Access Toolbar, click the *Save* icon to save the drawing.

6. Use the ZOOM command to better view the table.

7. Use the ZOOM and PAN commands to better view the alignment labeling.

The labels are now tags. When you create a table, the labels change their display mode to tag mode.

This completes alignment annotation. The next unit reviews how to create new entities from an existing alignment.

EXERCISE 5-6

After completing this exercise, you will:

- Be able to create points by intersection with an alignment.
- Be able to create points from the geometry with an alignment.
- Be able to import points with station and offset values.
- Be able to create an alignment ROW.

Exercise Setup

This exercise continues with the previous exercise's drawing. If you did not complete the previous exercise, browse to the Chapter 05 folder of the CD that accompanies this textbook and open the *Chapter 05 – Unit 6.dwg* file.

1. If not open, open the previous exercise's drawing or browse to the Chapter 05 folder of the CD that accompanies this textbook and open the *Chapter 05 – Unit 6* drawing.

Alignment Points — Station/Offset

Civil 3D creates points whose locations are critical centerline points, reference station and offset, measure lengths along it, or divide it into equal-length segments.

Setting point by Station and Offset is similar to labeling alignment stations and offsets.

1. On the Ribbon, click the ***View*** tab. On the Views panel's left, select and restore the named view **Proposed Starting Point**. You may need to scroll up the list by selecting the up arrow.

2. On the Ribbon, click the ***Home*** tab. On the Create Ground Data panel, click the Points drop-list arrow and on the shortcut menu, select POINT CREATION TOOLS.

3. On the toolbar, click (chevron on the toolbar's right side) the Expand the Create Points dialog.

4. Expand the Points Creation section, set the values in Table 5.1.

TABLE 5.1

Section	Setting	Value
Points Creation	Prompt For Elevations	None
Points Creation	Prompt For Descriptions	Automatic – Object
Points Creation	Default Description	CL

Your toolbar should look like Figure 5.22.

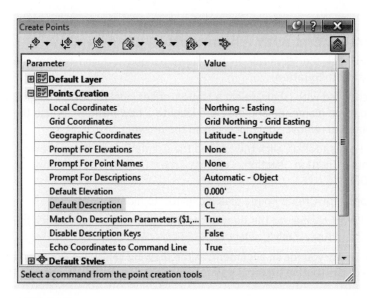

FIGURE 5.22

5. Click the Collapse the Create Points dialog chevron to close it.

6. In the Create Points toolbar, click the ***Alignment*** icon stack drop-list arrow, and from the list select AT GEOMETRY POINTS.

7. In the command line, the routine prompts you to select an alignment; select the Rosewood alignment.

8. In the command line, the routine prompts you for the starting and ending stations; set the starting station to 0+00, press ENTER, set the ending station to 1+00, and press enter twice to exit the routine.

The routine creates points representing the alignment's curve RPs, PCs, and PTs.

Import from File

You can import points from a file that references an alignment's stations and offsets.

The file for this exercise is in this textbook's CD, Chapter 05 folder.

1. From the Create Points toolbar, to the right of the Alignment icon stack, click the drop-list arrow, and from the list select IMPORT FROM FILE.

2. In the Select File dialog box, browse to the CD that accompanies this textbook, and from the Chapter 05 folder, open the *Station Offset.txt* file.

3. In the command line, the routine prompts you for a file format number (you may have to press F2 to see the format list). For the file format enter **1** (Station, Offset), and press ENTER to continue.

4. In the command line, the routine prompts you for a delimiter; enter **2** (comma), and press ENTER twice to use a comma and accept the default invalid station/offset indicator.

5. In the command line, the routine prompts you to select an alignment; in the drawing, select the Rosewood alignment.

The points reference stations 1+00 to 2+00.

6. Use the ZOOM command to better view the points in the area of stations 1+00 and 2+00.

7. At Civil 3D's top left, Quick Access Toolbar, click the **Save** icon to save the drawing.

Points from Intersections with Alignments

1. On the Ribbon, click the **View** tab. In the Views panel select and restore the named view **Station Equation**.

2. If the area is crowded with labels, select a station/offset label to display its grip; select the grip (cyan diamond), and drag the label to a new position, or use the ERASE command to remove some labels from the drawing.

3. In the Create Points toolbar, click the Expand the Create Points dialog chevron. In the Points Creation section, change the Default Description to **INTERSECTION**.

4. Click the Collapse the Create Points dialog chevron to hide the panel.

5. In the Create Points toolbar, to the right of the Intersection icon stack, click the drop-list arrow and from the list, select DIRECTION/ALIGNMENT.

6. The routine prompts you for an alignment; in the drawing select the Rosewood alignment. Next, the routine prompts you for an offset; for the offset, enter **-25** and press ENTER.

7. In the command line, the routine prompts you for a direction starting point; in the drawing, select a point near the Phase 1 line.

8. After selecting the starting point, the routine prompts you for a second point to set a direction (a drag helps you define the direction). In the drawing, select a second point to define a direction that intersects the Rosewood alignment.

9. In the command line, the routine prompts you for an offset; for the offset enter **0** (zero) and press ENTER. The routine marks a point that represents the inter-

section of an offset from the alignment and the direction from the Phase 1 line. Press ENTER to exit the routine.

10. In the Create Points toolbar, click the toolbar's red **X** to close it.

11. At Civil 3D's top left, Quick Access Toolbar, click the **Save** icon to save the drawing.

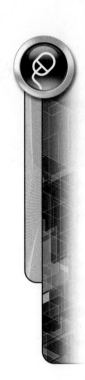

Creating a ROW

The Rosewood alignment divides the property into two new parcels. In each subdivision, there is a buffer that extends to either side of the alignment. This buffer is the ROW parcel. Civil 3D creates a ROW from all of the defined centerlines of a drawing.

1. Use the ZOOM EXTENTS command to view the entire drawing.

2. On the Ribbon, click the Home tab. On the Create Design Panel, click Parcel's drop-list arrow and from the shortcut menu select CREATE RIGHT OF WAY.

3. In the command line, the routine prompts you to select the parcels that are adjacent to the alignment. In the drawing, select the parcel labels to the east and west of the Rosewood alignment, and press ENTER to display the Create Right Of Way dialog box.

4. In the Create Parcel Right of Way section, set the Offset From Alignment value to **33** and set the cleanup methods for Cleanup at Parcel Boundaries and Alignment Intersections to **None** (no fillet or chamfer) (see Figure 5.23).

5. Click **OK** to create the ROW.

6. At Civil 3D's top left, Quick Access Toolbar, click the **Save** icon to save the drawing.

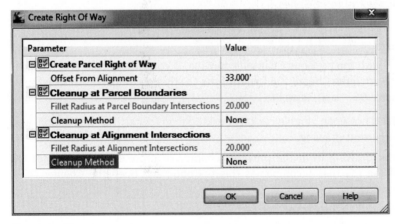

FIGURE 5.23

This completes the alignment exercises.

CHAPTER 6

Profile Views and Profiles

After completing this exercise, you will:

- Be familiar with Profile and Profile View Settings.
- Be familiar with Profile View Styles.
- Be familiar with Profile View bands.
- Be familiar with Profile Styles.

Exercise Setup

This exercise starts with the previous Chapter's drawing. If you did not complete the previous exercise, browse to the Chapter 06 folder of the CD that accompanies this textbook and open the *Chapter 06 – Unit 1.dwg* file.

1. If **Civil 3D** is not open, double-click its desktop icon to start the application, and close the current drawing.

2. Open your drawing from the previous chapter or browse to the Chapter 06 folder of the CD that accompanies this textbook and open the *Chapter 06 – Unit 1* drawing.

3. If necessary, click the **Prospector** tab.

4. In Prospector, select the **Points** heading, press the right mouse button, and from the shortcut menu, select EDIT POINTS....

5. In the Point Editor vista, click the **Point Number** heading and sort the point list.

6. Select point number 3, hold down SHIFT, and select the last point number.

7. With the points highlighted, press the right mouse button, from the shortcut menu, select DELETE..., and then click **Yes** in the Are you sure dialog box.

8. Close the Panorama.

9. In Layer Properties Manager, freeze the layer **V-NODE**, and click **X** to exit the Layer Properties Manager palette.

10. At Civil 3D's top left, Quick Access Toolbar, click the **Save** icon to save the drawing. If you are in the *Chapter 06 – Unit 1* drawing, from the Menu browse, select **Save as**..., and save the drawing with the name, *Alignments*.

Edit Drawing Settings

Edit Drawing Settings affects all styles and settings below it (see Figure 6.1).

1. Click the *Settings* tab.
2. At the Settings' top, select the drawing name, press the right mouse button, and from the shortcut menu, select EDIT DRAWING SETTINGS....
3. Click the *Object Layers* tab.
4. In Object Layers, scroll to Profile, Profile-Labeling, Profile View, and Profile View-Labeling, set their modifier to **Suffix**, and for their value enter, **–***.

The **–*** (dash followed by an asterisk) appends the layer names with the object's name.

5. Click the *Abbreviations* tab.
6. Collapse the Alignment and Superelevation sections to view the Profile section and review their values.
7. Click the **Ambient** *Settings* tab.
8. Expand the Grade/Slope and Station sections to review their values.
9. Click **OK** exiting the dialog box.

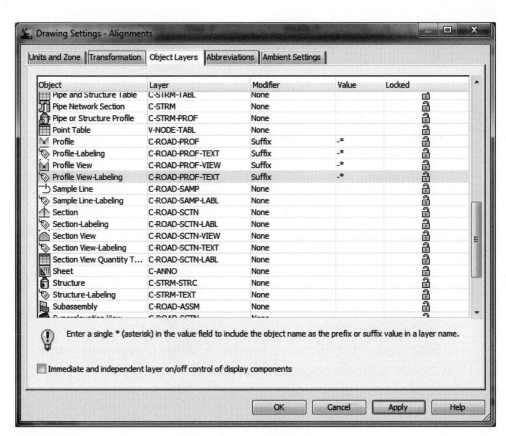

FIGURE 6.1

Profile View – Edit Feature Settings

Profile View's Edit Feature Settings set initial styles, band set assignments, naming formats, and other profile view creation settings (see Figure 6.2).

1. In Settings, click the **Profile View** heading, press the right mouse button, and from the shortcut menu, select EDIT FEATURE SETTINGS....

2. In Edit Feature Settings, expand Default Styles, Default Name Format, Profile View Creation, Split Profile View Options, Stacked Profile View Options, Default Projection Label Placement, and review their values.

3. Click **OK** to exit the dialog box.

Edit Feature Settings - Profile View

Property	Value	Override	Child Override	Lock
Default Styles				
Marker Style	Basic		⬇	🔒
Cut Area Shape Style	Basic			🔒
Fill Area Shape Style	Basic			🔒
Multiple Boundary Area Shape Style	Basic			🔒
Profile View Style	Profile View			🔒
Profile View Band Set	EG-FG Elevations and ...			🔒
Profile Label Set	_No Labels			🔒
First Split View Style	Profile View			🔒
Intermediate Split View Style	Profile View			🔒
Last Split View Style	Profile View			🔒
Top Stack View Style	Profile View			🔒
Middle Stack View Style	Profile View			🔒
Bottom Stack View Style	Profile View			🔒
Projection Style	Standard			🔒
Profile Station and Elevation Label ...	Station and Elevation			🔒
Profile Depth Label Style	Depth			🔒
Profile Projection Label Style	Standard			🔒

OK　　Cancel　　Apply　　Help

FIGURE 6.2

Profile – Edit Feature Settings

Profile's Edit Feature Settings dialog box sets initial styles, label assignments, and other critical profile values (see Figure 6.3).

1. In Settings, click the **Profile** heading, press the right mouse button, and from the menu, select EDIT FEATURE SETTINGS....

2. In Edit Feature Settings, expand each Profile section and review its settings.

3. Click **OK** to exit the dialog box.

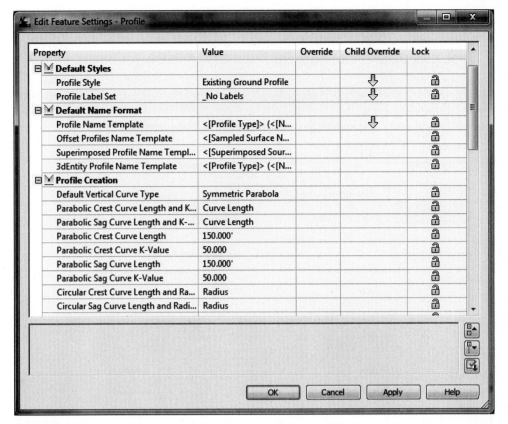

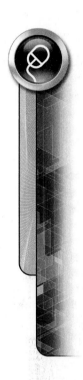

FIGURE 6.3

Profile View Style – Full Grid

Profiles appear within a profile view. A profile view has vertical lines to represent horizontal alignment stations and horizontal lines to represent elevations along the alignment. A Profile View Style defines the grid and its annotation (see Figures 6.4, 6.5, and 6.6).

1. In Settings, expand the Profile View branch until you are viewing Profile View Styles' list of styles.
2. From the list, select **Full Grid**, press the right mouse button, and from the shortcut menu, select EDIT....
3. Click the ***Graph*** tab.

This panel controls a profile view's vertical exaggeration and direction.

4. Click the ***Grid*** tab to review its settings.

This panel controls if the grid is clipped, and if there are any extra vertical or horizontal grid segments.

5. Click the ***Title Annotation*** tab to review its settings.

This panel controls the title, the title's location (left half), Axis Title, and location (right half).

6. Click the Horizontal Axes tab.

This tab controls the stationing annotation and interval.

7. In Major tick details, to the right of the Tick label text, and click the ***Text Component Editor*** icon.

The Text Component Editor opens and contains the label's format.

8. Click ***OK*** to exit the Text Component Editor.
9. Click the ***Vertical Axes*** tab.

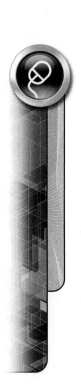

This tab controls the elevation annotation and interval.

10. Click the **Display** tab to review its component list and display values.

11. Click **OK** to exit the dialog box.

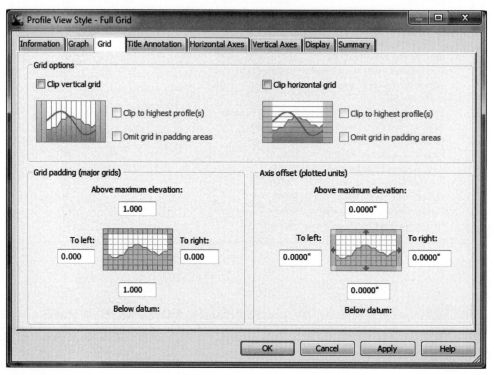

FIGURE 6.4

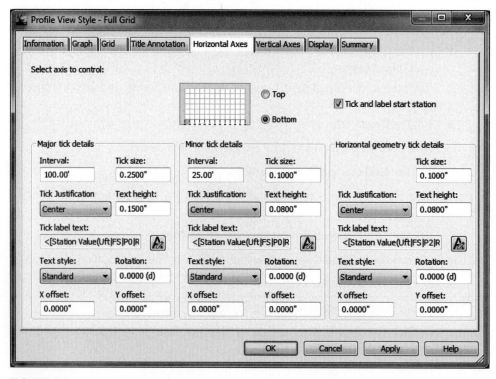

FIGURE 6.5

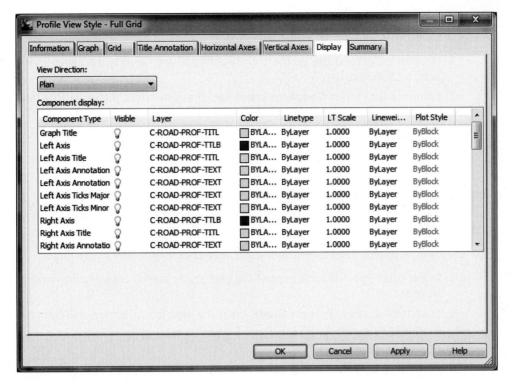

FIGURE 6.6

Profile View – Band Set

Civil 3D bands place information above or below a profile view. A band can contain profile data, horizontal and vertical geometry, superelevation, section sample line data, and pipe network values.

1. In Settings, expand the Profile View branch until you are viewing the Band Styles heading and its styles list.

2. In the Band Styles branch, expand Band Sets, from the styles list select **Profile Data with Geometry and Superelevation**, press the right mouse button, and from the shortcut menu, select EDIT….

3. Click the *Bands* tab to view its list of band types for the top and bottom of the profile view and label styles. Note the styles referenced in this style.

4. Click **OK** to exit the Band Set – Profile Data with Geometry and Superelevation dialog box.

Profile – Existing Ground

Existing Ground is a style that affects the display of elevations along an alignment. This style assigns each component a layer and properties.

1. In Settings, expand the Profile branch until you are viewing the Profile Styles list.

2. From the styles list, select **Existing Ground Profile**, press the right mouse button, and from the shortcut menu, select EDIT….

3. Click the *Display* tab to review the component layer and property assignments.

4. Click **OK** to exit the dialog box.

Profile – Label Sets

A Profile Label Set is an alias for a collection of label styles.

1. In Settings, expand the Profile branch until you are viewing the Label Styles' branch Label Sets list.

2. From the list, select **Complete Label Set**, press the right mouse button, and from the shortcut menu, select Edit....

3. If necessary, click the *Labels* tab and note the label types and label styles list.

4. Click *OK* to exit the dialog box.

Profile Label Set Styles – Line

Label sets annotate a profile view's profile and are used for design alignment labeling, and not for surface profile labeling.

The line styles label a profile tangent with a grade percent format (see Figure 6.7).

1. In the Label Styles branch, expand the Line styles branch until you are viewing its styles list.

2. From the list, select **Percent Grade**, press the right mouse button, and from the shortcut menu, select EDIT....

3. Click the *Layout* tab.

4. In the Text section, click the Contents' value cell to display an editing ellipsis.

5. Click the ellipsis to display the Text Component Editor, and in the Editor, review the tangent label's format.

6. At the dialog box's top left, click Properties' drop-list arrow to view all of the alignment and profile properties available to this label type.

7. Click *OK* to exit the dialog boxes and return to the command line.

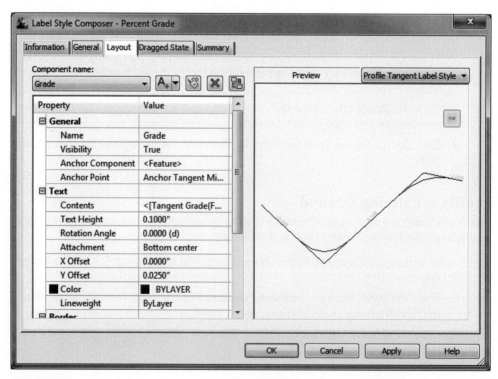

FIGURE 6.7

Profile Label Set Styles – Curve

Curve places a multi-component label on a vertical curve. This label is the most complex profile label. Its components range from arrow blocks to formatted text (see Figure 6.8).

1. In Profile, Label Styles, expand the Curve branch until you are viewing its styles list.

2. From the list of styles, select **Crest and Sag**, press the right mouse button, and from the shortcut menu, select EDIT....

3. If necessary, click the *Layout* tab.

4. In the top-left of the Layout panel, click the Component name's drop-list arrow to view its label components.

5. From the list, select **PVI Sta and Elev**.

6. In the Text section, click the Contents' value cell to display an editing ellipsis.

7. Click the ellipsis to display the Text Component Editor.

In Text Component Editor, review the component's format.

8. This label component has four properties.

9. At the top left, click the Properties drop-list arrow to view all of this label type's label properties.

10. Click **OK** to exit the dialog boxes and return to the command line.

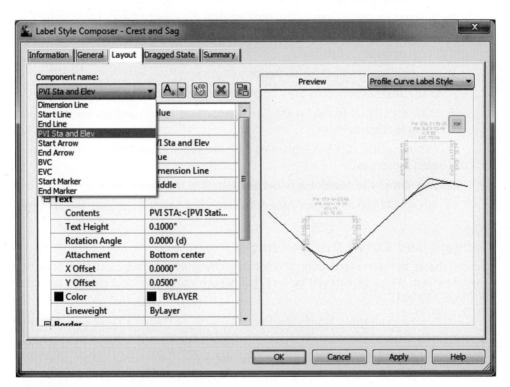

FIGURE 6.8

Profile Label Set Styles – Station and Elevation

Station and Label labels show profile spot elevation within a profile view. If you are moving the label from its original location, its dragged state settings define how it is displayed.

1. In Settings, expand the Profile View branch until you are viewing Label Styles' Station Elevation labels list.

2. From the list, select **Station and Elevation**, press the right mouse button, and from the shortcut menu, select EDIT....

3. If necessary, click the **Layout** tab.

4. If necessary, at its top right, change the Preview to Station Elevation Label Style.

5. In the Text section, click the Contents' value cell to display an editing ellipsis.

6. Click the ellipsis to view the Text Component Editor.

7. At the top left, click the Properties drop-list arrow to view all of this label type's label properties.

8. Click **OK** to exit the dialog boxes and return to the command line.

Profile Label Set Styles – Depth

Depth labels place labels between two user-selected points, which results in a label with a slope or grade between the selected points.

1. Expand the Profile View branch until you are viewing Label Styles' Depth styles list.

2. From the styles list, select **Depth**, press the right mouse button, and from the shortcut menu, select EDIT....

3. Click the **Layout** tab.

4. If necessary, at the panel's top right, change Preview to Depth Label Style.

5. In Layout, at the top, click the Component name drop-list arrow to view the label's components list.

6. Set the Component name to **Depth**.

7. In the Text section, click the Contents' value cell to display an editing ellipsis.

8. Click the ellipsis to display the Text Component Editor, and review the format string for Depth.

9. At the top left, click the Properties drop-list arrow to view all of this label type's label properties.

10. Click **OK** to exit the dialog boxes and return to the command line.

11. At Civil 3D' top left, Quick Access Toolbar, click **Save** icon to save the drawing.

Tangent and Curve Design Checks

Design checks are like expressions and identify any alignment segment that does not pass their test. All tangents must be less than 5 percent, and vertical curves cannot be less than 350 feet.

1. Expand the Profile branch until you are viewing Design Checks and its check type list.

2. From the check type list, select **Line**, press the right mouse button, and from the shortcut menu, select NEW....

3. For the check name, enter **DuPage 5% Grade**.

4. In the New Design Check dialog box, in the lower-middle right, click the **Insert Property** icon (first large button on the right), and from the list of properties, select **Tangent Grade**.

5. In the dialog box, click the less than sign (<) and then click **5**.

Your check should look like the left side of Figure 6.9.

6. Click **OK** to create the check.

7. From the check type list select *Curve*, press the right mouse button, and from the shortcut menu, select NEW....

8. For the check name, enter **Minimum Curve Length**.

9. In the New Design Check dialog box, in the lower-middle right click the **Insert Property** icon, and from the list of properties, select **Profile Curve Length**.

10. In the dialog box, click the less than sign (<) and then click **145**.

Your check should look like the right side of Figure 6.9.

11. Click **OK** to create the check.

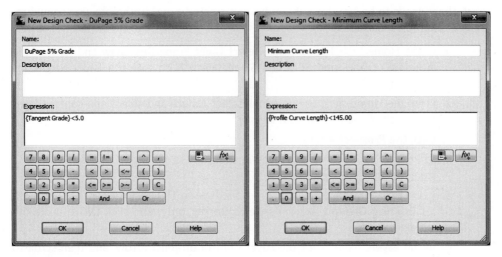
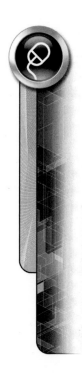

FIGURE 6.9

Creating a Vertical Design Check Set

1. Select the **Design Check Sets** heading, press the right mouse button, and from the shortcut menu, select NEW....

2. Click the Information tab for the name, enter **DuPage Vertical Checks**.

3. Click the **Design Checks** tab.

4. At the panel's top left, click the Type drop-list arrow, set it to **Line**, set the Line check to **DuPage 5% Grade**, and click **Add>>** to add it to the list.

5. At the panel's top left, click the types drop-list arrow, set it to **Curve**, set the Curve check to **Minimum Curve Length**, and click **Add>>** to add it to the list.

6. Click **OK** to create the Design Check Set.

7. At Civil 3D's upper left, Quick Access Toolbar, click the **Save** icon to save the drawing.

This ends the Profile View and Profile styles and settings review. The next unit covers creating an existing ground profile and a profile view.

EXERCISE 6-2

After completing this exercise, you will:

- Be able to create a Profile.
- Be able to create a Profile View.
- Be able to change an assigned Profile View.
- Be able to modify Profile View annotation.

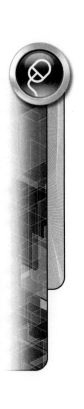

Exercise Setup

This exercise continues with the previous Unit's exercise drawing. If you did not complete the previous exercise, browse to the Chapter 06 folder of the CD that accompanies this textbook and open the *Chapter 06 – Unit 2.dwg* file and start the exercise.

1. If not open, open the previous exercise's drawing or browse to the Chapter 06 folder of the CD that accompanies this textbook's and open the *Chapter 06 – Unit 2* drawing. If using the *Chapter 06 – Unit 2* drawing, from the Menu browse select Save as… and save the drawing as *Alignments*.

Create an EG Surface

First, there needs to be a surface. If you haven't created a surface before, you should review Chapter 04 – Surfaces.

1. In the Home tab's Layers panel, select Layer Properties, thaw and turn on the layers **3EXCONT** and **3EXCONT5**, and click **X** to exit the Layer Properties Manager.

2. From the Home tab's Layers panel, select *Isolate* and select one contour layer from each of the two contour layers.

3. Click the *Prospector* tab.

4. In Prospector, click the **Surfaces** heading, press the right mouse button, and from the shortcut menu, select CREATE SURFACE….

5. In the Create Surface dialog box, set the type to **TIN surface**, for the Name, enter **Existing Ground**, set the style to **Border only**, and click **OK** to make the surface.

6. In the Layers panel, click the Layer Control drop-list arrow, and then scroll to and unlock the **C-TOPO-Existing Ground** layer.

The C-TOPO-Existing Ground layer contains the surface object.

7. In Prospector, expand Surfaces, then expand Existing Ground until you are viewing its Definition branch and its data list.

8. In Existing Ground's Definition list, select **Contours**, press the right mouse button, and from the shortcut menu, select ADD….

9. In the Add Contour Data dialog box, for the description enter **Aerial Contours** and set the Weeding and Supplementing values to match those in Figure 6.10.

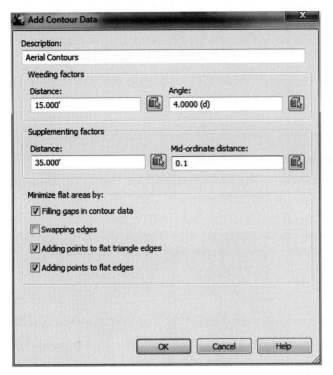

FIGURE 6.10

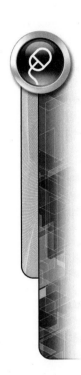

10. Click **OK**, in the drawing select the contours, and then press ENTER.

11. From the Layers panel, click the **Unisolate** icon to restore layer visibility.

12. In Layers panel, Layer Properties Manager, freeze the layers **3EXCONT** and **3EXCONT5**, thaw the layer **Boundary**, and click the **X** to exit.

13. In Existing Ground's Definition branch, select **Boundaries**, press the right mouse button, and from the shortcut menu, select ADD....

14. In the Add Boundaries dialog box, for the description enter **Outer**, set the type to **Outer**, toggle **OFF** Non-destructive breakline, and click **OK** to continue.

15. In the drawing, select the polyline boundary to add the boundary to the surface.

16. Reopen Layer Properties Manager, freeze the layer **Boundary**, and click the **X** to exit the palette.

17. If you still do not see the surface boundary, in the command line, enter REA and press ENTER.

18. Use the ZOOM and PAN commands to place the site at the drawing's left side.

19. At Civil 3D's top left, Quick Access Toolbar, click the **Save** icon to save the drawing.

Create a Surface Profile

After creating a surface, next you sample the surface along the alignment to create the existing profile.

1. From the Home tab, on the Create Design panel, click the Profile icon, select CREATE SURFACE PROFILE.

The drawing has three alignments.

2. In the upper left of the Create Profile from Surface dialog box, set the alignment to **Rosewood – (1)**, and in the upper right select the surface **Existing Ground**.

Next, you decide the station sample range. By default, sampling is from the alignment's beginning to end. If you want a right and left profile, toggle ON Offsets and enter their values.

If you decide to sample the alignments' entire lengths, do not adjust their values.

3. If it is not already off, toggle **OFF** offset sampling.

4. In the dialog box's middle right, click **Add>>** to place the current values in the Profile List.

If you click OK to exit, you next need to select from the Profile & Sections Views panel, Profile View's Create Profile View command to create a profile view with the current profile. A second option is to continue to Create Profile View by clicking Draw in Profile view.

Create a Profile View

Create Profile View reads the sampled data and displays the Create Profile View Wizard.

1. At the bottom of the Create Profile from Surface dialog box, click **Draw in Profile View** to display the Create Profile View wizard.

2. Using the values in Figures 6.11, 6.12, 6.13, 6.14, and 6.15, click **Next** until you are in the last panel, click **Create Profile View**, and in the drawing to the right of the site, select a point to locate the profile view.

3. At Civil 3D's upper left, Quick Access Toolbar, click the **Save** icon to save the drawing.

4. Use the ZOOM and PAN commands to better view the profile.

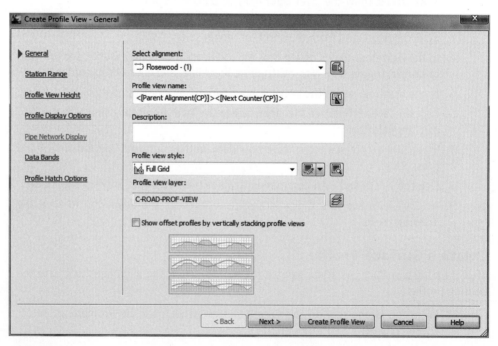

FIGURE 6.11

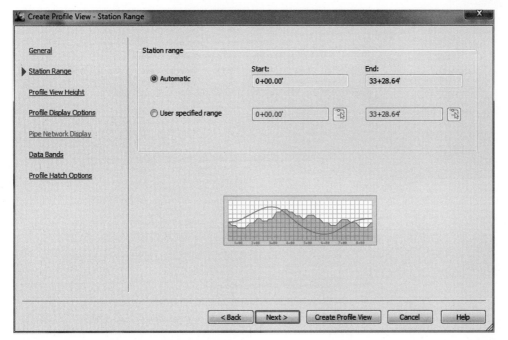

FIGURE 6.12

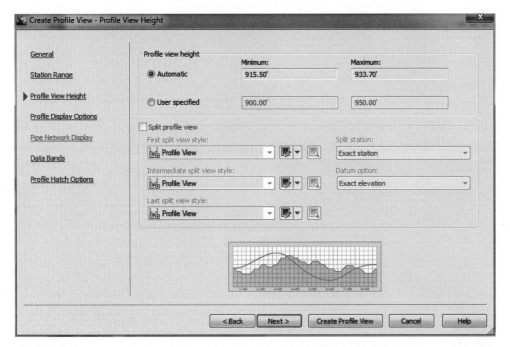

FIGURE 6.13

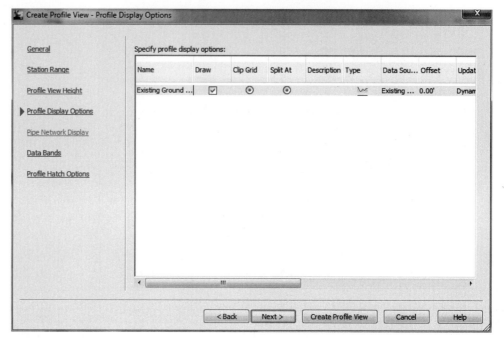

FIGURE 6.14

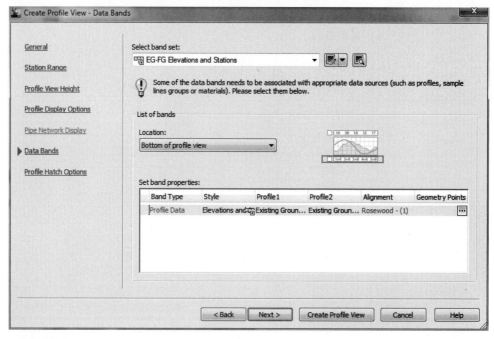

FIGURE 6.15

Change Profile View Styles

A Profile View style defines the view's grid and annotation. Changing the profile view style changes the view's display and shape.

1. In the drawing, select the profile view, press the right mouse button, and from the shortcut menu, select PROFILE VIEW PROPERTIES....

2. In the Profile View Properties dialog box, click the **Information** tab, click Object Style's drop-list arrow, and from the styles list, select **Major Grids**.

3. Click **OK** to exit the Profile View Properties dialog box.

4. Repeat the previous three steps, but set the Profile View Style to **Full Grid**.

Modify a Profile View Style

Modifying a profile view style changes how it is displayed.

1. Click the **Settings** tab.

2. In Settings, expand the Profile View branch until you are viewing the Profile View Styles list.

3. From the list, select **Full Grid**, press the right mouse button, and from the shortcut menu, select EDIT....

4. In the Profile View Style dialog box, click the **Grid** tab.

5. In Grid Options, toggle **ON** both Clip Vertical Grid and Clip Horizontal Grid.

6. Click **OK** to view the profile view changes.

7. In the drawing, select the profile view, press the right mouse button, and from the shortcut menu, select EDIT PROFILE VIEW STYLE....

8. In the Profile View Style dialog box, click the Grid tab and toggle **OFF** both Clip Horizontal Grid and Clip Vertical Grid.

9. Click **OK** to make the profile view changes.

10. At Civil 3D's top left, Quick Access Toolbar, click the **Save** icon to save the drawing.

This completes the exercise on creating a profile and its view.

The next unit describes how to create a proposed vertical alignment for Rosewood.

EXERCISE 6-3

After completing this exercise, you will:

- Be familiar with the Create By Layout Command Settings.
- Be able to create a Profile using the Profile Layout Tools toolbar.

Exercise Setup

This exercise continues with the previous Unit's exercise drawing. If you did not complete the previous exercise, browse to the Chapter 06 folder of the CD that accompanies this textbook and open the *Chapter 06 – Unit 3.dwg* file.

1. If not open, open the previous exercise's drawing or browse to the Chapter 06 folder of the CD that accompanies this textbook and open the *Chapter 06 – Unit 3* drawing. If using the *Chapter 06 – Unit 2* drawing, from the Menu browse select Save as... and save the drawing as *Alignments*.

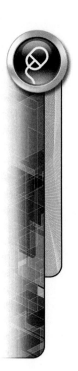

Edit Feature Settings

The Profile Edit Feature Settings dialog box settings affect vertical curves.

1. If necessary, click the **Settings** tab.

2. In Settings, click the *Profile* heading, press the right mouse button, and from the shortcut menu, select EDIT FEATURE SETTINGS....

3. Expand the Profile Creation section and review its values.

The first value sets the default vertical curve type. The remaining values are vertical curve type settings.

4. If necessary, set the Default Vertical Curve Type to **Symmetric Parabola**.

5. If necessary, set the parabolic Sag and Crest criteria to **Curve Length**.

6. Your values should match those in Figure 6.16.

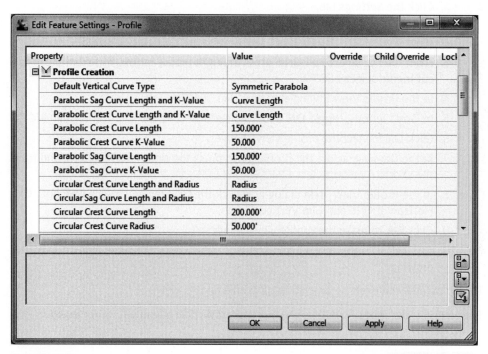

FIGURE 6.16

7. Click **OK** to set the values and to exit the dialog box.

Drafting a Proposed Centerline

Creating a vertical design is not always a simple single pass process. The vertical design may have grade, curve length, K Value, or distance between restrictions that affect the design. In addition to these design issues, you may have to design a road that balances or meets some targeted earthworks amount.

1. Use the ZOOM and PAN commands to view the profile view.

2. On the Ribbon, click the **View** tab. From the Views panel, select NAMED VIEWS, and create a New view, naming it **Profile**.

3. At Civil 3D's top left, Quick Access Toolbar, click the **Save** icon to save the drawing.

Make sure that the Transparent Commands toolbar is visible.

4. On the Ribbon, click the *Home* tab. From the Create Design panel, click the *Profile* icon, select PROFILE CREATION TOOLS and in the drawing, select the Rosewood profile view.

The Create Profile – Draw New dialog box opens.

5. In the dialog box, for the name replace <[Profile Type] > with **_Rosewood Preliminary -**, give the profile a short description, set the Profile Style to **Layout**, and set the Label Set to **Complete Label Set**.

Your dialog box should look similar to Figure 6.17.

6. Click the Design Criteria tab, toggle **ON** Use criteria-based design, toggle *OFF* Use design criteria file, toggle **ON** Use design check set, and change the set to **DuPage Vertical Checks**.

7. Click *OK* to exit the Create Profile – Draw New dialog box and to display the Profile Layout Tools toolbar.

The Rosewood vertical design criteria are the following:

No tangent grades over 5%.

No vertical curves less than 350.

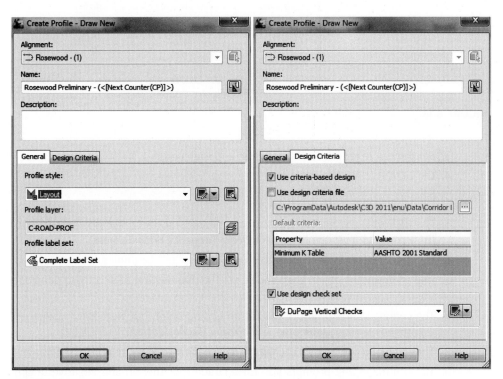
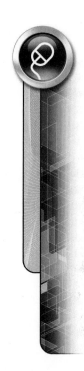

FIGURE 6.17

Drafting Tangent Segments with Vertical Curves

The vertical design uses the Draw Tangents With curves icon at the toolbar's left end. Table 6.1 has the tangents' From and To Stations and their grades. The first and last vertical alignment ends are Endpoint object snap selections of the existing ground profile.

1. On the Profile Layout Tools toolbar, in the first icon stack on the left, click its drop-list arrow and from the list, select Curve Settings....

2. In the Vertical Curve Settings dialog box, if necessary, at the top click the Select curve type drop-list arrow and select the **Parabolic curve** type from the list.

Your screen should look like Figure 6.18.

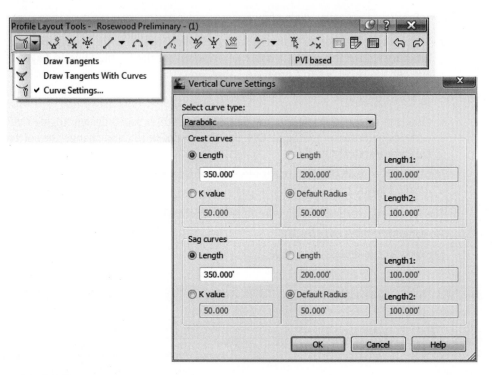

FIGURE 6.18

3. Set the Curve length to **350** for crest and sag curves.
4. Click **OK** to close the dialog box.

The command line prompts for a starting point.

5. On the Profile Layout Tools toolbar, the first icon stack on the left, click its drop-list arrow and from the list, select **Draw Tangents With Curves**.
6. Use the ZOOM command to better view the profile's beginning.
7. Use an **Endpoint** object snap and select the intersection of existing ground and the 0+00 profile station.
8. The command line prompts you for an ending point, but from the Transparent Commands toolbar, select the icon **Profile Grade Station**.
9. The command line prompt changes to Select a Profile View; select the profile view.
10. The command line prompts you for a Grade (positive up and negative down). For the grade, enter **1** and press ENTER.

The crosshairs display a 1 percent grade, a station jig, and a tooltip that reports a station and elevation.

11. In the profile view, select a point near station **900**, or in the command line enter **900**. If you are entering a command line value, you must press ENTER to assign the value.
12. Use the mouse wheel and pan and zoom to view the profile past station 900.

13. The command line prompts you for the next Grade. Set the grade to −3%, enter **−3** (and press ENTER), drag the cursor to 13+75, and select that point or enter **1375** (and press ENTER).

14. Use the mouse wheel and pan and zoom to view the profile past station 2000.

15. The command line prompts you for the next grade. Set the grade to 3%, enter **3**, and select a point near 28+50 or enter **2850**.

16. Press ESC once to end the Transparent Command. The prompt changes to Specify End Point. Using an Endpoint object snap, select the intersection of the existing ground and the profile's last vertical grid line, and press ENTER to complete the profile.

17. Close the Profile Layout Tools toolbar.

TABLE 6.1

From Station	To Station	Grade
Endpoint	9+00	1%
9+00	13+75	−3%
13+75	28+50	3%
28+50	Endpoint	No specific grade

Your profile should look similar to Figure 6.19.

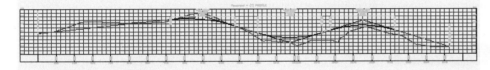

FIGURE 6.19

Assigning the Profile to Profile2

The Profile2 assignment is done in Profile View Properties Band Set values.

1. Use the ZOOM and PAN commands to better view the band's Major and Minor Station annotation.

Currently, both values represent existing ground elevations.

2. In the drawing, select the Profile View, press the right mouse button, and from the shortcut menu, select PROFILE VIEW PROPERTIES....

3. Click the *Bands* tab.

4. In the Bands panel, scroll it completely to the right to view the Profile1 and Profile2 assignments.

5. Under the Profile2 entry, click in the cell to display the drop-list, and from the list select **_Rosewood Preliminary - (1)**.

6. Click **OK** to exit the dialog box.

7. At Civil 3D's top left, Quick Access Toolbar, click the **Save** icon to save the drawing.

8. Close the drawing.

Create a Vertical Alignment from Cogo Points

This section uses the same drawing from the previous Chapter's Unit 2 exercise. The file was saved as *Alignments from Points*.

1. Open the *Alignments from Points* drawing. If you did not do this exercise, on the CD that accompanies this book, in the Chapter 06 folder is the *Alignments* from Points drawing.

2. Use the Zoom extents command to view the entire drawing.

3. From the Home tab, the Profile and Section View panel, select Create Profile View to display the Profile View wizard.

4. At the wizard's bottom left, click Create Profile View and in the drawing to the alignment's east, place the profile view.

5. From the Home tab, the Create Design panel, select the Profile icon to display a command list.

6. From the Profile command list, select Create Best Fit Profile.

7. The routine prompts you to select a profile view. Select the just-created profile view; the Create Best Fit Profile dialog box displays.

8. At the dialog box's top, set the Input type to Cogo Points, the Path 1 Point group to Crown, and click OK to define the profile. The Best Fit Report displays.

9. Review the Best Fit report and then dismiss it.

10. Use the Pan and Zoom commands to better view the new profile.

11. After reviewing the profile, close and Save the drawing.

This ends the drafting vertical alignment tangents and vertical curves exercise. Next, you focus on evaluating and editing vertical alignments.

EXERCISE 6-4

After completing this exercise, you will:

- Be familiar with profile reports.
- Be able to create and view a vertical alignment report.
- Be able to graphically edit a vertical alignment.
- Be able to edit the vertical alignment in a grid view or sub-entity editor.

Exercise Setup

This exercise continues with the previous Unit's exercise drawing. If you did not complete the previous exercise, browse to the Chapter 06 folder of the CD that accompanies this textbook and open the *Chapter 06 – Unit 4.dwg file.*

1. Open the drawing from the previous exercise or browse to the Chapter 06 folder of the CD that accompanies this textbook and open the *Chapter 06 – Unit 4* drawing. If using the *Chapter 06 – Unit 2* drawing, from the Menu browse select Save as... and save the drawing as *Alignments*.

Profile Review

The first profile report is a tangents and curves review.

1. Click the Toolbox tab. If necessary, click the **View** tab. At the Palettes panel's right, select the **Toolbox** icon displaying the Toolbox tab on the Toolspace.

2. In Toolbox expand the Reports Manager's Profile section, and from the list, select PVI STATION AND CURVE REPORT.

3. After selecting the report, press the right mouse button, and from the shortcut menu, select EXECUTE....

4. In the Create Reports dialog box, click **Create Report** to create the report.

The report is displayed.

5. Review the report, close Internet Explorer, and click **Done**.

Inquiry Tool

Inquiry Tool reports profile and profile view values in a panel.

1. On the Ribbon, click the **Analyze** tab. On the Inquiry panel, select INQUIRY TOOL.

2. At the palette's top, click the Inquiry type drop-list arrow, expand the Profile View section, and from the inquiries list, select PROFILE VIEW STATION AND ELEVATION AT POINT.

3. In the drawing, select a few points within the profile view and review the Inquiry results in the tool palette.

4. At the palette's top, click the Inquiry type drop-list arrow, expand the Profile section to change the Inquiry type to **Profile Station and Elevation at Point**. In the Select Profile dialog box, select the profile _Rosewood Preliminary - (1), and click **OK**.

5. In the profile view, select some points and review their values in the Inquiry Tool palette.

6. Close the Inquiry Tool palette by clicking its **X**.

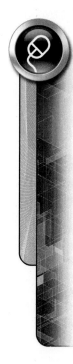

Graphically Editing a Proposed Profile

You can graphically edit a vertical alignment.

1. At Civil 3D's top left, Quick Access Toolbar, click the **Save** icon to save the drawing.

2. Use the ZOOM and PAN commands to better view the first PVI.

3. Select the profile _Rosewood Preliminary - (1) to display its grips.

4. Adjust the vertical curve by selecting its round grip and relocating it, and then press ESC.

5. Adjust the PVI by selecting one tangent arrow grip and moving its location, and then press ESC.

6. Adjust the PVI by selecting the opposite tangent arrow and moving its location, and then press ESC.

7. Adjust the PVI by selecting it and moving it, and then press ESC.

Sub-Entity Editor Dialog Box

When making any adjustments in an editor, all associated changes accommodate the edit.

1. In the drawing, select the _Rosewood Preliminary - (1) profile, press the right mouse button, and from the shortcut menu, select EDIT PROFILE GEOMETRY....

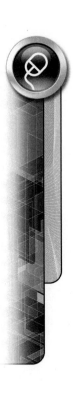

The Profile Layout Tools toolbar is displayed.

2. On the toolbar's right side, click the ***Profile Layout Parameters*** icon to display the Profile Layout Parameters dialog box.

3. If necessary, use the ZOOM and/or PAN commands until you view the first PVI.

4. From the toolbar, click the ***Select PVI*** icon, and in the profile view, select a point near the first PVI.

5. In the Editor, change the PVI Station to **8+50** and change its elevation to **933.50.**

6. Use the ZOOM and/or PAN commands to view the second PVI.

7. Select near the second PVI.

8. Change its PVI Station to **13+75** and raise its elevation to **916.50.**

9. Use the ZOOM and/or PAN commands until you are viewing the last PVI.

10. Select near the last PVI.

11. In the Profile Layout Parameters dialog box, change the PVI's location to **28+50** and change its Elevation to **926.5.**

12. Click the **X** in the Profile Layout Parameters dialog box to close it.

Profile Grid View Vista

Profile Grid View displays all of the profile values in a single Panorama vista.

1. In the Profile Layout Tools toolbar, on its right side, click the ***Profile Grid View*** icon.

2. Scroll through the values and note which values can be edited.

3. Close the Panorama and close the Profile Layout Tools toolbar.

4. At Civil 3D's top left, Quick Access Toolbar, click the ***Save*** icon to save the drawing.

Your profile and profile view should look similar to Figure 6.20.

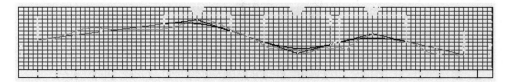

FIGURE 6.20

This completes the vertical alignment evaluation and editing exercise. Editing is done by two methods: graphically manipulating the alignment's grips; or by editing the vertical design in the profile editors.

EXERCISE 6-5

After completing this exercise, you will:

- Be able to label a profile view.
- Be able to apply different profile label styles.
- Project a Watermain into a Profile

Exercise Setup

This exercise continues with the previous Unit's exercise drawing. If you did not complete the previous exercise, browse to the Chapter 06 folder of the CD that accompanies this textbook and open the *Chapter 06 – Unit 5.dwg* file.

1. If not open, open the previous exercise's drawing or browse to the Chapter 06 folder of the CD that accompanies this textbook and open the *Chapter 06 Unit 5* drawing. If using the *Chapter 06 – Unit 2* drawing, from the Menu browse select Save as... and save the drawing as *Alignments*.

Spot Profile Elevations – Station and Elevations

The Add Labels dialog box creates two types of spot profile labels: a station and elevation and depth label.

1. Click Ribbon's **Annotate** tab. On the Labels & Tables panel's left, click the Add Labels drop-list arrow to display a shortcut menu. Place your cursor over the menu's Profile View entry and in the flyout select ADD PROFILE VIEW LABELS.

The Add Labels dialog box opens.

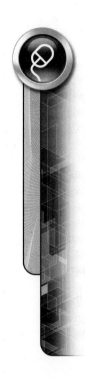

2. In the Add Labels dialog box, if necessary, change the Label Type to **Station Elevation** and click **Add**.

3. The command line prompts you for a Select a Profile View. In the drawing, select the **Rosewood Profile View**.

4. The command line prompts you for a Station. Using the station jig, select a point in the profile view or enter a Station value and press ENTER.

5. The station jig freezes and a second one appears prompting you for an elevation. Select a point in the profile view or enter an elevation and press ENTER.

6. Place two more labels in the profile view.

7. Press ENTER to exit the command.

8. Use the ZOOM command to better view the labels.

Depth

1. In the Add Labels dialog box, change the Label Type to **Depth** and click **Add**.

2. The command line prompts you to Select a Profile View. In the drawing, select the **Rosewood Profile View**.

3. The command line prompts you to Select First Point. In the profile view, select a point. A jig appears connecting the cursor to the selected point.

4. The command line prompts you to Select a Second Point. In the profile view, select a second point.

The routine draws a line between the two selected points, labels the line with a distance, and exits to the command line.

5. Create two more Depth labels, and ZOOM in to better view the labels.

6. Click the Add Labels' **Close** button.

7. In Civil 3D's upper left, Quick Access Toolbar, click **Save** icon to save the drawing.

Changing Profile Labels

Label sets apply label types and their styles to a profile. Edit Labels... changes or assigns new labels to an existing profile.

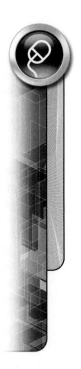

1. Use ZOOM command and zoom in to better view the first vertical curve.

2. In the profile view, select the **_Rosewood Preliminary - (1)** profile, press the right mouse button, and from the shortcut menu, select EDIT LABELS....

3. In the Profile Labels dialog box, review the currently assigned labels.

4. Delete all of the labels by selecting each type from the list and clicking the red **X**.

5. In the panel's upper left, change the label type to **Grade Breaks**, set the style to **Station over Elevation**, and click **Add>>**.

6. In the panel's upper left, change the label type to **Crest Curves**, set the style to **Crest Only**, and click **Add>>**.

7. For Crest Curves, click in its Dim anchor opt cell and change it to **Graph view top**.

8. Click **OK** to label the profile.

9. In the profile view, select the **_Rosewood Preliminary - (1)** profile, press the right mouse button, and from the shortcut menu, select EDIT LABELS....

10. In the Profile Labels dialog box, review the currently assigned labels.

11. At the panel's bottom, click IMPORT LABEL SET....

12. In the Select Style Set dialog box, click the drop-list arrow, from the list select **Complete Label Set**, and click **OK** to return to the Profile Labels dialog box.

13. Click **OK** to assign the label set to the profile.

14. At Civil 3D's top left, Quick Access Toolbar, click **Save** icon to save the drawing.

Projecting a Watermain to a Profile View

1. Select Ribbon's **Home** tab. On the Layers panel, select Layer Properties, locate and turn on the layer **C-TOPO-FEAT-Watermain**, and exit the Layer Properties Manager.

2. In the command line, enter REA (regenall) to view the watermain.

The watermain is adjacent to the Rosewood alignment.

3. Use ZOOM and PAN to better view the watermain.

4. Select the Ribbon's Modify tab and then on the Profile & Section Views panel, click the **Profile View** icon.

5. On the Profile View tab's, Launch Pad panel's right, select Project Objects To Profile View.

6. The command line prompts you to select an object to project. In the drawing, select the watermain near the Rosewood alignment and press ENTER.

7. The command line prompts you to select a profile view. Select the Rosewood profile view and the Project Objects To Profile View dialog box displays.

8. In the Project Objects To Profile View dialog box, set the style to Proposed Watermain, Elevation Options to Use Object, and click OK to project the watermain to the profile view.

9. Select Ribbon's **Home** tab. On the Layers panel, select Layer Properties, locate and turn OFF the layer **C-TOPO-FEAT-WATERMAIN**, and exit the Layer Properties Manager.

10. In the drawing, select the profile view, press the right mouse button, and from the shortcut menu, select **Profile View Properties....**

11. In Profile View Properties, click the Elevations tab, toggle on User specified height, change it to 940, and click OK to exit the Profile View Properties.

12. Review the projected watermain.

13. Select the profile view, right click, and from the shortcut menu, select **Profile View Properties…**.

14. In Profile View Properties, click the Projections tab, toggle off Feature Lines, and click **OK** to exit the Profile View Properties dialog box.

15. At Civil 3D's top left, Quick Access Toolbar, click *Save* icon to save the drawing.

This completes the Profile labels exercise.

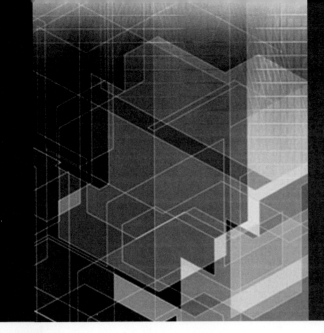

EXERCISE 7-1

After completing this exercise you will:

- Be familiar with Marker, Link, and Shape styles.
- Be familiar with the settings for the CreateSimpleCorridor command.

Exercise Setup

This exercise starts with the previous chapter's exercise drawing. If you did not complete the previous chapter's exercise, browse to the Chapter 07 folder of the CD that accompanies this textbook and open the *Chapter 07 – Unit 1.dwg* file.

1. If you are not in **Civil 3D**, double-click its desktop icon to start the application.

2. When you are at the command prompt, close the open drawing and do not save it.

3. Open the drawing from the previous exercise's chapter or browse to the CD that accompanies this textbook and open the *Chapter 07 – Unit 1* drawing. If you use the *Chapter 07 – Unit 1* drawing, use Save as to save it as the drawing *Alignments*.

Edit Drawing Settings

The assembly, corridor, and corridor section layers need a suffix and a dash asterisk (–*).

1. Click the **Settings** tab.

2. At Settings' top, select the drawing name, press the right mouse button, and from the shortcut menu, select EDIT DRAWING SETTINGS….

3. Click the **Object Layers** tab.

4. In Object Layers, change the Modifier for Assembly, Corridor, and Corridor Section to **Suffix**, and change their Value to **–*** (a dash followed by an asterisk).

5. Click **OK** to set the values and exit the dialog box.

Subassembly — Edit Feature Settings

It is important that each subassembly name include its side. In Civil 3D 2010, this is a default value. In this exercise this value has not been set. You will change the Subassembly feature settings' value to use the side as a part of the subassembly name. The side property makes it easier to correctly assign controlling alignments and profiles to complicated corridors.

1. In Settings, from the headings list, select **Subassembly**, right mouse click, and from the shortcut menu, select EDIT FEATURE SETTINGS….

2. Expand the Subassembly Name Templates section.

3. Click the Create from Macro's value cell that is displaying an ellipsis.

4. Click the ellipsis to display the Name Template dialog box.

5. Click in the Name cell, after Macro Short Name's dash, add two spaces, and an additional dash.

6. Place the cursor between the two dashes, click the Properties field's drop-list arrow, from the list select **Subassembly Side**, and click *Insert* to add the side property to a Subassembly's name.

7. Click **OK** until the dialog boxes are closed.

Point Code Styles

Points, links, and shapes styles define their symbols, layers, and visibility.

1. In Settings, expand General and Multipurpose Styles until you are viewing the Marker Styles list. From the list, select **Crown**, press the right mouse button, and from the shortcut menu, select EDIT….

2. If necessary, in the Marker Style dialog box, click the *Marker* tab.

Marker defines what shape a roadway crown marker displays.

3. Click the *Display* tab.

The Display tab defines marker's visibility, layer, and properties.

4. Click **OK** to exit the dialog box.

5. In Settings, collapse the Marker Styles branch.

Link Styles

A link style defines a link's visibility, layer, and layer properties.

1. In Settings, General, Multipurpose Styles branch, expand Link Styles until you view its styles list. From the list, select **Pave1**, press the right mouse button, and from the shortcut menu, select EDIT….

2. Click the *Display* tab to review its values.

3. Click **OK** to exit the dialog box.

4. In Settings, collapse the Link Styles branch.

Shape Styles

A shape style defines visibility, outline and fill layers, their properties, and the fill pattern.

1. In Settings' Multipurpose Styles branch, expand Shape Styles until you view its styles list.

2. From the list, select **Pave1**, press the right mouse button, and from the shortcut menu, select EDIT….

3. In the Shape Style dialog box, click the *Display* tab to review its contents.

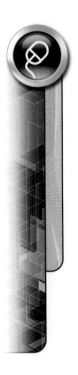

4. Click **OK** to exit the dialog box.

5. In Settings, collapse the General branch.

Create Simple Corridor

CreateSimpleCorridor assigns several default object styles and corridor parameters.

1. In Settings, expand the Corridor branch until you view the Commands list.

2. From the list, select **CreateSimpleCorridor**, press the right mouse button, and from the shortcut menu, select EDIT COMMAND SETTINGS....

3. Expand the Assembly Insertion Defaults section.

These values set the corridor section interval (every 25 feet), what critical geometry points to include (horizontal, vertical, and superelevation), and how often to sample a vertical curve.

4. Collapse the Assembly Insertion Defaults section and expand the Default Styles section.

All styles in this section are from Corridor's Edit Feature Settings.

5. Collapse the Default Styles section and expand the Default Name Format section.

The Default Name Format section sets a corridor naming for corridors, surfaces, and feature lines.

6. Collapse the Default Name Format section and expand the Region Highlight Graphics section.

The Region Highlight Graphics section sets a corridor region's appearance and display.

7. Click **OK** to close the dialog box.

Imperial Subassembly Catalogs

The Roadway Catalogs provide content for the Civil 3D Imperial and Metric subassembly palettes.

1. On the Ribbon, click the View tab. In the Palettes panel's middle right, click the **Content Browser** icon.

2. The Imperial and Metric Catalogs are displayed.

3. In the catalog library, select the Corridor Modeling Catalogs (Imperial) icon.

4. Select and review each catalog's contents.

5. Close Corridor Modeling Catalogs dialog box.

6. At Civil 3D's top left, Quick Access Toolbar, click the **Save** icon to save the drawing.

The Imperial Roadway Palette

Imperial Roadway is a multi-tabbed palette containing subassemblies that address several road design issues. Each tab represents the subassemblies categories.

1. In the Palette panel, to the left of the Content Browser icon, select the **Tool Palettes** icon.

2. Click each palette tab to review each subassembly collection.

This completes the exercise that reviews corridor settings, styles, catalogs, and palettes. The next unit creates an assembly by using various subassemblies.

EXERCISE 7-2

After completing this exercise, you will:

- Be able to create an assembly.
- Be familiar with various subassemblies.
- Be able to select and attach a subassembly to an assembly.
- Be able to edit a subassembly's properties.
- Be able to review an assembly's properties.

Exercise Setup

This exercise continues with the previous exercise's drawing. If you did not complete the previous exercise, browse to the Chapter 07 folder of the CD that accompanies this textbook and open the *Chapter 07 – Unit 2.dwg* file.

1. If not open, open the previous exercise's drawing, or browse to the Chapter 07 folder of the CD that accompanies this textbook and open the *Chapter 07 – Unit 2* drawing. If you use the *Chapter 07 – Unit 2* drawing, use Save as to save it as the drawing *Alignments*.

Create the Assembly

Creating an assembly requires naming it, assigning styles, and placing it in the drawing. After placing the assembly in the drawing, you next add the appropriate subassemblies.

1. In the Home tab's, Create Design panel, click the **Assembly** icon and from the drop-list, select CREATE ASSEMBLY.

2. In the Create Assembly dialog box, for the assembly name replace "Assembly" with **Rosewood**, leave the Assembly Style as **Basic** and the Code Set Style to **All Codes**, and click **OK** to close the dialog box (see Figure 7.1).

3. The command line prompts you for the assembly's location. Select a point just to the left of Rosewood Profile View's lower-left corner.

The assembly is a vertical line with connection symbols at its midpoint.

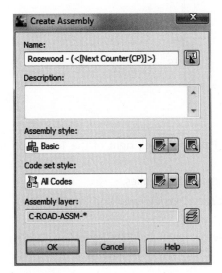

FIGURE 7.1

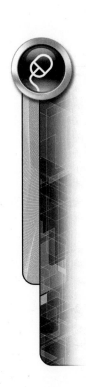

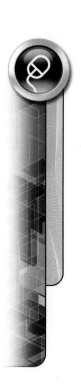

Add Subassemblies — Travel Lanes

Subassemblies create the roadway section and are from the Civil 3D – Imperial palette.

Travel lanes are 12 feet wide, have four materials with varying thicknesses, and have a –2 percent cross slope. You adjust these parameters before attaching the subassembly to the assembly (see Figure 7.2).

Most subassemblies have right- and left-side property. When you attach a subassembly to the assembly or to an inner subassembly, take care to set the correct side parameter and to be able to view the correct attachment code.

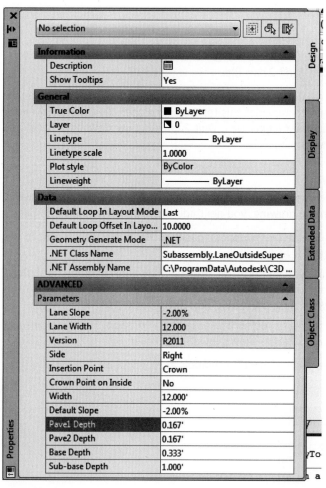

FIGURE 7.2

1. If necessary, click the View tab. On the Palettes panel, right of the Toolspace icon, select the **Tool Palettes** icon.

2. Click the **Imperial – Lanes** tab to display its subassemblies.

3. From the Imperial – Lanes palette, select **LaneOutsideSuper** and the Properties palette displays (see Figure 7.2).

4. In Properties, set the Side to **Right**, change the Pave1 and Pave2 depths to **0.167**, and in the drawing, select the assembly to attach the subassembly.

The subassembly attaches the right travelway to the assembly.

5. In the Properties, change the Side to **Left**, and in the drawing, select the assembly to attach the subassembly to its left side.

6. Press ENTER twice to end the routine.

7. Select the ***Prospector*** tab.

8. In Prospector, expand the Subassemblies branch, select the first subassembly from the list, press the right mouse button, and from the shortcut menu, select PROPERTIES….

The Information tab displays a name with the side parameter (see Figure 7.3).

9. Click the ***Parameters*** tab to review its values.

10. Click **OK** to exit the dialog box.

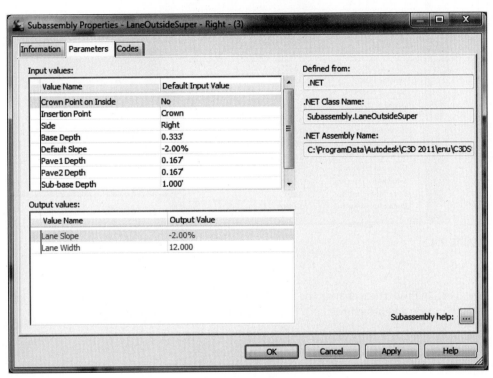

FIGURE 7.3

Add Subassemblies — Curbs

The Curb subassembly attaches to the top of the pavement's red ringlet on the left side.

1. Click the ***Imperial – Curbs*** tab, and from the panel click ***UrbanCurbGutterGeneral***.

2. In Properties, change the Side property to **Right.** In the drawing, pan to the assembly's right side, and attach the curb to the travelway's top code (red circulate) (see Figure 7.4).

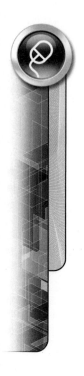

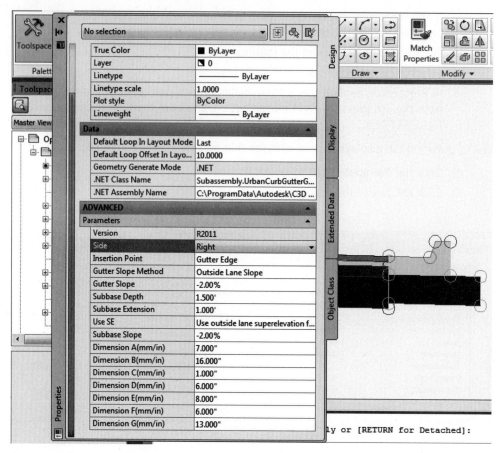

FIGURE 7.4

3. In Properties, change the Side to **Left**. In the drawing, pan to the assembly's left side and attach the curb to the top outside travelway red ringlet.

4. Press ENTER twice to end the routine.

5. Your subassembly should look like Figure 7.5.

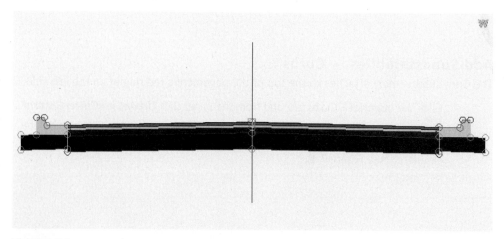

FIGURE 7.5

Add Subassemblies — Ditch and Daylight Slopes

BasicSideSlopeCutDitch daylights to a surface. When attaching the daylight subassembly to the curb's outside back edge, it shows as a sideways "V." When processed, it creates the expected daylight.

1. In the Civil 3D – Imperial tool palette, click the ***Imperial – Basic*** tab, and from the palette, select ***BasicSideSlopeCutDitch***.

2. In the Properties palette, review its settings, change the side to **Right**, and in the drawing, attach the subassembly to the curb's upper-right side.

3. In Properties, change the Side to **Left**, and in the drawing, attach the subassembly to the back of the curb's upper-left side.

4. Press ENTER twice to exit the command.

Your subassembly should look like Figure 7.6.

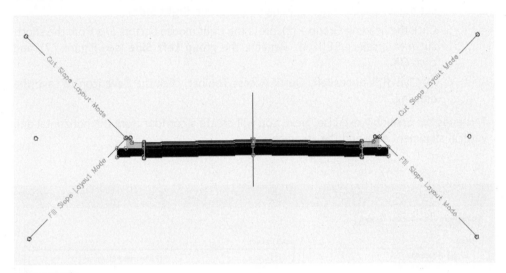

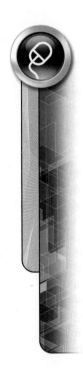

FIGURE 7.6

5. At Civil 3D's upper left, Quick Access Toolbar, click the ***Save*** icon to save the drawing.

Subassembly Properties

The curb's Subbase depth does not match the travelway's depth.

1. If necessary, select the ***Prospector*** tab.

2. If necessary, expand the Subassemblies branch until you view the subassembly list.

3. From the list, select **UrbanCurbGutterGeneral – Right**, press the right mouse button, and from the shortcut menu, select PROPERTIES….

4. In the Subassembly Properties dialog box, select the ***Parameters*** tab, and in the lower right, select the ellipsis for Subassembly Help.

In Help, the Subbase depth represents edge-of-travelway depth. This value needs to be set to 1.6666.

5. Close Help to return to the Subassembly Properties dialog box.

6. Scroll down the Default Input Values list, locate the value for **Subbase Depth**, and change its value to **1.6666**.

7. Click **OK** to exit the dialog box.

8. Repeat Steps 3 through 7 and update the values for **UrbanCurbGutterGeneral – Left**.

9. At Civil 3D's upper left, Quick Access Toolbar, click the **Save** icon to save the drawing.

Assembly Properties

Assembly Properties' Construction tab lists the assembly's subassemblies and their side.

1. In Prospector, expand the Assemblies branch.

2. From the list, click **Rosewood – (1)**, press the right mouse button, and from the shortcut menu, select PROPERTIES....

3. In the Properties dialog box, click the **Construction** tab.

4. Click the heading Group – (1), press the right mouse button, and from the shortcut menu, select RENAME. Rename the group **Right Side**.

5. Click the heading Group – (2), press the right mouse button, and from the shortcut menu, select RENAME. Rename the group **Left Side** (see Figure 7.7), and click **OK**.

6. At Civil 3D's upper left, Quick Access Toolbar, click the **Save** icon to save the drawing.

This ends the assembly exercise. Next, you will create a corridor from the horizontal and vertical alignment data and the assembly.

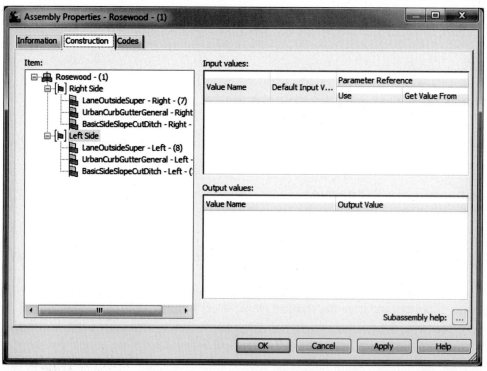

FIGURE 7.7

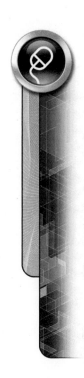

EXERCISE 7-3

After completing this exercise, you will:

- Be able to create a simple corridor.
- Be familiar with corridor properties.

Exercise Setup

This exercise continues with the previous exercise's drawing. If you did not complete the previous exercise, browse to the Chapter 07 folder of the CD that accompanies this text-book and open the *Chapter 07 – Unit 3.dwg* file.

1. If not open, open the previous exercise's drawing or browse to the Chapter 07 folder of the CD that accompanies this textbook and open the *Chapter 07 – Unit 3* drawing. If you use the *Chapter 07 – Unit 3* drawing, use Save as to save it as the drawing *Alignments*.

Create Simple Corridor

1. If necessary, click the **Home** tab. At the Create Design panel's middle, click the **Corridor** icon and from the drop-list, select CREATE SIMPLE CORRIDOR.

2. In the Create Simple Corridor dialog box, leave the counter and replace "Corridor" with **Rosewood**. Enter a short description, and click **OK** to begin identifying the corridor components.

3. The command line prompts you for an alignment. If you are able to select the alignment from the drawing, select it or right mouse click, from the list select **Rosewood – (1)**, and click **OK** to continue.

4. The command line prompts you for a profile. If you are selecting the profile from the drawing, select it or right mouse click, from the list select **_Rosewood Preliminary – (1)**, and click **OK** to continue.

5. The command line prompts you for an assembly. If you are selecting the assembly from the drawing, select it or right mouse click, from the list select **Rosewood – (1)**, and click **OK** to continue.

6. The Target Mapping dialog box opens. For Surfaces, Object Name click in the cell <Click here to set all>. The Pick a Surface dialog box opens. From the list, select the **Existing Ground** surface, and click **OK** to specify the surface.

7. Click **OK** to build the corridor.

8. An Events Panorama displays. Close the Events Panorama.

The corridor builds and is displayed as a mesh.

9. At Civil 3D's top left, Quick Access Toolbar, click the **Save** icon to save the drawing.

Viewing a Corridor

The corridor is a 3D model and can be viewed with Object Viewer or with 3D Orbit.

1. Use the ZOOM and PAN commands to view the site's corridor.

2. In the drawing, select any corridor segment, press the right mouse button, and from the shortcut menu, select OBJECT VIEWER….

3. In Object Viewer, begin viewing by clicking and holding the left mouse button down in the area of curve 2, and slowly moving the cursor toward the center of the object viewer. This tilts the roadway toward your point of view.

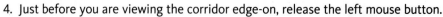

4. Just before you are viewing the corridor edge-on, release the left mouse button.

5. After viewing the corridor, select the Object Viewer's **X** to close it.

Corridor Properties

Corridor properties range from alignments, parameters, codes, surfaces, boundaries, slope patterns, and feature lines.

1. In the drawing, select any corridor segment, press the right mouse button, and from the shortcut menu, select CORRIDOR PROPERTIES....

2. In the Corridor Properties dialog box, click the ***Parameters*** tab.

Parameters list the alignment, profile, and assembly assignments for the corridor. You change the name of any of these elements by clicking the named element and selecting a new alignment, profile, etc., from a list in a dialog box.

3. Click **Set All Targets** at the panel's top right.

This panel reports the Right and Left target surface (existing ground) and any alignment or profile assignment.

4. Click ***OK*** to exit the Target Mapping dialog box and return to the Corridor Properties dialog box.

5. Select the ***Codes*** tab.

Codes lists all corridor links, points, and shapes.

6. Click the ***Feature Lines*** tab.

Feature lines are threads that pass through the assembly point codes. Each code has a name and a specific function. You can import feature lines and use them to design a surface or use them as grading objects.

7. Click the ***Surfaces*** tab.

This panel creates surfaces from feature lines and links.

8. Click the ***Boundaries*** tab.

This panel defines surface boundary control.

9. Click the ***Slope Patterns*** tab.

If you want to include a corridor slope pattern, define it here. You identify where it occurs in the corridor and what pattern to use.

10. Click ***Cancel*** to exit the Corridor Properties dialog box.

Feature Lines

The All Codes style assigns each feature line a style.

1. On the Ribbon, click the ***View*** tab and in the Views panel, select and restore the named view **Proposed Starting Point**.

2. Use the ZOOM and PAN commands to better view the corridor (see Figure 7.8).

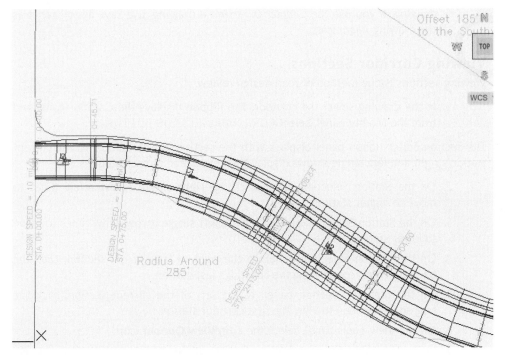
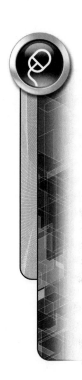

FIGURE 7.8

3. In the drawing, select any corridor segment, press the right mouse button, and from the shortcut menu, select CORRIDOR PROPERTIES….

4. In the Corridor Properties dialog box, click the **_Feature Lines_** tab to review its contents.

This panel lists all corridor feature lines. Each has a name, is visible, and has styles.

5. Click **_OK_** to exit the Corridor Properties dialog box.

6. At Civil 3D's top left, Quick Access Toolbar, click the **_Save_** icon to save the drawing.

This ends the corridor exercise on creating and viewing its properties. Once you have a corridor, you can create new objects from it. The next unit reviews and edits a corridor's values.

EXERCISE 7-4

After completing this exercise, you will:

- Be able to use the View/Edit Section Editor.
- Be familiar with section editing.

Exercise Setup

This exercise continues with the previous exercise's drawing. If you did not complete the previous exercise, browse to the Chapter 07 folder of the CD that accompanies this textbook and open the *Chapter 07 – Unit 4.dwg* file.

1. If not open, open the previous exercise's drawing, or browse to the Chapter 07 folder of the CD that accompanies this textbook and open the *Chapter 07 – Unit 4*

drawing. If you use the *Chapter 07 – Unit 4* drawing, use Save as to save it as the drawing *Alignments*.

Viewing Corridor Sections

Viewing sections is one method of road design review.

1. In the drawing select the corridor. The Ribbon displays the Corridor panel and from the Modify panel, select CORRIDOR SECTION EDITOR.

The Section Editor ribbon panel displays with the section. In the Ribbon, to the current station's right and left, single arrows display the next ahead or back stations.

2. On the Station Selection panel, click the **right single arrow** a few times to move to higher stations.

3. On the Station Selection panel, click the **left single arrow** a few times to move to lower stations.

4. On the Station Selection panel, to the right of the current section, click the **barred-arrow** icon to view the corridor's last station.

5. On the Station Selection panel, to the left of the current section, click the **barred-arrow** icon to view the first corridor station.

6. On the View Tools panel, select the *Edit/View Options* icon.

7. In the View/Edit Corridor Section Options dialog box, View/Edit Options section, set the Default View Scale to **2.0**, adjust any colors, Default Styles section, set the Code Set Style to **All Codes with No Shading**, and click *OK* to return to the section view.

Editing Sections

Edit each subassembly's parameters in an Parameters Panorama. Any changed value is applied to the current section or to a range of stations.

1. In the Station Selection panel, in the panel's middle, click the stations drop-list arrow and select station **9+25**.

2. With 9+25 as the current station and in the Corridor Edit Tools panel, click the **Parameter Editor** icon to display the Corridor Parameters palette. You may need to expand the palette to view its contents.

3. In Corridor Parameters palette, locate the Right Lane's Default Slope entry. Click in the Value cell for slope and change it to **6** (percent).

This changes the right lane slope to 6 percent. After you make the change, a check appears in the override box, an information icon appears, and the section responds by raising its right pavement to the 6 percent slope.

4. In Corridor Parameters, click the **X** to close the palette.

5. In the Corridor Edit Tools panel, at its middle right, select *Apply to a Station Range*.

6. In the Apply to a Range of Stations dialog box, set the station range from **925** to **1350** and click *OK* to apply the change.

7. In the Station Selection panel, to the right of the current station, click the **single arrow** and view the changes to station **14+00**.

8. To remove the change and restore the original value, in the Station Selection panel, click the stations drop-list arrow, and select station **9+25** from the list.

9. In the Corridor Edit Tools panel, click the *Parameter Editor* icon. In the Right Lane Default Slope entry, **uncheck** the override (the value returns to −2%), and close the Parameter Editor.

10. In the Corridor Edit Tools panel, click the ***Apply to a Station Range*** icon, apply the change to the full station range, and click ***OK*** to exit the Apply to a Range of Stations dialog box.

11. Verify the change by reviewing the stations between 9+50 and 13+50.

12. At the Section Editor tab's right, click the **X** to close the panel.

This exercise ends corridor review and editing. The Corridor Section Editor tweaks values for any corridor parameter. After each edit, a section reacts to the change. Changes can apply to the current station or to a range of stations.

EXERCISE 7-5

After completing this exercise, you will:

- Be able to create 3D polylines from a corridor.
- Be able to create feature lines from a corridor.
- Be able to create points from a corridor.
- Be able to build corridor surfaces.
- Be able to add a corridor surface boundary.
- Be able to add a corridor slope pattern.

Exercise Setup

This exercise continues with the previous exercise's drawing. If you did not complete the previous exercise, browse to the Chapter 07 folder of the CD that accompanies this textbook and open the drawing *Chapter 07 – Unit 5.dwg* file.

1. If not open, open the previous exercise's drawing or browse to the Chapter 07 folder of the CD that accompanies this textbook and open the *Chapter 07 – Unit 5* drawing. If you use the *Chapter 07 – Unit 5* drawing, use Save as to save it as the drawing *Alignments*.

Polyline

You can export a 3D polyline by selecting a corridor feature line.

1. In Ribbon's Home tab, open the Layer Properties Manager, create a new layer, and name it **3D poly**. Make it the current layer, assign it a color, and click the **X** to exit the palette.

2. If necessary, from Ribbon's View tab, Views panel, select and restore the named view **Proposed Starting Point**.

The Daylight feature line represents the intersection of the ditch slope out to an intersection with Existing Ground.

3. Click the ***Modify*** tab, on the ***Design*** panel, click the ***Corridor*** icon. From the ***Corridor*** tab, click the Launch Pad panel's drop-list arrow, and from the shortcut menu, select POLYLINE FROM CORRIDOR.

4. In the drawing, select the corridor's north **Daylight Line** (yellow), the Select a Feature Line dialog box displays, select **Daylight**, click ***OK*** to create the polyline, and press ENTER to exit the routine.

5. Select the new object, press the right mouse button, and from the shortcut menu, select PROPERTIES….

The new object is a 3D polyline with daylight feature elevations.

6. Click the **X** to exit the Properties palette.

7. Erase the just-created 3D polyline.

Feature Lines

Feature Lines are similar to 3D polylines. They have varying elevations at each vertex, but feature lines are custom Civil 3D grading objects. The Feature Line export routine works exactly like the Polyline from Corridor command.

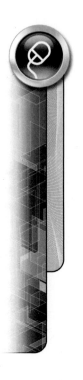

1. From the Launch Pad panel, at its top right, select FEATURE LINES FROM CORRIDOR.

2. Select the corridor's north daylight line, if the Select a Feature Line dialog box displays, select **Daylight**, click **OK** to display the Create Feature Line from Corridor dialog box.

3. Change the style to Corridor Daylight, toggle **OFF** the Smoothing option, click **OK**, to continue, and press ENTER to exit.

4. Select the new object, press the right mouse button, and from the shortcut menu, select PROPERTIES....

The object is listed as an Auto Corridor Feature Line. The feature line is on C-TOPO-FEAT, the default feature line layer.

5. Click **X** to close the Properties palette.

6. Erase the just-created feature line.

Points

Points that represent corridor elevations are critical to placing a design in the field. The point numbers should be offset from other existing points and they should have their own point group.

1. Click the **Prospector** tab.

2. In Prospector, select the **Points** heading, press the right mouse button, and from the shortcut menu, select CREATE ...to display the Create Points toolbar.

3. At the toolbar's right, click the Expand the Create Points dialog chevron.

4. Expand the Point Identity section, change the **Next Point Number** to **10000**, and press ENTER.

5. At the toolbar's right, click the Collapse the Create Points dialog chevron.

6. Close the Create Points toolbar.

7. Click the Ribbon's **Corridor** tab, click the Launch Pad panel's drop-list arrow, and select POINTS FROM CORRIDOR.

8. The command line prompts you to identify a corridor. In the drawing, select a Rosewood – (1) corridor segment.

9. In the Create COGO Points dialog box, set the station range to the entire corridor length, for the point group name enter **Design Points**, and export the following point codes:

   Back_Curb (Back of Curb)

   Daylight

   ETW

   Flowline_Gutter

10. Click **OK** to create the points and their point group.

The points are displayed and are in the Design Points point group.

11. In Prospector, expand Point Groups, select the **Design Points** point group, press the right mouse button, and from the shortcut menu, select EDIT POINTS….

12. After reviewing the points, click the Panorama's **X** to hide it.

13. In the point groups list, select the **Design Points** point group, press the right mouse button, and from the shortcut menu, select DELETE POINTS….

14. Click **OK** in the Are you sure? dialog box to delete the points.

15. With the **Design Points** point group still highlighted, press the right mouse button, and from the shortcut menu, select DELETE… to delete the point group.

16. Click **Yes** in the Are you sure? dialog box.

Slope Patterns

When preparing for a submission or a presentation, you may want to show slope patterns along the corridor's path.

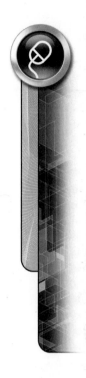

1. If necessary, on the Ribbon, click the **View** tab. In the Views panel, select and restore the named view **Proposed Starting Point**.

2. If necessary, use the PAN and ZOOM commands until you are viewing the corridor cut area (stations 1+25 to 2+50).

3. Use the ZOOM command to better view the northern corridor section. You will be identifying the **Ditch_Out** and **Daylight_Cut** feature lines around station 2+50. Use the tooltip to identify the location of feature lines.

4. In the drawing, select any corridor segment, press the right mouse button, and from the shortcut menu, select CORRIDOR PROPERTIES….

5. Click the **Slope Patterns** tab.

6. At the dialog box's top, click **Add Slope Pattern >>**.

7. The command line prompts you for the first feature line. On the corridor's north side, select the **Ditch_Out** line just north of the Ditch_In feature line.

8. The command line prompts you for the second feature line. On the corridor's north side, select the **Daylight_Cut**. The Select a Feature Line dialog box opens. Select **Daylight_Cut** and click **OK**.

The Corridor Properties dialog box again opens and lists the first and second feature lines.

9. In the dialog box, in the Slope Pattern Style column, click the **Slope Pattern Style** icon, and in the Pick Style dialog box, select the style, **Slope Schemes**. Click **OK** to return to the Corridor Properties dialog box.

10. Again, at the panel's top, click **Add Slope Pattern >>**.

You may have to pan to the corridor's southern side to select the next two lines.

11. In the drawing on the corridor's southern side, select the same two lines, **Ditch_Out** and **Daylight_Cut**.

The Corridor Properties dialog box again opens, listing the first and second feature lines.

12. For the new entry, click the **Slope Pattern Style** icon, in the Pick Style dialog box select the style, **Slope Schemes**, and click **OK** to close it and return to Corridor Properties dialog box.

13. Click **OK** to exit the dialog box and assign the slope patterns to the corridor.

14. Use the ZOOM and PAN commands to better view the pattern.

15. At Civil 3D's top left, Quick Access Toolbar, click the **Save** icon to save the drawing.

Corridor Surfaces

Calculating corridor earthworks compares existing ground and the corridor datum elevations.

1. In the drawing, select any corridor segment, press the right mouse button, and from the shortcut menu, select CORRIDOR PROPERTIES....

2. Click the *Surfaces* tab.

This tab defines the surface names and assigns their data.

3. At the dialog box's top left, click the *Create a Corridor Surface* icon.

A surface entry appears.

4. Click in the Name column and change the surface name to **Rosewood – Top**.

5. In the dialog box top center, Add Data area, set the Data Type to **Links**. Set Specify Code to **Top**, and click the + (plus sign) to assign the link data.

6. Assign Rosewood – Top the surface style **Contours 1' and 5' (Design)**.

7. Click in the Overhang Correction column and select **Top Links** (see Figure 7.9).

8. To make a second surface, at the dialog box's top left, click the **Create a Corridor Surface** icon.

9. Click in the second surface's Name column and change the surface name to **Rosewood – Datum**. At the top center of the dialog box, Add Data area and set the Data Type to Links. Set Specify Code to **Datum** and click the + (plus sign) to add the link data.

10. Assign Rosewood – Datum the surface style **_No Display**.

11. Change Overhang Correction to **Bottom Links**.

12. Click the *Boundaries* tab.

13. Select **Rosewood – Top**, press the right mouse button, and in the Add Automatically flyout menu from the boundaries list select **Daylight**. Make sure the Use Type is **Outside Boundary** (see Figure 7.10).

14. Repeat the previous step and add the same boundary to **Rosewood – Datum**.

15. Click **OK** to create the surfaces and exit the dialog box.

16. Place your cursor over the corridor and review the corridor's station and surface elevations.

17. At Civil 3D's top left, Quick Access Toolbar, click the *Save* icon to save the drawing.

Any corridor changes automatically update the corridor surfaces.

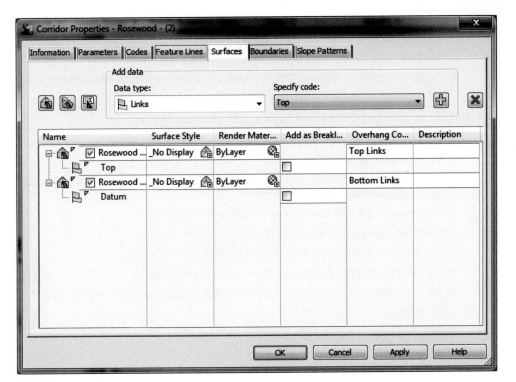

FIGURE 7.9

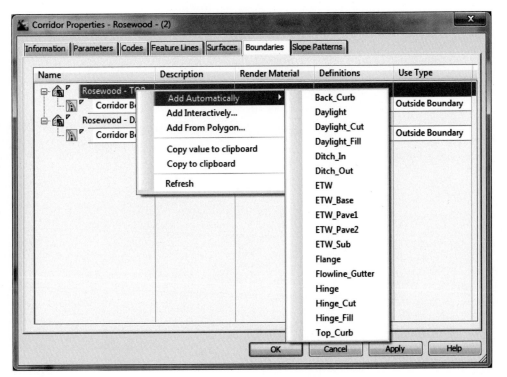

FIGURE 7.10

Calculate an Earthworks Volume

1. If necessary, click the Ribbon's **Modify** tab. On the Ground Data panel, click the **Surface** icon. In the Surface tab's Analyze panel, select VOLUMES.

2. In the Composite Volumes vista's upper left, click the **Create new volume entry** icon.

Clicking in the Base and Comparison Surface cells displays a surface drop-list.

3. In Base Surface, click twice in <select surface>, and from the surfaces list, select **Existing Ground**.

4. In Comparison Surface, click in <select surface>, and from the surfaces list, select **Rosewood – Datum**.

5. If necessary, click in the Cut cell to calculate a volume.

6. Click the Panorama's **X** to hide it.

7. Click the Tool Palettes's **X** to close the panel.

8. At the Surface tab's right, click the **X** to close the panel.

9. At Civil 3D's top left, Quick Access Toolbar, click the **Save** icon to save the drawing.

CHAPTER

8

Cross-Sections and Volumes

After completing this exercise, you will:

- Be familiar with Edit Drawing Settings.
- Be familiar with Sample line, Section, and Section View Edit Feature Settings.
- Be familiar with sample lines and sections Label Styles.
- Be familiar with sample lines and sections Edit Command Settings.

Exercise Setup

This exercise starts with the previous chapter's exercise drawing. The drawing contains an alignment, profile, assembly, and corridor. If you did not complete the Chapter 7 exercises, browse to the Chapter 08 folder of the CD that accompanies this textbook and open the *Chapter 08 – Unit 1.dwg* file.

1. If you are not in **Civil 3D**, double-click the **Civil 3D** desktop icon to start the application.
2. When you are at the command line, close the open drawing and **do not save it**.
3. At Civil 3D's top left, Quick Access Toolbar, click the Open icon to open the exercise drawing from the previous chapter or browse to the Chapter 08 folder of the CD that accompanies this textbook and open the *Chapter 08 – Unit 1* drawing. If using the chapter file, from the Menu Browser select Save As... and save the drawing as *Alignments*.

Edit Drawing Settings

Edit Drawing Settings affect sample lines, sections, and section views.

1. Click the **Settings** tab.
2. At Settings' top, click the drawing name, press the right mouse button, and from the shortcut menu, select EDIT DRAWING SETTINGS….
3. Select the **Object Layers** tab.

If you have multiple sample line groups and section objects, their layers entries should have a modifier and a value (see Figure 8.1).

4. In the Object Layers panel, change the modifier for **Sample Line, Sample Line-Labeling, Section, Section-Labeling, Section View, Section View-Labeling, Section View Quantity Takeoff Table**, and **Sheet** to **Suffix**. For the value, type (–*) (a dash followed by an asterisk).

5. Click **OK** to exit the dialog box.

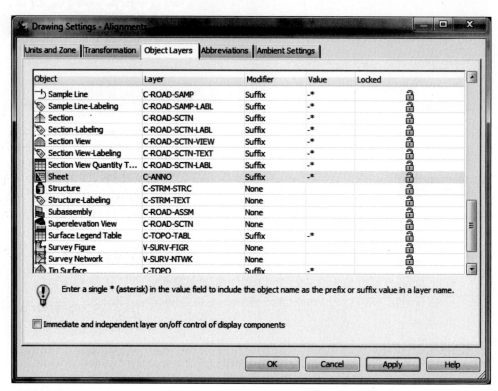

FIGURE 8.1

Edit Feature Settings — Sample Line

Sample Line's Edit Feature Settings dialog box sets default styles and name formats (see Figure 8.2).

1. In Settings, select the Sample Line heading, press the right mouse button, and from the shortcut menu, select EDIT FEATURE SETTINGS....

2. In Edit Feature Settings, expand the Default Styles and Default Name Format sections to review their values.

3. Click **OK** to close the dialog box.

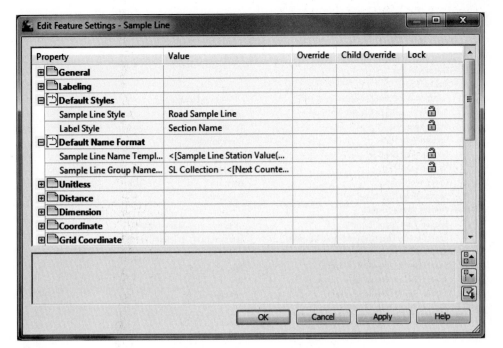

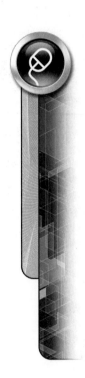

FIGURE 8.2

Edit Feature Settings — Section View

Section View's Edit Feature Settings assign default view styles, its section label sets, plotting styles, label styles, and default name formats (see Figure 8.3).

1. In Settings, select the Section View heading, press the right mouse button, and from the shortcut menu, select EDIT FEATURE SETTINGS….

2. In Edit Feature Settings, expand the Default Styles, Default Name Format, Section View Creation, and Default Projection Label Placement sections to review their values.

3. Click **OK** to close the dialog box.

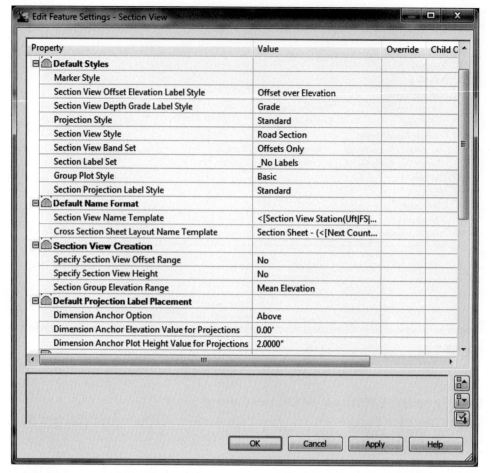

FIGURE 8.3

Edit Feature Settings — Section

Section's Edit Feature Settings set the default style and naming format for a section (see Figure 8.4).

1. In Settings, select the Section heading, press the right mouse button, and from the shortcut menu, select EDIT FEATURE SETTINGS….

2. In Edit Feature Settings, expand the Default Styles and Default Name Format sections to review their values.

3. Click **OK** to close the dialog box.

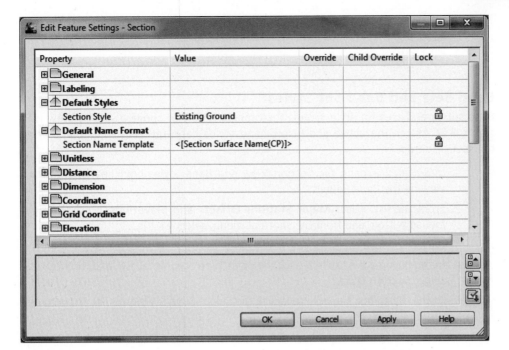

FIGURE 8.4

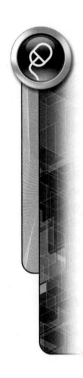

Object and Label Styles — Sample Line

Sample lines, section views, and section styles specifically label their critical values. Section view band sets and section label sets are aliases for a collection of styles with specific purposes.

The sample line object style sets its drawing layers. The sample line label styles affect its appearance and labeling.

1. In Settings, expand the Sample Line's Label Styles branch until you view its styles list.
2. From the styles list, select Section Name and Marks, press the right mouse button, and from the shortcut menu, select EDIT….
3. If necessary, click the **Layout** tab.
4. At Layout's top, click the Component name drop-list arrow to display the label component list. Select each one and review each component's values.
5. From the Component name drop-list, select **Sample Name**. In its Text section, click in the Contents value cell to display an ellipsis, and click the ellipsis to display the Text Component Editor.
6. Review this component's format string.
7. Click **Cancel** until you have returned to the command line.

Object and Label Styles — Section View

Section view is a grid that encloses the section station and elevations. The view styles provide basic station and elevation annotation, grid, exaggeration, and title content.

1. In Settings, expand the Section View branch until you view Section View Styles' styles list.
2. From the list, select **Road Section**, press the right mouse button, and from the shortcut menu, select EDIT….

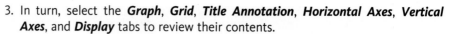

3. In turn, select the **Graph**, **Grid**, **Title Annotation**, **Horizontal Axes**, **Vertical Axes**, and **Display** tabs to review their contents.

4. Click **OK** to exit the Road Section view style.

Band Sets and Band Set Styles

Band sets are aliases for style groups that appear below or above a section view (see Figure 8.5). Assign the above and below band styles here, rather than when creating the section views.

1. In Settings, expand the Section View and Band Styles branches until you are viewing the Band Sets list.

2. From the list, select **Major Stations Offsets and Elevations**, press the right mouse button, and from the shortcut menu, select EDIT....

3. Click the **Bands** tab.

4. In the dialog box's top left, click the Band Type drop-list arrow, and from the list, select **Section Data**.

5. At the dialog box's middle top, click the Select Band Style drop-list arrow to display the styles list.

6. In the dialog box's top left, click the Band Type drop-list arrow and from the list, select **Section Segment**.

7. At the dialog box's top middle, click the Select Band Style drop-list arrow to display a styles list.

8. Click **Cancel** to close the dialog box.

The band set styles are below the Band Set heading, Section Data, and Section Segment headings list.

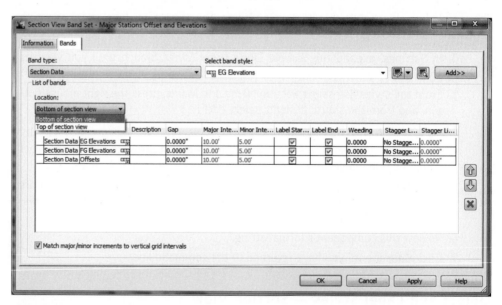

FIGURE 8.5

Section Data

1. In Settings, expand the Section View, Band Styles branch until you view the Section Data's styles list.

2. From the Section Data styles list, select **Offsets**, press the right mouse button, and from the shortcut menu, select EDIT....

3. Click the **Band Details** tab.

The panel's left side defines the band's title. At the top right, Labels and Ticks sets their size. At the center are the label's annotation locations. The Compose Label... button displays the Label Style Composer dialog box.

4. In the Band Details top middle, from the At: area, select **Centerline** and click **Compose Label...** to display the Label Style Composer.

5. In the Label Style Composer's Text section, click in the Contents value cell to display an ellipsis, and then click the ellipsis to display the Text Component Editor.

6. At the Text Component Editor's left side, click the Properties drop-list arrow to view the properties list.

7. Click **Cancel** until you have returned to the command line.

Section Segment

This band label type annotates section segment lengths and grades.

1. In Settings, expand the Section View, Band Styles branch until you view the Section Segment styles list.

2. From the Section Segment styles list, select **Segment Length**, press the right mouse button, and from the shortcut menu, select EDIT... .

3. If necessary, click the **Band Details** tab.

The panel's left side defines the band's title. At the center, the Labels and Ticks area lists all label types and tick sizes and locations. Compose Label... displays the Label Style Composer dialog box.

4. In Band Details' top middle, in the At: area, select **Segment Labels** and click **Compose Label...** to display the Label Style Composer.

5. In Label Style Composer's Text section, click in the Contents' value cell displaying an ellipsis, and then click the ellipsis to display the Text Component Editor.

This band label annotates the surface segment lengths.

6. In the Text Component Editor, at its top left, click the Properties drop-list arrow to view the label's list properties.

7. Click **Cancel** until you have returned to the command line.

Section Labels — Section Label Set

Section labels annotate surface elevations and stations within a section view.

Section label sets are aliases containing one or more style types: Major and Minor Offsets; Grade Breaks; and Segments style. A set also specifies the label's location and a weeding factor. Weeding removes overlapping labels.

1. In Settings, expand Section's Label Styles branches until you view the Label Sets' styles list.

2. From the styles list, select **FG Sections Labels**, press the right mouse button, and from the shortcut menu, select EDIT... .

3. If necessary, click the **Labels** tab.

4. At the dialog box's top left, click Type's drop-list arrow to view the style types list.

5. From the types list, select **Segments**.

6. In the dialog box's middle top, click the Section Segment Label Style's drop-list arrow to view its styles.

7. Click **Cancel** to return to the command line.

Section Labels — Major and Minor Offset

Major and Minor Offset label styles annotate offsets and elevations. A section view defines the major and minor styles intervals.

1. In Settings, expand the Section branch until you view the Major and Minor Offset styles list.

2. From the Major Offset styles list, select **Offset and Elevation**, press the right mouse button, and from the shortcut menu, select EDIT….

3. Click the **Layout** tab.

4. In the Layout tab, in the Text section, click in the Contents' value cell to display an ellipsis, and then click the ellipsis to display the Text Component Editor.

The label at the major station intervals annotates the section view.

5. In the Text Component Editor, at its top left, click the Properties drop-list arrow to view the label's list properties.

6. Click **Cancel** until you return to the command line.

Section Labels — Grade Break

Grade Break styles label a surface or design section grade break using EG or FG. The labeling appears in the section view.

1. In Settings, expand the Section branch until you view the Grade Break styles list.

2. From the styles list, select **FG Section Offset and Elevation**, press the right mouse button, and from the shortcut menu, select EDIT….

3. If necessary, click the **Layout** tab.

4. Set the Component name to txtOffset. In its Text section, click in the Contents' value cell to display an ellipsis, and then click the ellipsis to view its format string.

5. Click **Cancel** until you return to the command line.

Section Labels — Segment

This label type annotates a surface section segment length and cross slope.

1. In Settings, expand the Section branch until you view the Segment styles list.

2. From the styles list, select the **Percent Grade**, press the right mouse button, and from the shortcut menu, select EDIT….

3. Click the **Layout** tab.

4. In the Text section, click in the Contents' value cell to display an ellipsis, and then click the ellipsis to display its label format string.

5. In the Text Component Editor, at its top left, click the Properties drop-list arrow to view the label's list properties.

6. Click **Cancel** until you return to the command line….

Command — Create Sample Lines

Commands create sample lines, section views, and sections default settings in addition to those set by Edit Feature Settings.

The Create Sample Line command samples a corridor at an interval, at specific stations, or at a station range.

1. In Settings, expand the Sample Line branch until you view the Commands list.

2. From the list, select **CreateSampleLines**, press the right mouse button, and from the shortcut menu, select EDIT COMMAND SETTINGS….

3. Expand the Default Swath Widths section.

A swath width is the distance sampled to the roadway's left and right.

4. Expand the Sampling Increments section.

This section sets the initial corridor sampling frequency.

5. Expand the Additional Sample Controls section.

This section sets the sampling of additional critical corridor points.

6. Expand the Miscellaneous section.

If Lock To Station is true, if any change occurs to the alignment geometry or properties, then the sections resample.

7. Expand the Default Styles and Default Name Format sections.

Sample Line Edit Feature Settings sets these styles, and you can change them here.

8. Click **Cancel** to return to the command line.

Command — Create Section View

Create Section View (single section) and Create Multiple Section Views (multiple sections) commands use the same settings.

1. In Settings, expand the Section View branch until you view the Commands list.

2. From the list, select **CreateSectionView**, press the right mouse button, and from the shortcut menu, select EDIT COMMAND SETTINGS….

3. Expand the Table Creation, Default Styles, Default Name Format, Section View Creation, and Default Projection Label Placement sections and review their values.

Table Creation controls the format and behavior of quantity takeoff tables. The Default Styles lists the style for a new section view. The Default Name Format specifies a new view's naming format. Section View Creation governs prompting for Offset Range and Height, and how to display the section group's elevation range. Default Projection Label Placement controls the placement of labels for objects projected to section views.

4. Click **Cancel** to return to the command line.

5. At Civil 3D's top left, Quick Access Toolbar, click the **Save** icon to save the drawing.

This ends the exercise that reviews objects, styles, and commands that affect sample lines, section views, and sections. The next unit reviews creating section sample lines.

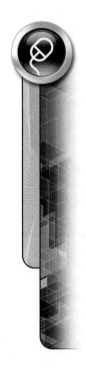

EXERCISE 8-2

After completing this exercise, you will:

- Be able to create sample lines.
- Be able to create sample lines by different methods.
- Be familiar with the grid view of a section.
- Be able to view sample line properties.

Drawing Setup

This exercise continues with the previous exercise's drawing. If you did not complete the previous exercise, browse to the Chapter 08 folder of the CD that accompanies this text-book and open the *Chapter 08 – Unit 2* drawing.

1. If not open, open the previous exercise's drawing or browse to the Chapter 08 folder of the CD that accompanies this textbook and open the *Chapter 08 – Unit 2* drawing. If using the chapter file, from the Menu Browser select Save As... and save the drawing as *Alignments*.

Create Sample Lines

Create Sample Lines displays a Create Sample Line Group dialog box. This dialog box identifies what components to include in the sampling. After the sample component has been identified, the Sample Line Tools toolbar is displayed. The Sample Line methods icon stacks sets and changes the sampling methods.

1. On the Ribbon, click the **Home** tab. On the Profile & Section Views panel, click the **Sample Lines** icon.

2. The command line prompts you for an alignment. If you are able to select the alignment from the drawing, select it or right mouse click, from the list select **Rosewood – (1)**, and click **OK** to continue displaying the Create Sample Line Group dialog box.

3. In the dialog box, toggle off Rosewood – Top, change the styles for Rosewood – TOP and DATUM to **Finished Ground**, and click **OK** to display the Sample Line Tools toolbar with a jig attached to the alignment that reports its current station.

4. Move your cursor around the corridor to view its station reporting. If necessary, zoom or pan to better view the corridor.

5. In the Sample Line Tools toolbar, to the Sample Line Group name's right, click the Sample line creation method's icon stack drop-list arrow, and from the list, select FROM CORRIDOR STATIONS.

The Create Sample Lines dialog box opens and lists the beginning and ending stations and the left and right swath width.

6. Click **OK** to create the sample lines.

The sample lines appear, and creation mode returns to At a Station.

7. Without changing the creation mode, in the drawing, select a couple of stations and, when you are prompted, for both swath widths enter **50**.

These stations are added to the sample line list's end.

Reviewing Sample Line Data

1. On the Sample Line Tools toolbar's right, click the **SampleLine Entity View** icon to display the Edit Sample Line dialog box editor.

2. In the Sample Line Tools toolbar to the left of the SampleLine Entity View icon, click the **Select/Edit Sample Line** icon, and in the drawing, select a sample line.

The sample line's properties are displayed in the editor. You can change the station name, swath width, or review its station and elevations (center, left, and right).

3. Click the **Next Vertex** or **Previous Vertex** to review the section values.

4. Click the editor's red **X** to close it.

5. Press ENTER to exit the Create Sample Lines command.

6. At Civil 3D's upper left, Quick Access Toolbar, click the *Save* icon to save the drawing.

Sample Line Group Properties

Each SLG has properties: Sample Line Properties include sample line data, what has been sampled, what section views use the section's data, and material list data.

1. Click the *Prospector* tab.
2. In Prospector, expand the Sites branch until you are viewing the Rosewood – (1), Sample Line Groups' SL Collection – 1 branch.
3. Select the SL Collection – 1 heading, press the right mouse button, and from the shortcut menu, select PROPERTIES….
4. Click the *Sample Lines* tab to review its information.

This lists the group's sample lines, their layer, style, and swath width offset.

5. Click the *Sections* tab.

This displays the sample groups sample components list.

6. Click *OK* to exit the dialog box.

Sample Line Properties

1. If necessary, in Prospector, expand the Site1 branch until you are viewing the Sample Lines heading under SL Collection – 1.
2. Select the Sample Line heading to display, in Preview, the sample line list.
3. Scroll the sample line list until you locate line **2+75**, select it, press the right mouse button, and from the shortcut menu, select ZOOM TO.
4. In Preview, select **2+75**, press the right mouse button, and from the shortcut menu, select PROPERTIES….
5. Click the *Sample Line Data* tab.

This panel's content lists the section's location, number, alignment, and if the sample line is station locked.

6. Click the *Sections* tab.

This panel lists the section's sampled components.

7. Click *Cancel* to exit the dialog box.
8. At Civil 3D's upper left, Quick Access Toolbar, click the *Save* icon to save the drawing.

This ends the unit on creating sample lines and their properties. Next, you will create section views with sections.

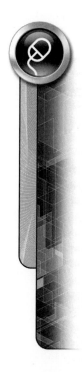

EXERCISE 8-3

After completing this exercise, you will:

- Be familiar with the settings for paper styles.
- Be familiar with the plot page and plot all styles.
- Be able to adjust and plot pages of sections.
- Be able to plot all of the sections.
- Be familiar with section view properties.
- Be familiar with section properties.

Exercise Setup

This exercise continues with the previous exercise's drawing. If you did not complete the previous exercise, browse to the Chapter 08 folder of the CD that accompanies this textbook and open the *Chapter 08 – Unit 3.dwg* file.

The next step is to create a section sheet. After reviewing the section sheet, you will be defining the section annotation.

1. If not open, open the previous exercise's drawing or browse to the Chapter 08 folder of the CD that accompanies this textbook and open the *Chapter 08 – Unit 3* drawing. If using the chapter file, from the Menu Browser select Save As... and save the drawing as *Alignments*.

Review Page Sheet Style

When you plot with a page group style, the style defines the sheet size, its printable area, and its grid spacing. The Model or Layout space default layout must make the sheet style size definition (see Figure 8.6).

1. Click the **Settings** tab.
2. In Settings, expand the Section View branch until you view the Sheet Styles' styles list.
3. From the list, select **Sheet Size – D (24×36)**, press the right mouse button, and from the shortcut menu, select EDIT....
4. Click the **Sheet** tab.

The default model plot sheet size must set the size for this style.

5. Click **Cancel** to exit the dialog box.

Plot By Page

Plot By Page plots sections as an array. Currently, the array is four sections wide and has as many rows as necessary to plot all of the sections.

1. In Settings, expand the Section View branch until you view the Group Plot Styles' styles list.
2. From the Styles list, select **Plot By Page**, press the right mouse button, and from the shortcut menu, select EDIT....
3. Click the **Array** tab.
4. In the Array tab, set the Row and Column spacing to 1.00".
5. Click the Plot Area Tab. When you are done, click **OK** to exit the dialog box.
6. Use the PAN and ZOOM commands to move the site west.
7. At Civil 3D's upper left, Quick Access Toolbar, click the **Save** icon to save the drawing.

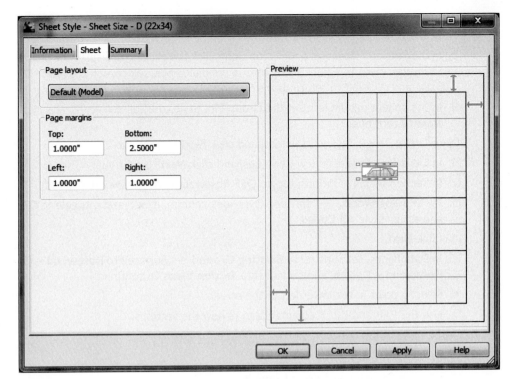

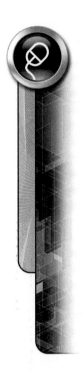

FIGURE 8.6

Set Model Space Plotting

The model space paper size must match the section view sheet size.

1. Click the Ribbon's Output tab, from the Plot panel, select `PAGE SETUP MANAGER`.

2. With Model highlighted, click `MODIFY...` to display the Plot Setup – Model dialog box.

3. Click the Paper size's drop-list arrow, and from the list of paper sizes, select **Arch expand D (36.00✕24.00 inches)**.

4. Click **OK** and then click Close to return to the command line.

5. At Civil 3D's top left, Quick Access Toolbar, click the **Save** icon to save the drawing.

Creating Section Views

1. Click the **Prospector** tab.

2. In Prospector, expand the Corridors branch.

3. Pan the drawing so you have empty space for the sections.

4. If the Rosewood corridor is out-of-date, select Rosewood – (1), press the right mouse button, and from the shortcut menu, select `REBUILD`.

5. If necessary, set the annotation scale to **1″ = 20′**,

6. Click the **Modify** tab, on the Profile & Section Views panel, and click the **Sample Line** icon. From the Ribbon's Sample Line tab's Launch Pad panel, select the **Create Section View** icon displaying a shortcut menu and from the shortcut menu, select `CREATE MULTIPLE SECTION VIEWS` to display the Create Multiple Section Views wizard.

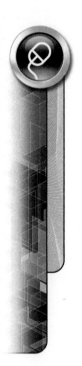

7. In the General panel's lower left, make sure the section view style is set to Road Section and then click Next.

8. In the Section Placement panel's top right, click the ellipsis, to display the Select Layout as Sheet Template dialog box. In the dialog box, select Arch D Section 20 Scale and click OK to return to the Section Placement panel.

9. In the panel's middle left, set the Group Plot Style to PLOT BY PAGE, and click **Next** to continue.

10. In Offset Range, review its values, and click **Next** to continue.

11. In Elevations Range, review its values, and click **Next** to continue.

12. In Section Display Options, toggle **OFF** Rosewood – (1) Rosewood – Datum.

13. For the Rosewood – (1) corridor Overrides column, if necessary, toggle it **ON** and assign style **All Codes**.

14. Click **Next**.

15. In Data Bands, set Surface1 to **Existing Ground**, set Surface2 to **Rosewood – (1) Rosewood – Datum**, and click **Create Section Views** to continue.

16. Select a point in the lower left of the screen.

17. Use the PAN and ZOOM commands to review the sections.

The sections sit properly in the sheet. The Plot by Page settings are correct for your section sheets.

18. Close the drawing and **do not save it**.

This concludes creating section views by page. The key to successful sheets is setting the Model Space sheet size to the desired sheet style size.

Next is defining is the cross-section's corridor section labels from the Multipurpose style, All Codes.

EXERCISE 8-4

After completing this exercise, you will:

- Be familiar with the code set label styles.
- Be familiar with the code label styles.
- Be able to assign a code label style for a code label set.

Exercise Setup

This exercise continues with the previous exercise's drawing. If you did not complete the previous exercise, browse to the Chapter 08 folder of the CD that accompanies this textbook and open the *Chapter 08 – Unit 4.dwg* file.

1. Open the previous exercise's drawing or browse to the Chapter 08 folder of the CD that accompanies this textbook and open the *Chapter 08 – Unit 4* drawing. If using the chapter file, from the Menu Browser select Save As... and save the drawing as *Alignments*.

Set Model Space Plotting

The model space paper size must match the section view sheet size.

1. Click the Ribbon's Output tab, from the Plot panel, select PAGE SETUP MANAGER.

2. With Model highlighted, click MODIFY... to display the Plot Setup – Model dialog box.

3. Click the Paper size's drop-list arrow, and from the list of paper sizes, select **Arch expand D (36.00✕24.00 inches)**.

4. Click *OK* and then click Close to return to the command line.

Review All Codes Code Set Style

All Codes Set style contains all possible subassembly codes, links, and shapes.

1. Click the *Settings* tab.

2. In Settings, expand the General, Multipurpose Styles branch until you are viewing the Code Set Styles' style list.

3. From the styles list, select **All Codes**, press the right mouse button, and from the shortcut menu, select EDIT....

4. If necessary, click the *Codes* tab.

5. Review the Link label style assignments.

The only assigned label style is Daylight (Steep Grades).

6. Review the Point section.

7. In the Point section, for BackCurb and Daylight do not have an assigned label style.

8. In their Label Style column assign the **Offset Elevation** style, and click *OK* until you return to the command line.

Modifying Labels

The Daylight slope and Offset and Elevation labels need to be adjusted.

1. Expand the General, Labels Styles branch until you view the Link styles list.

2. From the styles list, select **Steep Grades**, press the right mouse button, and from the shortcut menu, select EDIT....

3. If necessary, click the *Layout* tab.

4. In the Label Style Composer – Steep Grades, Text section, click in the Text Height value cell, change the size to **0.03**, and click *OK* to exit the dialog box.

5. Expand the Label Style's Marker branch to display its styles list.

6. From the styles list, select **Offset Elevation**, press the right mouse button, and from the shortcut menu, select EDIT....

7. In the Label Style Composer – Offset Elevation, Text section, click in the Text Height value cell, and change the size to **0.03**.

8. Click the Component name drop-list arrow, and select Point Code from the list. Set its size to **0.03**. Scroll to the Border section, set Background Mask to **False**, and click *OK* to exit the dialog box.

9. At Civil 3D's top left, Quick Access Toolbar, click the *Save* icon to save the drawing.

Create Multiple Section Views

It is time to review the sections with their new labels.

1. Click the *Prospector* tab.

2. In Prospector, expand the Corridors branch.

3. If the Rosewood corridor is out-of-date, select **Rosewood – (1)**, press the right mouse button, and from the shortcut menu, select REBUILD.

4. Use the ZOOM EXTENTS command to view an open area of the drawing.

5. If necessary, set the Annotation scale to **1" = 20**.

6. On the Sample Line tab, the Launch Pad panel, select **Create Section View** displaying a shortcut menu and from the shortcut menu, select CREATE MULTIPLE SECTION VIEWS to display the Create Multiple Section Views wizard. If the Sample Line tab, is not displayed on the Ribbon, click the **Modify** tab and on the Profile & Section Views panel, click the **Sample Line** icon.

7. After reviewing the values in the General panel, click next to display the Section Placement panel.

8. In the panel's middle, click the browse button to make sure the template for section sheet is set to ARCH D – Section 20 scale.

9. In the Group Plot Style section change it to **Plot By Page**, and click **Next** to continue.

10. In Offset Range, review its values, and click **Next** to continue.

11. In Elevations Range, review its values, and click **Next** to continue.

12. In Section Display Options, toggle **OFF** Rosewood – (1) Rosewood – Datum.

13. For the Rosewood – (1) corridor Style column, assign the style **All Codes**.

14. Click **Next**.

15. In Data Bands, set Surface1 to **Existing Ground**, set Surface2 to **Rosewood – (1) Datum**, and click **Create Section Views** to continue.

16. Select a point in the lower left of the screen.

17. Use the PAN and ZOOM commands to review the sections.

18. Close the drawing and **do not save it**.

This concludes this unit's discussion on labeling corridor components. The next unit covers calculating earthwork and material quantities and then adding the volumes as section view tables.

EXERCISE 8-5

After completing this exercise, you will:

- Be familiar with the quantity criteria takeoff.
- Be able to create a quantity takeoff earthwork and material report.
- Be able to create a quantity takeoff table.

Exercise Setup

This exercise continues with the previous exercise's drawing. If you did not complete the previous exercise, browse to the Chapter 08 folder of the CD that accompanies this textbook and open the *Chapter 08 – Unit 5.dwg* file.

1. Open the previous exercise's drawing or browse to the Chapter 08 folder of the CD that accompanies this textbook and open the *Chapter 08 – Unit 5* drawing. If using the chapter file, from the Menu Browser select Save As... and save the drawing as *Alignments*.

Material List — Criteria Styles

First, you define a cut and fill and then an earthworks material list. Sections use cut and fill data for hatching an assembly's cut and fill areas and earthworks is used as a section label or a detailed earthworks report.

1. Click the *Settings* tab.
2. Expand the Quantity Takeoff branch until you are viewing the Quantity Takeoff Criteria list.
3. From the list, select **Cut and Fill**, press the right mouse button, and from the shortcut menu, select EDIT….
4. Click the *Material List* tab.
5. Expand the Material Name sections and review the values.

The Condition settings are confusing. The settings do not define the surfaces' relative positions, but do define the areas that are cut and fill. For example, the area below EG and above Datum creates a cut area. The area below Datum and above EG creates a fill area.

6. Click *OK* to exit the dialog box.
7. From the Quantity Takeoff Criteria list, select **Earthworks**, press the right mouse button, and from the shortcut menu, select EDIT….
8. Click the *Material List* tab.
9. Expand the Material Name and review the values.

The criteria compare a base to a comparison surface, just like the Ribbon's Modify, Surface panel's, Volumes routine.

10. Click *OK* to exit the dialog box.

The assembly has five potential volume materials: Pave1, Pave2, Base, Subbase, and Curb.

11. From the Quantity Takeoff Criteria list, select **Material List**, press the right mouse button, and from the shortcut menu, select EDIT….
12. Click the *Material List* tab.
13. Expand the Material Name sections and review the values.

The materials list contains three materials: Pavement, Base, and SubBase. The list needs two more materials: Binder and Curb.

14. In the dialog box's top left, make a new material entry by clicking *Add New Material*.
15. Click in the new Material's Name cell, and for its name enter **Binder**.
16. Click in Binder's Quantity Type cell, and from the drop-list select **Structures**.
17. At the Panel's top center, change the Data Type to **Corridor Shape**. Adjacent to the Data Type, click Select Corridor Shape's drop-list arrow, from the list select **Pave2**, and click the plus sign (+), to add Pave2 to Binder.
18. Repeat Steps 14–17 and add **Curb** to the material list. The Curb material is a **structure** using the corridor shape of **Curb**.
19. For the Pavement entry, click its Pavement shape (Pavement Material) and click the red **X** to delete it.
20. With Pavement still highlighted, at the top center, change the Data Type to **Corridor Shape**, from the Select Corridor Shape list select **Pave1**, and click the plus sign (+), to add Pave1 as Pavement's data type.
21. Change the Shape Style for Binder to **Pave2**, and the Shape Style for Curb to **Curb**.
22. Your values should match those in Table 8.1 and Figure 8.7.

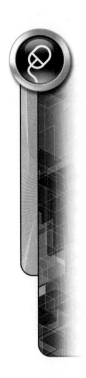

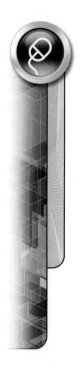

TABLE 8.1

Material Name	Quantity Type	Data Type	Corridor Shape and Style
Pavement	Structures	Corridor Shape	Pave
SubBase	Structures	Corridor Shape	Subbase
Base	Structures	Corridor Shape	Base
Binder	Structures	Corridor Shape	Pave
Curb	Structures	Corridor Shape	Curb

23. Click **OK** to exit the Quantity Takeoff Criteria – Material List dialog box.

24. At Civil 3D's top left, Quick Access Toolbar, click the **Save** icon to save the drawing.

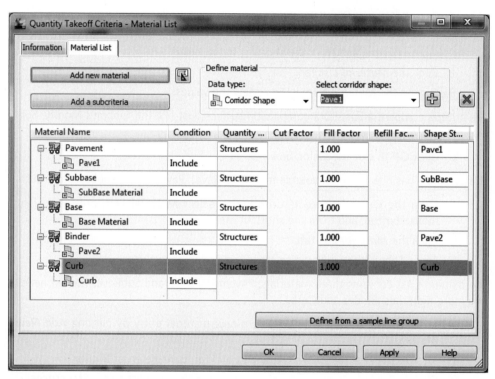

FIGURE 8.7

Sample Line Properties Material Lists — Cut and Fill

1. Use the ZOOM and PAN commands to view the corridor and its section lines.

2. Click the **Prospector** tab.

3. In Prospector, expand the Sites branch until you are viewing the Rosewood – (1) Sample Line Group groups list.

4. From the sample line list, select **SL Collection – 1**, press the right mouse button, and from the shortcut menu, select PROPERTIES….

5. Click the **Material List** tab.

6. At the panel's lower right, click Import Another Criteria to display the Select a Quantity Takeoff Criteria dialog box. From its criteria list, select **Cut and Fill**, and click *OK* to display the Compute Materials dialog box.

7. In the dialog box's top center, in EG's Object Name cell, click (<Click here to set all>), and from the surface list, select **Existing Ground**.

8. Repeat the previous step for the Datum entry and select **Rosewood – (2) Rosewood – Datum**.

9. Click *OK* to compute the materials and return to the Sample Line Group Properties dialog box.

Sample Line Properties Material Lists — Earthworks

1. At the panel's lower right, click *Import Another Criteria* to display the Select a Quantity Takeoff Criteria dialog box. Click the drop-list arrow, from its list of criteria, select **Earthworks**, and click *OK*.

2. The Compute Materials dialog box opens. In the top center, in Existing Ground's Object Name cell, click (<Click here>), and from the surface list, select **Existing Ground**.

3. In Datum's Object Name cell, click (<Click here>), and from the surface list, select **Rosewood – (2) Rosewood – Datum**.

4. Click *OK* until you return to the command line.

5. At Civil 3D's top left, Quick Access Toolbar, click the *Save* icon to save the drawing.

Material List — Sample Line Panel

1. From the Sample Line tab's Launch Pad panel, select COMPUTE MATERIALS. If the Sample Line tab, is not displayed on the Ribbon, click the Modify tab and on the Profile & Section Views panel, click the Sample Line icon.

2. In the Select a Sample Line Group dialog box, set the alignment to **Rosewood – (1)**, the Sample line group to **SL Collection – 1**, and click *OK* to continue.

3. In the Edit Material List's bottom right, click *Import Another Criteria* button.

4. In the Select a Quantity Takeoff Criteria dialog box, click the drop-list arrow, select **Material List**, and click *OK*.

5. Using Figure 8.8 as a guide, set the object types in Compute Materials. Click Map objects with same name to fill in some of the values.

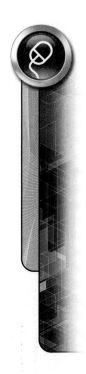

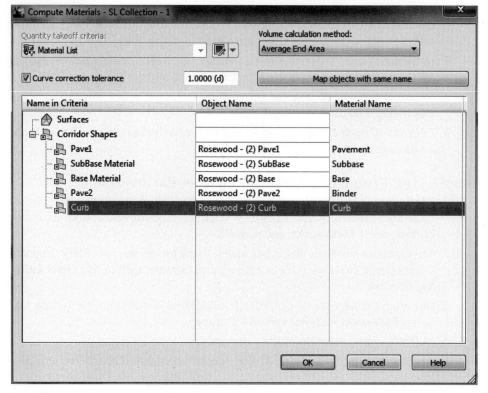

FIGURE 8.8

6. Click **OK** twice to add the materials to the list and to exit the dialog boxes.

7. At Civil 3D's top left, Quick Access Toolbar, click the **Save** icon to save the drawing.

Calculating Earthwork Volumes

1. From the Sample Line tab's Launch Pad panel, select GENERATE VOLUME REPORT.

The Report Quantities dialog box opens.

2. In the Report Quantities dialog box, set the Alignment to **Rosewood – (1)**.

3. In the Report Quantities dialog box, set the Material List to **Material List – (2)**.

4. In the Report Quantities dialog box, to the right of Select a style sheet, click its icon, in the Select Style Sheet dialog box, select **Earthwork.xsl**, and click the **Open** button.

5. Click **OK** to create an earthwork volume report.

6. If you get a scripts warning, click **Yes** to continue.

7. Review the values, and after reviewing them, click the Internet Explorer **X** to close the report window.

Calculating Material Volumes

1. From the Sample Line tab's Launch Pad panel, select GENERATE VOLUME REPORT.

The Report Quantities dialog box opens.

2. In the Report Quantities dialog box, set the Alignment to **Rosewood – (1)**.

3. In the Report Quantities dialog box, set the Material List to **Material List – (3)**.

4. In the Report Quantities dialog box, to Select a style sheet's right, click its icon, in the Select Style Sheet dialog box, select **Select Material.xsl**, and click **Open**.

5. Click **OK** to create a materials volume report.

6. If you get a scripts warning, click **Yes** to continue.

7. Review the values, and after reviewing them, click the Internet Explorer **X** to close the report.

Creating a Volume Table

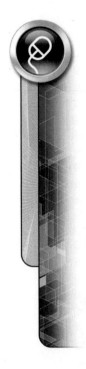

1. Use the ZOOM EXTENTS command to view an open area of the drawing.

2. From the Sample Line tab's Labels & Tables panel, click the Add Tables icon, and from the drop-list, select TOTAL VOLUME.

3. In the Create Total Volume Table dialog box, set the Alignment to **Rosewood – (1)**, set the Material List to **Material List – (1)**, and click **OK**.

4. The command line prompts you for a table location. In the drawing, select a point below the profile view to locate the table.

5. Use the ZOOM and PAN commands to better view the table.

6. At Civil 3D's top left, Quick Access Toolbar, click the **Save** icon to save the drawing.

Create a Mass Haul Diagram

1. Use the PAN and ZOOM commands to view an open area in the drawing.

2. From the Sample Line tab's Launch Pad panel, select CREATE MASS HAUL DIAGRAM.

3. The Create Mass Haul Diagram wizard opens. Review its contents and click **Next**.

4. Set the Material list to **Material List – (1)**, set the Material to display as mass haul to **Total Volume**, and click **Next**.

5. Review the values in the Balancing Options panel, and, finally, click **Create Diagram**.

6. The command line prompts you for a diagram location. In the drawing, select a point to locate the Mass Haul View.

7. Use the ZOOM and PAN commands to better view the graph.

8. At Civil 3D's top left, Quick Access Toolbar, click the **Save** icon to save the drawing.

Set Model Space Plotting

The model space paper size must match the section view sheet size.

1. Click the Ribbon's Output tab, from the Plot panel, select PAGE SETUP MANAGER.

2. With Model highlighted, click MODIFY... to display the Plot Setup – Model dialog box.

3. Click the Paper size's drop-list arrow, and from the list of paper sizes, select **Arch expand D (36.00✕24.00 inches)**.

4. Click **OK** and then click Close to return to the command line.

5. If necessary, set the Annotation scale to **1" = 20'**.

Creating Sections with Volume tables

1. Use the PAN and ZOOM commands to view an open area in the drawing.

2. From the Sample Line tab's Launch Pad panel, click the **Create Section View** icon and from the drop-list, select CREATE MULTIPLE SECTION VIEWS.

3. In Create Multiple Section Views – General, make sure the Section View style is set to Road Section and click Next.

4. In the Section Placement panel, click the Template for cross-section sheet, select AECH D Section 20 Scale and click OK to return to the Sections Wizard.

5. In the panel's middle left, change the Group plot style to **Plot by Page**, and click **Next**.

6. In Offset Range, click **Next**.

7. In Elevation Range, click **Next**.

8. In Section Display Options, toggle OFF **Rosewood – (1) Rosewood – Datum, Earthworks, Pavement, Subbase, Base, Binder,** and **Curb**.

9. For the Rosewood – (1) corridor Style column assign **All Codes**.

10. Click **Next**.

11. In Data Bands, set Surface1 to **Existing Ground**, and click **Next**.

12. In Section View Tables, set Type to **Total Volume**, set Table Style to **Standard**, and click **Add>>**.

13. In Section View Tables, Position of Tables, set X-Offset to **0.5**, and click **Create Section Views**.

14. Select a point in the lower left of the screen.

15. Use the PAN and ZOOM commands to review the sections and tables.

16. At the Sample Line tab's right, click the **X** to close the panel.

17. At Civil 3D's top left, Quick Access Toolbar, click the **Save** icon to save the drawing.

Creating a Cross-Section Sheets

If you want to create cross-section views in a layout, use the Output tab's Create Section Sheets. You must have sections present in the drawing to use this command.

1. Click the Output tab and in the Plan Production panel, select Create Section Sheets to display the Create Section Sheets dialog box.

2. Match your settings to those in Figure 8.9 and when correct, click Create Sheets to create the section layouts.

A dialog box displays stating the drawing has to be saved to continue.

3. Click OK to continue.

The Sheet Set Manager displays.

4. In Sheet Manager, select a few sheets and review their contents.

5. Close Sheet Set Manager and save the drawing.

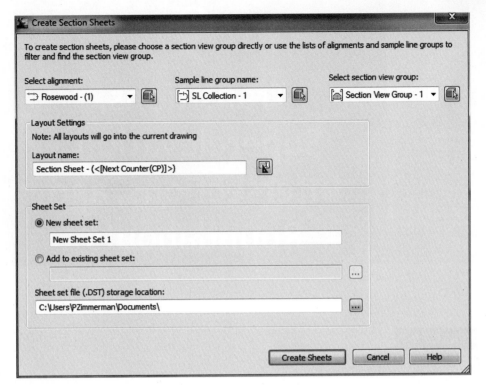

FIGURE 8.9

Creating a Cross-Sections Report

1. If necessary, from the Home tab's Palettes panel, click the Toolbox icon to display the Toolbox tab.

2. Click the Toolbox tab.

3. In Toolbox, expand the Reports Manager section and then expand the Corridor section.

4. From Corridor's report list select Daylight Line, right click, and select **Execute** to display the Daylight Line Report application.

5. The report should include SL Collection-1 and then click Create Report.

The Report lists the daylight points, their station and offset, coordinates, and elevations.

6. Close the report and the report generator.

7. From Corridor's report list select Slope Stake Report, right click, and select **Execute** to display the Slope Stake Report application.

8. Set the report values Corridor to **Rosewood – (2)**, Alignment to **Rosewood – (1)**, Sample Line group to **SL Collection-1**, set the Corridor Link to **Datum**, and click the plus sign (+) to add the values to the List of Corridors. When set, click Create Report.

The Report lists each datum point, their station and offset, slope, and elevation.

9. Close the report and the report generator.

10. Save the drawing and then close it.

This ends the review of corridor section and volumes. Sample Line Groups are the basis for cross-sections, their annotation, and volume reports. Material lists compute section data for reports or tables that display volume values.

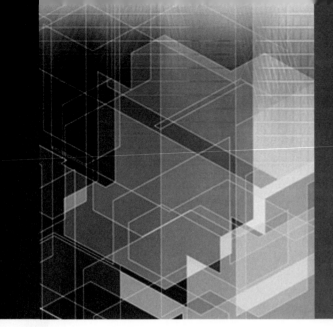

Transitions, Superelevation, Intersections, and Roundabouts

EXERCISE 9-1

After completing this exercise, you will:

- Be familiar with creating a transition assembly.
- Be familiar with the BasicLane transition subassembly.
- Be able to transition a lane using an offset horizontal and vertical alignment.
- Create a widening.

Exercise Setup

This exercise uses the *Chapter 09 - Unit 1.dwg* file. To find this file, browse to the CD's Chapter 09 folder that accompanies this textbook and open the *Chapter 09 - Unit 1.dwg* file.

1. If you are not in **Civil 3D**, double-click the **Civil 3D** desktop icon to start the application.

2. When you are at the command line, CLOSE the open drawing and **do not save it**.

3. At Civil 3D's top left, Quick Access toolbar, click the Open icon, browse to the Chapter 09 folder of the CD that accompanies this textbook, and open the *Chapter 09 - Unit 1* drawing.

4. At Civil 3D's top left, click Civil 3D's drop-list arrow, from the Application Menu, highlight SAVE AS, from the flyout menu, select AUTOCAD DRAWING, browse to the Civil 3D Projects folder, for the drawing name enter **Basic Transition**, and click **Save** to save the file.

5. If necessary, click Ribbon's Home tab, in the Palettes panel, click the **Tool Palettes** icon to display the Civil 3D Imperial tool palette.

Add Side to Subassembly Name

1. Click the **Settings** tab.

2. In Settings, select *Subassembly*, press the right mouse button, and from the list, select EDIT FEATURE SETTINGS….

3. Expand the Subassembly Name Templates section, click the **Create From Macro** value cell to display an ellipsis, and click the ellipsis to open the Name Template dialog box.

4. Place the cursor after the dash between the two name values, press the Spacebar **twice**, add a dash (-), and press the back arrow to place the cursor between the two spaces.

5. Click the Property Fields drop-list arrow, and from the list, select **Subassembly Side**.

6. Click *Insert* to place the Subassembly side in the name template (see Figure 9.1).

7. Click **OK** until you have returned to the command line.

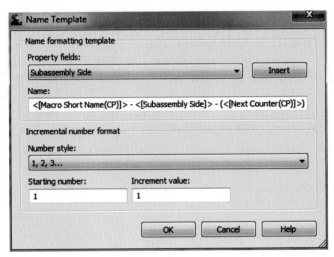
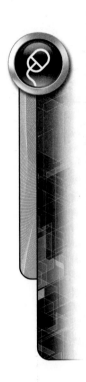

FIGURE 9.1

Create a Basic Lane Transition Assembly — Senon

This assembly uses the basic lane transition, basic curb and gutter, and urban sidewalk subassemblies (see Figure 9.2).

FIGURE 9.2

1. From Ribbon's Home tab, Create Design panel, click the Assembly icon, and from the shortcut menu, select CREATE ASSEMBLY.

2. In the Create Assembly dialog box, for the assembly name, enter **Senon**, leaving the counter. Click **OK**, and in the drawing, select a point to the profile's left.

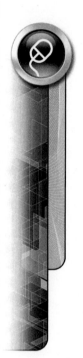

Basic Lane — Right Side and Basic Lane Transition — Left Side

1. On the tool palette, select the *Imperial - Basic* tab, and from the palette, select *BasicLane* to display its properties dialog box.

2. In Properties, the Parameters area, if necessary, change Side to **Right**, set the lane width to **14** feet, in the drawing select the assembly to attach the subassembly, and press ENTER twice to exit.

3. On the tool palette, select the *Imperial - Basic* tab, and from the palette, select *BasicLaneTransition* to display its properties dialog box.

4. In Properties, the Parameters area, set the lane width to **14** feet and set the Side to **Left** and the Transition property to **Hold grade, change offset**. In the drawing, attach the subassembly by selecting the assembly and press ENTER twice to exit the routine.

Curb, Gutter, and Sidewalk

1. Repeat the process of attaching subassemblies and add to the right and left sides a *BasicCurbAndGutter*. The curbs attach to the BasicLaneTransition's outside top.

2. Repeat attaching the subassemblies and add a **BasicSidewalk** to the right and left sides. The sidewalks attach to BasicCurbAndGutter's outside top.

3. At Civil 3D's top left, Quick Access toolbar, click the *Save* icon to save the drawing.

Your assembly should now look like Figure 9.2.

Assembly Properties

1. Click the *Prospector* tab.

2. In Prospector, expand the Assemblies branch. From the list, select **Senon - (1)**, press the right mouse button, and from the shortcut menu, select PROPERTIES….

3. Click the *Construction* tab.

There should be two groups: one (right side) and two (left side) of the assembly.

4. In the dialog box, select the Group - (1) heading that represents the assembly's right side, and press the right mouse button. From the shortcut menu, select *RENAME*, and rename the group **Non-Transitioning Right**.

5. In the dialog box, select the Group - (2) heading that represents the assembly's left side, and press the right mouse button. From the shortcut menu, select *RENAME*, and rename the group **Transitioning Left**.

Your Assembly Properties dialog box should now look like Figure 9.3.

6. Click *OK* to exit the Assembly Properties dialog box.

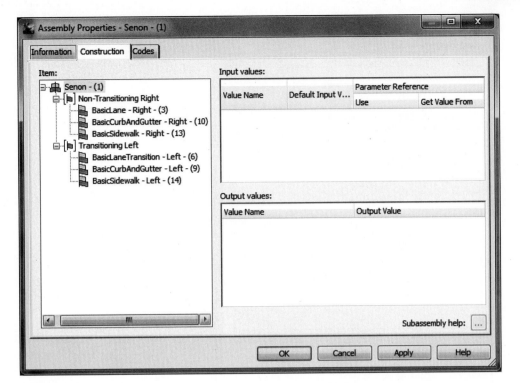

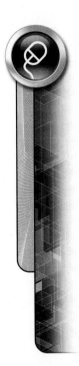

FIGURE 9.3

Create a Simple Corridor

1. From Ribbon's Home tab, in the Create Design panel, click the ***Corridor*** icon, and from the shortcut menu, select CREATE SIMPLE CORRIDOR to open the Create Simple Corridor dialog box.

2. In Create Simple Corridor, keeping the counter for the corridor, for the name enter **Senon**, and click ***OK*** to continue.

3. The command line prompts you for an alignment. Press ENTER key, from the list select **Senon-CL - (1) (1)**, and click ***OK*** to continue.

4. The command line prompts you for a profile. Press the right mouse button, click the drop-list arrow, and from the list, select **Senon Centerline (1)**. Click ***OK*** to continue.

5. The command line prompts you for an assembly. Press ENTER, click the drop-list arrow, and from the list, select **Senon - (1)**. Click ***OK*** to continue.

The Target Mapping dialog box opens.

6. In the dialog box, for the Transitioning Left assembly group, set the Width or Offset Targets section by clicking in the Transition Alignment's Object Name cell (<None>). A Set Width Or Offset Target dialog box opens. In the Select Alignments list, select **Senon-Left Transition - (1) (1)**, click ***Add >>***, and, making sure the alignment appears in Selected Entities to target, click ***OK*** to return to the Target Mapping dialog box (see Figure 9.4).

You do not need to set a profile because the transitioning subassembly parameter (Hold grade, change offset) sets the edge of the travelway's elevation.

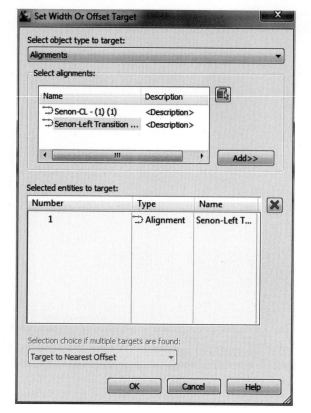

FIGURE 9.4

7. Click **OK** to create the corridor.

8. If an event panorama is displayed, close it.

9. At Civil 3D's top left, click the **Save** icon to save the drawing.

Review Corridor

1. Use the ZOOM and PAN commands to view the new corridor.

The section frequency does not create a good transition representation in the knuckle. The section frequency needs to be increased, and the station range should be reduced.

2. In the drawing, select the corridor, press the right mouse button, and from the shortcut menu, select CORRIDOR PROPERTIES....

3. In Corridor Properties, select the **Parameters** tab, and for Region (1) of RG-Senon - (1) - (1), change its range Start Station to 1 + 75 (**175**), and its End Station to 3 + 25 (**325**) (see Figure 9.5).

You could graphically select start and end stations by clicking the pick station icon in each cell, and, in the drawing, by selecting a station along the corridor.

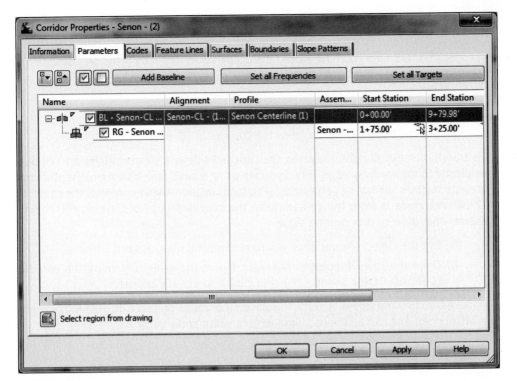

FIGURE 9.5

4. For Region (1), in the Frequency cell, click the ellipsis to open the Frequency to Apply Assemblies dialog box.

5. In the Apply Assembly section, change each Along entry value to **5** (see Figure 9.6).

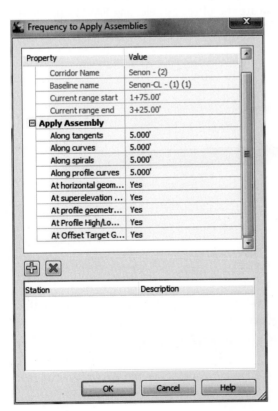

FIGURE 9.6

You can graphically select a frequency by clicking the ellipsis in a value cell and selecting a distance from the drawing.

6. Click **OK** until you have exited the dialog boxes and rebuilt the corridor.
7. At Civil 3D's top left, Quick Access toolbar, click the **Save** icon to save the drawing.
8. Use the ZOOM and PAN commands to better view the knuckle.
9. Use the Object Viewer to view the corridor in 3D.

The transition looks all right. However, the curb and sidewalk subassemblies are not perpendicular to the roadway edge. When you are using a basic lane transition, the curb and sidewalk sections will not be perpendicular to the transition alignment. Also, the constant −2 percent grade is from the centerline to the knuckle's edge-of-travelway. This subassembly has little control over its slope.

10. Exit the Object Viewer after you have reviewed the corridor.
11. Open the Layer Properties Manager, freeze the layer containing the corridor **(C-ROAD-CORR-Senon - (1))**, and click **X** to exit the palette.

To make the curb and sidewalk perpendicular to the transition curve, the assembly needs an offset point and two additional alignments: a width and a vertical.

Assembly with Offset

Previously, an offset alignment defined the transitional control point for the BasicLane-Transition (the edge-of-travelway) and all of the subassemblies attached to that point. Making the curb and sidewalk perpendicular to the edge-of-travelway requires an assembly offset and an alignment to control its horizontal location and a profile to control its elevations.

Figure 9.7 shows an assembly using an offset (the vertical line between the edge-of-travelway and the curb flange). You locate the offset by selecting PAVE1's upper-left endpoint. The left curb subassembly attaches to the OFFSET, not the assembly. The sidewalk attaches to the top back-of-curb.

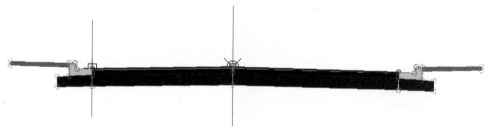

FIGURE 9.7

Create New Assembly with Offset

1. If necessary, click the **Prospector** tab.
2. Expand Assemblies, from the list select Senon - (1), click the right mouse button, and from the shortcut menu, select ZOOM TO.
3. PAN the assembly so there is a clear working space beneath the previous assembly.
4. From Ribbon's Home tab, Create Design panel, click the **Assembly** icon, and from the shortcut menu, select CREATE ASSEMBLY.

5. In the Create Assembly dialog box, for the assembly name enter **Senon**, leaving the counter. Click **OK**, and just below the Senon - (1) assembly, select a point.

Create the Assembly's Right Side

The right side uses a transitional lane subassembly, but there will be no attached alignment.

1. If necessary, click Ribbon's Home tab, in the Palettes panel, click the Tool Palettes icon to display the Civil 3D Imperial tool palette.

2. Click the **Lanes** tab. From the tool palette, select **LaneOutsideSuper**.

3. In Properties, in the Parameters area, set the Side to **Right**, change the width to **14** feet, and in the drawing, select the assembly.

4. Press ENTER twice to exit the routine.

5. On the palette, click the **Curbs** tab and select **UrbanCurbGutterGeneral**.

6. In Properties, the Parameters area, set the Side to **Right**. In the drawing, attach the curb and gutter to the right lane's upper-right outside marker (edge-of-travelway).

7. Press ENTER twice to exit the routine.

8. From the same palette tab, select **UrbanSidewalk**.

9. In Properties, the Parameters area, set the Side to **Right**. In the drawing, attach the sidewalk to the curb and gutter's upper-right marker (back-of-curb), and press ENTER twice to exit the routine.

10. At Civil 3D's top left, Quick Access toolbar, click the **Save** icon to save the drawing.

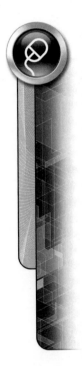

Create Assembly Left Side

1. Click the **Lanes** tab. From the palette, select **LaneOutsideSuper**.

2. In the Properties palette, the Parameters area, change the Side to **Left**, set the width to **14** feet, and attach it to the assembly. Press ENTER twice to exit the routine.

Add Assembly Offset

1. Use the ZOOM and PAN commands to view the assembly's left half.

2. In the drawing, select the assembly (red vertical line), press the right mouse button, and from the shortcut menu, select ADD OFFSET….

3. Use the ZOOM command to better view PAVE1's upper-left vertex and using the Endpoint object snap, select its upper endpoint to set the offset (see Figure 9.7).

An offset vertical line appears at the selected point.

Attach the Curb and Sidewalk

The left side curb subassembly must attach to the offset, not the assembly. The curb and its attached subassemblies are perpendicular to the offset's (transition) alignment.

1. Click the **Curbs** tab. From the tool palette, select **UrbanCurbGutterGeneral**.

2. If necessary, in Properties, Advanced Parameters, change the Side to **Left**. In the drawing, attach the curb and gutter to the offset by selecting the **offset** (left vertical line). Press ENTER twice to exit the routine.

3. From the same palette, select **UrbanSidewalk**. If necessary, in Properties, Advanced Parameters, change the Side to **Left**. Select the curb's upper-left marker (back-of-curb) and press ENTER twice to exit the routine.

4. At Civil 3D's top left, Quick Access toolbar, click the **Save** icon to save the drawing.

5. Click the Subassembly tool palette's **X** to close it.

Edit Assembly Properties

1. In Prospector, expand Assemblies, select **Senon - (2)**, press the right mouse button, and from the shortcut menu, select PROPERTIES....

2. In the dialog box, click the **Construction** tab.

Construction indicates the UrbanCurbandGutterGeneral - Left and UrbanSidewalk - Left subassemblies attach to the offset, not to the main assembly (see Figure 9.8).

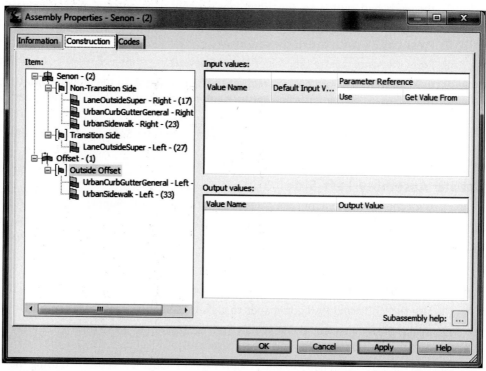

FIGURE 9.8

3. Rename the subassembly groups using Figure 9.8 as a guide.

4. In Construction, click the right side subassemblies group name, press the right mouse button, from the shortcut menu, select RENAME, and for the group name enter **Non-Transition Side**.

5. In Construction, click the left side of the LaneOutsideSuper group name, press the right mouse button, from the shortcut menu, select RENAME, and for the group name enter **Transition Side**.

6. In Construction, under Offset - (1) containing the Left UrbanCurbGeneral - Left and UrbanSidewalk - Left subassemblies, click the group name, press the right mouse button, from the shortcut menu, select RENAME, and for the group name enter **Outside Offset**.

7. Click **OK** to exit Assembly Properties.

8. At Civil 3D's top left, Quick Access toolbar, click the **Save** icon to save the drawing.

Create a Corridor

1. In Ribbon's Home tab, Create Design panel, click the Corridor icon, and from the shortcut menu, select CREATE CORRIDOR.

2. The command line prompts you for an alignment name. Press the right mouse button, and from the alignment list, select **Senon-CL - (1) (1)**, and click **OK** to continue.

3. The command line prompts you for a profile name. Press the right mouse button, click the drop-list arrow, and from the profile list, select **Senon Centerline (1)**. Click **OK** to continue.

4. The command line prompts you for an assembly name. Press the right mouse button, click the drop-list arrow, and from the assembly list, select **Senon - (2)**. Click **OK** to continue, and the Create Corridor dialog box opens.

5. Replace "Corridor" with **Senon**, leaving the counter, in the Corridor Name area at the top left of the dialog box.

The first items to set are the corridor stations for Region 1 (see Figure 9.9).

6. In the dialog box, click in the Region (1) RG – Senon - (2) – (2) Start Station cell and set it to **125**.

7. In the dialog box, click in the Region (1) End Station cell and set it to **375**.

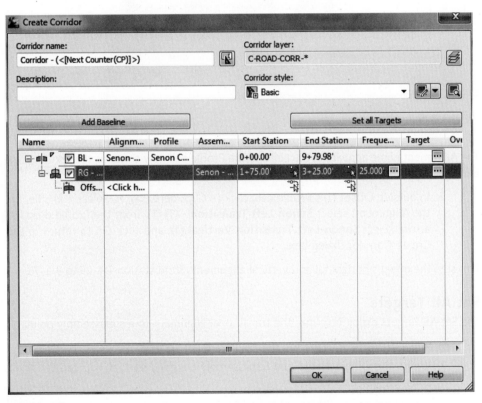

FIGURE 9.9

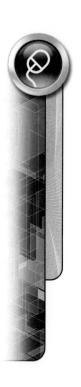

Set Offset Alignment and Profile Names

1. In the Alignment cell for outside Offset (1), click (<Click here...>). In the Pick Horizontal Alignment dialog box, click the drop-list arrow, select **Senon-Left Transition - (1) (1)** from the list, and click **OK** to return to the Create Corridor dialog box (see Figure 9.10).

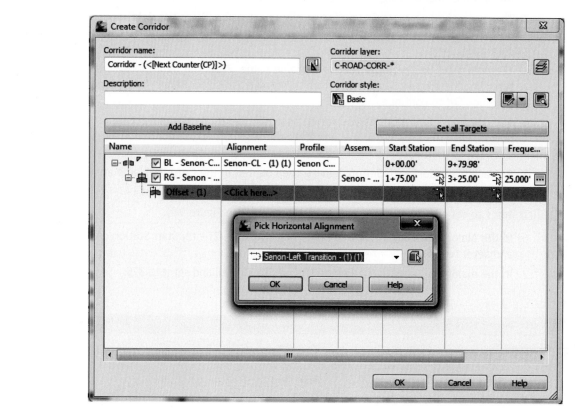

FIGURE 9.10

The Alignment name appears in the cell and the Profile cell now contains the text "<Click here...>".

2. In outside Offset (1)'s Profile cell, click (<Click here...>). In Select a Profile, for the Alignment, select **Senon-Left Transition - (1) (1)**, from the Profile drop-list arrow, select **Senon-Left Transition Vertical (1)**, and click **OK** to return to the Create Corridor dialog box.

This sets the offset's horizontal and vertical alignments from station 1 + 25 to 3 + 75.

Set All Targets

The Set All Targets dialog box links the transitional alignment to a subassembly point.

1. In the dialog box, at the top right, click **Set All Targets**.
2. Widen the Object Name and Subassembly columns so that you can see their complete names.

The right side width is the subassembly's width parameter. The left side's LaneOutsideSuper uses the transition alignment.

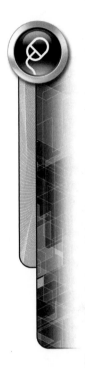

3. In the Width or Offset Targets (Alignments) section, LaneOutsideSuper – Left, Transition Side, click in the Width Alignment's Object Name cell (<None>). In the Set Width Or Offset Target dialog box, select **Senon-Left Transition - (1) (1)**, click **Add >>**, and click **OK** to return to the Target Mapping dialog box.

When you are using an assembly offset, a profile is required. The Profiles section sets edge-of-travelway elevation (see Figure 9.11).

4. In the Slope or Elevation Targets (Profiles) section, for LaneOutsideSuper – Left, click in the Outside Elevation Profile's Object Name cell (<None>), in the Set Slope Or Elevation Target dialog box, set the Alignment to **Senon-Left Transition - (1) (1)**, in Select Profiles, select **Senon-Left Transition Vertical (1)**, click **Add >>**, and click **OK** to return to the Target Mapping dialog box.

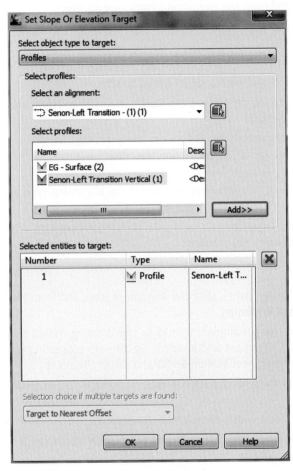

FIGURE 9.11

5. Click **OK** to exit the Target Mapping dialog box and return to the Create Corridor dialog box.

Change Region Frequency

1. In the Region (1) Frequency cell, click the ellipsis.

2. In Frequency to Apply Assemblies, change the Along Tangents and Along Curves frequency to **5**, and click **OK** to return to the Create Corridor dialog box.

3. Click **OK** to create the corridor.

4. At Civil 3D's top left, Quick Access toolbar, click the **Save** icon to save the drawing.

5. Use the ZOOM and PAN commands to view the new Corridor.

The curb and sidewalk subassemblies are perpendicular to the offset alignment's path.

6. Use the OBJECT VIEWER to view the corridor in 3D.

7. Exit the Object Viewer after you have reviewed the corridor.

8. At Civil 3D's upper left, Quick Access toolbar, click the Save icon to save the drawing.

Widening

Senon needs a widening alignment to create a corridor with a bus stop. The widening starts at 5 + 50 and ends at 6 + 25 and both the transition lengths are 55 feet.

1. In Prospector, expand the Site's Site1 branch until viewing the Senon alignments.

2. In the Centerline Alignment branch, select **Senon-Left Transition - (1) (1)**, right mouse click, and from the shortcut menu, select **Move to site**.

3. In the Move to Site dialog box, at the top set the Destination site to **None** and click **OK** to transfer the alignment.

4. In Prospector, expand the Alignments branch, select the alignment Senon-Left Transition - (1) (1), right mouse click, and from the shortcut menu, select PROPERTIES....

5. Click Senon-Left Transition - (1) (1)'s **Information** tab, set the alignment type to **Offset**, and click OK.

6. From the Ribbon's Home tab, Layers panel's right side, click the **Freeze** icon, and in the drawing, select the **Senon - (2)** corridor.

Define the Widening

1. On the Home tab, Create Design panel, click the Alignment icon, and from the shortcut menu, select **Create Widening**.

2. The command line prompts for an alignment, and in the drawing, select the Senon-Left Transition - (1) (1) alignment and a station jig displays prompting for the starting station. Move the jig near station **5+50** and select the point.

3. The routine now prompts for the ending station. Move the jig near station **6+25** and select the point.

4. The command line prompts for a width; press enter to accept **25**.

5. The command line prompts for a side; type in '**L**' and press enter to display the Offset Alignment Parameters palette.

6. In the palette, change the start station to 5 + 50 (**550**) and the end station to 6 + 25 (**625**).

As you edit the widening, the drawing contains interactive graphics showing the changes.

7. Change both transition lengths to **55**.

8. Close the Offset Alignment Parameters palette.

Create an Offset Alignment Profile for the Widening

The new offset alignment needs to have a vertical design profile. You can create a vertical design that is a superimposed copy Senon-Left Transition - (1) (1)'s vertical alignment.

1. If necessary, click the Home tab. In the Profile & Section Views panel, click the Profile View icon, and from the shortcut menu, select Create Profile View.

2. In the Create Profile View wizard, the General panel, set the alignment to **Senon-Left Transition - (1)(1)-Left 0.000**, the profile style to **Major Grids**, click Create Profile View, and in the drawing, select the profile view's location near the other profile views.

Superimpose Profile

1. In the Create Design panel, click the Profile icon, and from the shortcut menu, select CREATE SUPERIMPOSED PROFILE.

2. The command line prompts for a profile. In the Senon Left Transition profile view (the left of the two original profile views), select the **Senon-Left Transition Vertical (1)** vertical design.

3. The command line prompts for a profile view. In the drawing, select the new offset alignment's profile view.

4. The Superimpose Profile Options dialog box opens. Toggle **ON** select Start and End stations and click **OK**.

5. The profile transfers to the new profile view.

Create New Corridor

1. In Ribbon's Home tab, Create Design panel, click the Corridor icon, and from the shortcut menu, select CREATE CORRIDOR.

2. The command line prompts you for an alignment name. Press the right mouse button, and from the alignment list, select **Senon-CL - (1) (1)**, and click **OK** to continue.

3. The command line prompts you for a profile name. Press the right mouse button, click the drop-list arrow, and from the profile list, select **Senon Centerline (1)**. Click **OK** to continue.

4. The command line prompts you for an assembly name. Press the right mouse button, click the drop-list arrow, and from the assembly list, select **Senon - (2)**. Click **OK** to continue, and the Create Corridor dialog box opens.

5. In the dialog box's top left, replace "Corridor" with **Senon**, leaving the counter.

The first items to set are the corridor stations for Region 1.

6. In the dialog box, click in the Region (1) RG – Senon - (2) – (2) Start Station cell and set it to **125**.

7. In the dialog box, click in the Region (1) End Station cell and set it to **800**.

Set Offset Alignment and Profile Names

1. In the Alignment cell for outside Offset - (1), click (<Click here...>). In the Pick Horizontal Alignment dialog box, click the drop-list arrow, select **Senon-Left Transition - (1) (1)-Left 0.000** from the list, and click **OK** to return to the Create Corridor dialog box.

The Alignment name appears in the cell and the Profile cell now contains the text "<Click here...>".

2. In outside Offset - (1)'s Profile cell, click (<Click here...>). In Select a Profile, for the Alignment, select **Senon-Left Transition - (1) (1)-Left 0.000**, from the Profile drop-list arrow, select **Senon-Left Transition Vertical (1) - [Senon-Left Transition - (1) (1)] - (1),** and click **OK** to return to the Create Corridor dialog box.

This sets the offset's horizontal and vertical alignments from station 1+25 to 8+00.

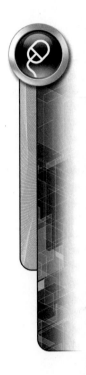

Set All Targets

The Set All Targets dialog box links the transitional alignment to a subassembly point.

1. In the dialog box, at the top right, click **Set All Targets**.

2. Widen the Object Name and Subassembly columns so you can see their complete names.

The right side width is the subassembly's width parameter. The left side's LaneOutside-Super uses the transition alignment.

3. In the Width or Offset Targets (Alignments) section, LaneOutsideSuper – Left, Transition Side, click in the Width Alignment's Object Name cell (<None>). In the Set Width Or Offset Target dialog box, select **Senon-Left Transition - (1) (1)-Left 0.000**, click **Add >>**, and click **OK** to return to the Target Mapping dialog box.

When you are using an assembly offset, a profile is required. The Profiles section sets edge-of-travelway elevation.

4. In the Slope or Elevation Targets (Profiles) section, for LaneOutsideSuper – Left, click in the Outside Elevation Profile's Object Name cell (<None>). In the Set Slope Or Elevation Target dialog box, set the Alignment to **Senon-Left Transition - (1) (1))-Left 0.000**, in Select Profiles, select **Senon-Left Transition Vertical (1) - [Senon-Left Transition - (1) (1)] - (1)**, click **Add >>**, and click **OK** to return the Target Mapping dialog box.

5. Click **OK** to exit the Target Mapping dialog box and return to Create Corridor.

Change Region Frequency

1. In the Region (1) Frequency cell, click the ellipsis.

2. In Frequency to Apply Assemblies, change the Along Tangents and Along Curves frequency to **5**, and click **OK** to return to the Create Corridor dialog box.

3. Click **OK** to create the corridor.

4. At Civil 3D's top left, Quick Access toolbar, click the **Save** icon to save the drawing.

5. Use the ZOOM and PAN commands to view the new Corridor.

6. Exit Civil 3D.

This completes the simple transitions and widening exercise.

EXERCISE 9-2

After completing this exercise, you will:

- Be able to create a cul-de-sac.
- Be familiar with the Create Corridor command.
- Be able to define corridor regions.
- Be able to create a corridor surface.
- Manually define a corridor surface boundary.

Exercise Setup

This exercise uses the *Chapter 09 - Unit 2.dwg* file. To find this file, browse to the CD that accompanies this textbook and open the *Chapter 09 - Unit 2.dwg* file.

1. If you are not in **Civil 3D**, double-click the **Civil 3D** desktop icon to start the application.

2. When you are at the command line, CLOSE the open drawing and **do not save it**.

3. At Civil 3D's top left, Quick Access toolbar, click the Open icon, browse to the Chapter 09 folder of the CD that accompanies this textbook, and open the *Chapter 09 - Unit 2* drawing.

4. At Civil 3D's top left, click Civil 3D's Menu Browse drop-list arrow, from the Application Menu, highlight SAVE AS, from the flyout menu, select AUTOCAD DRAWING, browse to the Civil 3D Project folder, for the name enter **Cul-de-sac**, and click **Save** to save the file.

There are three alignments: one entire roadway baseline (Lorraine - (1)) and two offset (transition) alignments for the cul-de-sac (Lorraine Left and Right Cul-de-sac – (1)). There are two roadway assemblies: one non-transitioning assembly applied to stations 0 + 00 to 5 + 25 and one transition assembly.

The two offset alignments and their profiles already exist. You would have had to calculate or determine their profile elevations so they initially match the main alignment's elevations and then reflect cul-de-sac drainage design. One way of developing a cul-de-sac design is to use an ETW feature line that determines the starting elevations. Creating Feature lines from a corridor are in the Ribbon's Modify, Corridors' panel: Feature Lines from Corridor. Alignments from Corridor also prompt to create a profile.

Create the Transition Assembly

Use Figure 9.12 as a guide when you are creating the transitional assembly.

1. If necessary, click the **Prospector** tab.

2. Expand the Assemblies branch. From the assembly list, select **Lorraine – No Transition – (1)**, press the right mouse button, and from the shortcut menu, select ZOOM TO.

3. From Ribbon's Home tab, Create Design panel, click the **Assembly** icon and from the shortcut menu, select CREATE ASSEMBLY. For the name, replace "Assembly" with **Lorraine – Transition**, keeping the counter. Click **OK**, and in the drawing, select a point to locate the assembly.

Right Side Transition Subassembly

1. If necessary, click Ribbon's **Home** tab. In the Palettes panel, click the Tool Palettes icon to display the Civil 3D Imperial tool palette.

2. On the tool palette, click the **Lanes** tab.

3. From the Imperial – Lanes palette, select **LaneOutsideSuper**.

4. In Properties, the Parameters area, set the Side to **Right**, and set the width to **12** feet. In the drawing, place the subassembly by selecting the assembly, and press ENTER twice to exit the routine.

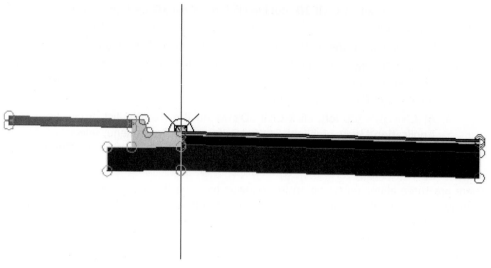

FIGURE 9.12

Left Side Curb and Sidewalk

1. On the tool palette, click the **Curbs** tab.

2. From the Imperial – Curbs palette, select **UrbanCurbGutterGeneral**.

3. In Properties, the Parameters area, set the Side to **Left**, and in the drawing, select the assembly, and press ENTER twice to exit the routine.

4. From the Imperial – Curbs palette, select **UrbanSidewalk**.

5. In Properties, the Parameters area, set the Side to **Left**, and in the drawing, select the upper back-of-curb marker to place the sidewalk, and then press ENTER twice to exit the routine.

6. Close the tool palette.

7. At Civil 3D's top left, Quick Access toolbar, click the **Save** icon to save the drawing.

Edit Assembly Properties

1. If necessary, expand the Prospector Assemblies branch until you view the assembly list.

2. From the list, select **Lorraine – Transition – (1)**, press the right mouse button, and from the shortcut menu, select PROPERTIES….

3. Click the **Construction** tab, and select and rename the group with right LaneOutsideSuper to **Transition Subassembly**.

4. Select and rename the group with Curb and Sidewalk to **Non-Transition Subassemblies**.

5. Click **OK** to exit the dialog box.

6. At Civil 3D's top left, Quick Access toolbar, click the **Save** icon to save the drawing.

Create the Non-Transition Corridor Region

The process starts with a simple corridor, and later adds to its complexity by editing its properties.

1. From Ribbon's Home tab, Create Design panel, click the Corridor icon, and from the shortcut menu, select CREATE SIMPLE CORRIDOR. For the name, replace "Corridor" with **Lorraine**, and click **OK** to continue creating the corridor.

2. The command line prompts you for an Alignment. Press the right mouse button, and from the list, select **Lorraine - (1)**. Click **OK** to continue.

3. The command line prompts you for a profile. Press the right mouse button, click the drop-list arrow, and from the list, select **Lorraine CL Vertical (1)**. Click **OK** to continue.

4. The command line prompts you for an assembly. Press the right mouse button, click the drop-list arrow, and from the list, select **Lorraine – No Transition – (1)**. Click **OK** to continue.

The Target Mapping dialog box opens. There are no target names to set at this time.

5. Click **OK** to create the corridor.

6. If the Event Viewer displays, close it by clicking the green check mark in the panorama's upper-right corner.

7. Use the ZOOM and PAN commands to view the corridor.

8. Use the ZOOM and PAN commands to view the cul-de-sac from station 5 + 25 to its end.

9. At Civil 3D's top left, Quick Access toolbar, click the **Save** icon to save the drawing.

The corridor goes from the beginning to the end (center of the cul-de-sac). The non-transitional corridor region should stop at station 5 + 25.

Edit the Corridor Properties

1. In the drawing, select the corridor, press the right mouse button, and from the shortcut menu, select CORRIDOR PROPERTIES....

2. If necessary, scroll right until you are viewing the beginning and ending stations. Click in the Region (1) RG – Lorraine – No Transition – (1) – (1) ending station cell and change its station to 5 + 25 (**525**).

3. Click **OK** to modify the corridor's ending station.

Define and Assign Baseline (2)

The corridor control around the cul-de-sac's northern edge passes to a second baseline (centerline) at station 5 + 25.01. This alignment's vertical setting controls the edge-of-travelway's vertical location. The Lorraine - (1) alignment stretches the pavement to create the cul-de-sac's paved area.

1. In the drawing, select the corridor, press the right mouse button, and from the shortcut menu, select CORRIDOR PROPERTIES....

2. In the Parameters panel, in the panel's upper center, click **Add Baseline** to create a new baseline entry in the Create Corridor Baseline dialog box.

3. In Create Corridor Baseline dialog box, click the Horizontal alignment drop-list arrow. From the list, select **Lorraine Left Cul-de-Sac – (1)**, and click **OK** to return to the Corridor Properties dialog box.

4. For Baseline (2) Lorraine Left Cul-de-Sac – (1) – (1), in its Profile cell, click (<Click here...>). In the Select a Profile dialog box, click the drop-list arrow. From the list, select **Lorraine Left Transition - Vertical (1)** and click **OK** to return to the Corridor Properties dialog box.

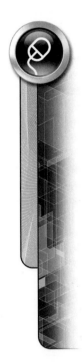

Create Baseline (2)'s Region 1

1. In the Corridor Properties dialog box, with the Baseline (2) highlighted, press the right mouse button, and from the shortcut menu, select ADD REGION... to open Create Corridor Region dialog box.

2. In Create Corridor Region, click the Assembly drop-list arrow. From the list, select **Lorraine – Transition – (1)**, and click **OK** to return to the Corridor Properties dialog box (see Figure 9.13).

The preceding steps will create Region (1) for Baseline (2) with the selected assembly.

3. Click **OK** to create the modified corridor.

4. Use the ZOOM and PAN commands to inspect the intersection of the two corridor regions.

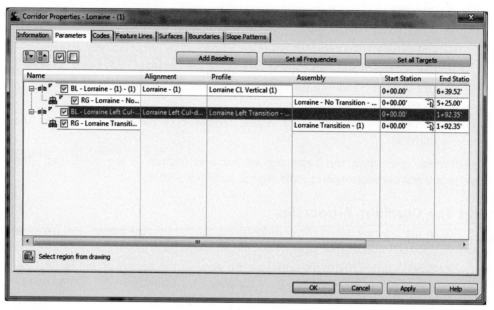

FIGURE 9.13

The new baseline creates sections that follow the new alignment and its vertical design. However, the right side of the assembly does not reach the centerline of the road or the cul-de-sac.

To create the correct width, in Target Mapping, assign the Lorraine - (1) centerline and its vertical as the lane width target alignment and profile.

Assign Transition Targets

1. In the drawing, select the corridor, press the right mouse button, and from the shortcut menu, select CORRIDOR PROPERTIES....

2. If necessary, click the *Parameters* tab.

3. If necessary, click Baseline (2)'s Region (1) entry to highlight it, scroll to the right, and in the Target column, click the ellipsis.

4. Under the Width or Offset Targets (Alignments) section, locate the Transition Subassembly's Width Alignment entry, and click in the Object Name cell (<None>). The Set Width Or Offset Target dialog box opens. In Select alignments, select **Lorraine - (1)**, click **Add >>**, and click **OK** to return to the Target Mapping dialog box (see Figure 9.14).

5. In the Slope or Elevation Targets (Profiles) section, for Transition Subassembly locate its Outside Elevation Profile entry. Click in its Object Name cell

(<None>) to open the Set Slope Or Elevation Target dialog box. With Alignment set to Lorraine - (1), from the Select Profile list, select **Lorraine CL Vertical (1)**, click **Add >>**, and click **OK** to return to the Target Mapping dialog box (see Figure 9.14).

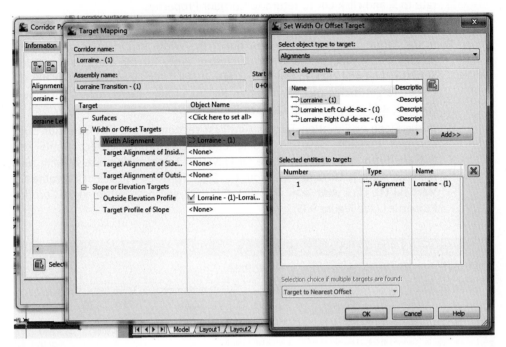
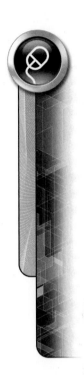

FIGURE 9.14

6. Click **OK** to exit the Target Mapping dialog box.

7. In Baseline (2), the Region (1)'s Frequency column, click the ellipsis, change the Along Tangents and Along Curves sampling rate to **5**, and click **OK** to return to the Corridor Properties dialog box.

8. Click **OK** to exit the dialog box and update the corridor.

9. Use the ZOOM and PAN commands to inspect both corridor baselines and their regions.

The new baseline and its region stretch the pavement to Lorraine - (1)'s centerline.

Define and Assign Baseline (3)

1. In the drawing, select the corridor, press the right mouse button, and from the shortcut menu, select CORRIDOR PROPERTIES….

2. If necessary, click the **Parameters** tab.

3. In the panel's upper center, click **Add Baseline** and create Baseline (3).

4. In Create Corridor Baseline, click the Horizontal Alignment drop-list arrow. From the list, select **Lorraine Right Cul-de-sac – (1)** and click **OK** to return to Corridor Properties.

5. For Baseline (3), click in its Profile cell (<Click here…>). In the Select a Profile dialog box, click the drop-list arrow, and from the list, select **Lorraine Right Transition Vertical (1)**. Click **OK** to return to Corridor Properties.

6. With Baseline (3) still highlighted, press the right mouse button, and from the shortcut menu, select ADD REGION….

7. In Create Corridor Region, click the drop-list arrow. From the list, select **Lorraine – Transition – (1)**, and click OK to return to Corridor Properties.

8. Expand Baseline (3) to view Region (1). Scroll the corridor panel to the right until you view the Frequency and Target columns.

9. In the Baseline (3) Region (1) Frequency cell, click the ellipsis. In the Frequency to Apply Assemblies, change the Along Tangents and Along Curves sampling rate to **5**, and click **OK** to return to Corridor Properties.

10. In the Baseline (3) Region (1) Target cell, click the ellipsis.

11. In the Target Mapping dialog box, under the Width or Offset Targets (Alignments) section, locate the Width Alignment entry for the Lorraine Transition Subassembly, and click in the Object Name cell (<None>). The Set Width Or Offset Target dialog box opens. From the list of alignments, select **Lorraine - (1)**, click **Add >>**, and click **OK** (see Figure 9.11).

12. In the Slope or Elevation Targets (Profiles) section, locate the Transition Subassembly Outside Elevation Profile entry. Click in its Object Name cell (<None>) to open the Set Slope Or Elevation Target dialog box. Making sure the Alignment is Lorraine - (1), from the Select profiles list, select **Lorraine CL Vertical (1)**, click **Add >>**, and click **OK** until you have returned to Corridor Properties (see Figure 9.15).

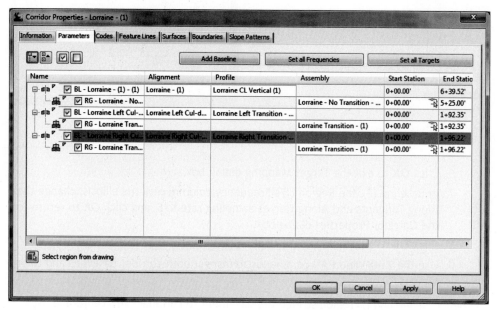

FIGURE 9.15

13. Click **OK** to exit and update the corridor.

14. Use the ZOOM and PAN commands to inspect the cul-de-sac.

15. At Civil 3D's top left, Quick Access toolbar, click the **Save** icon to save the drawing.

Corridor Surface

1. In the drawing, select the corridor, press the right mouse button, and from the shortcut menu, select CORRIDOR PROPERTIES….

2. Click the **Surfaces** tab.

3. In the panel's upper left, click the **Create a corridor surface** icon.

4. Rename the surface **Lorraine – Top**, assign it the style **Border & Triangles & Points**, and set Overhang correction to Top Links.

5. From the panel's top center, set the Data type to **Links**, from the list of links select **Top**, and click the plus sign (**+**) to add it as surface data.

Each Baseline contributes to an overall surface and their "extents" define the surface boundary.

6. Click the **Boundaries** tab.

7. In the panel, select Lorraine - (1) Surface - Top, right mouse click, and from the shortcut menu, select CORRIDOR EXTENTS AS OUTER BOUNDARY….

8. Click **OK** to build the corridor surface.

9. Select a surface triangle, and use the Object Viewer to review the surface.

10. At Civil 3D's top left, Quick Access toolbar, click the **Save** icon to save the drawing.

This ends the exercise on cul-de-sac transitions.

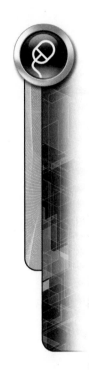

EXERCISE 9-3

After completing this exercise, you will:

- Be able to review Design Criteria values.
- Be able to use Design Criteria.
- Be able to superelevate a two-lane crowned road.
- Be able to superelevate a four-lane divided highway.

Exercise Setup

This exercise uses the *Chapter 09 - Unit 3.dwg* file. To find this file, browse to the CD that accompanies this textbook and open the *Chapter 09 - Unit 3.dwg* file.

1. If you are not in **Civil 3D**, double-click the **Civil 3D** desktop icon to start the application.

2. When you are at the command line, CLOSE the open drawing and **do not save it**.

3. At Civil 3D's top left, Quick Access toolbar, click the Open icon, browse to the Chapter 09 folder of the CD that accompanies this textbook, and open the *Chapter 09 - Unit 3* drawing.

4. At Civil 3D's top left, click Civil 3D's drop-list arrow, from Application Menu, highlight SAVE AS, from the flyout menu, select AUTOCAD DRAWING, browse to the Civil 3D Projects folder, for the drawing name enter **Superelevation**, and click **Save** to save the file.

Alignment — Edit Feature Settings

1. Click the **Settings** tab.

2. Select the Alignment heading, press the right mouse button, and from the shortcut menu, select EDIT FEATURE SETTINGS….

3. Expand the Superelevation Options section and review its values. See Figure 9.16.

4. Expand the Criteria-Based Design Options section. If necessary, using Figure 9.17 as a guide, set the values to match the figure.

5. Click **OK** to exit the dialog box.

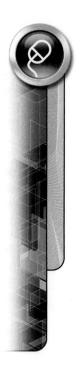

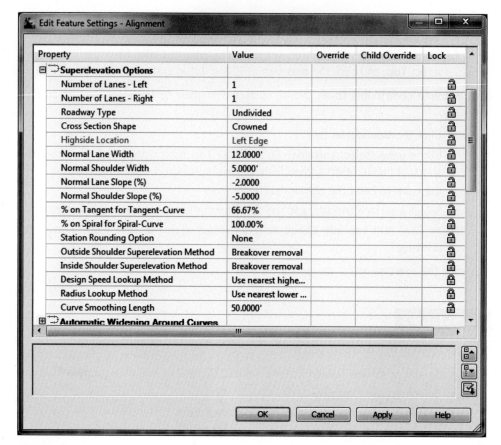

FIGURE 9.16

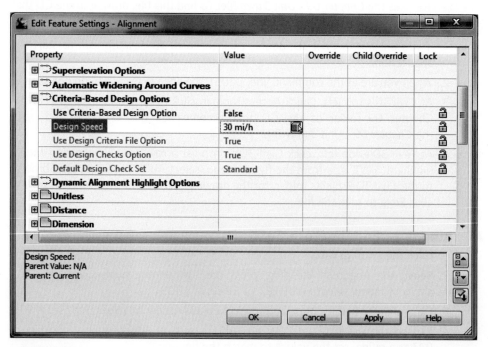

FIGURE 9.17

Import the Briarwood Alignment and Profile

The Briarwood Alignment and its vertical design profile are in a second LandXML file. When you import an alignment and profile, the alignment is drawn, but the profile data is read. You must create a profile view, sampling the EG surface, and when you draw the view, you must include the Briarwood vertical alignment.

1. In the Import panel, select LANDXML.
2. In the Import LandXML dialog box, browse to the Chapter 09 folder of the CD that accompanies this textbook, select the file *Briarwood.xml*, and click **Open**.
3. In the Import LandXML, click **OK** to import the file.
4. Use the ZOOM and PAN commands to better view the alignment (the surface's western center).

The Briarwood centerline is the small subdivision road at the surface's middle left.

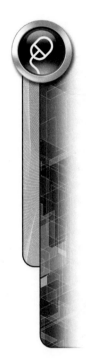

Assign Design Speeds

1. In the drawing, select the Briarwood alignment, press the right mouse button, and from the shortcut menu, select ALIGNMENT PROPERTIES....
2. Click the **Design Criteria** tab.
3. Click the **Add Design Speed** icon to set the first station and its design speed.
4. Click the **Add Design Speed** icon two more times.
5. Using Table 9.1 as a guide to set the following stations and their design speeds.

TABLE 9.1

Station	Speed
1 + 05	10
5 + 50	20
27 + 50	10

Assign Criteria

1. In the Design Criteria tab, in its upper right, toggle **ON** Use criteria-based design and Use design criteria file.
2. Click **OK**.

Set Superelevation Properties

The Superelevation editor uses the alignment's design speeds and criteria files to calculate critical superelevation stations.

1. On the Ribbon, click the Modify tab.
2. In Modify's Design panel, click the Alignment icon.
3. In the Alignment ribbon, the Modify panel, click the **Superelevation** icon to display a command list. From the command list, select Edit/Calculate Superelevation.
4. The routine prompts you to select an alignment. In the drawing, select the Briarwood alignment.

The alignment has no superelevation data.

5. The Edit Superelevation-No Data dialog box displays. Click the Calculate super-elevation now to display the Superelevation Wizard's Roadway type panel.

6. In the Roadway type panel, set the roadway type to Undivided Crowned, Pivot Center. See Figure 9.18. When set, click Next.

7. In the Lanes panel, at its top right, make sure Symmetrical Roadway is toggled on and the lane number is set to 1. Next make sure the Normal lane width is set to 12 and the Normal lane slope is set to -2 percent. See Figure 9.19. When set, click Next.

8. In the Shoulder Control panel, toggle on Calculate outside shoulders and set the Shoulder slope treatment to Breakover removal. Set the Normal shoulder width to 5 and the Normal shoulder slope to -5 percent. See Figure 9.20. When set, click Next.

9. In the Attainment panel, use Figure 9.21 to set the values and toggle on the required methods. When entering a curve, the cross slope will be two-thirds of the maximum superelevation value and the transitions will use a 50-foot radius curve.

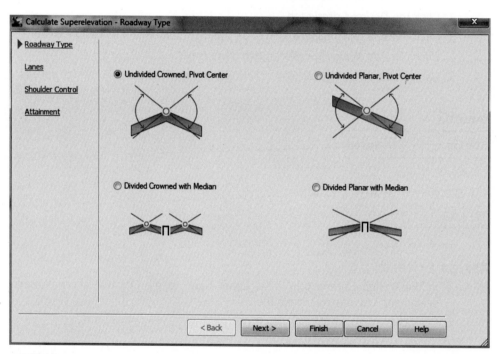

FIGURE 9.18

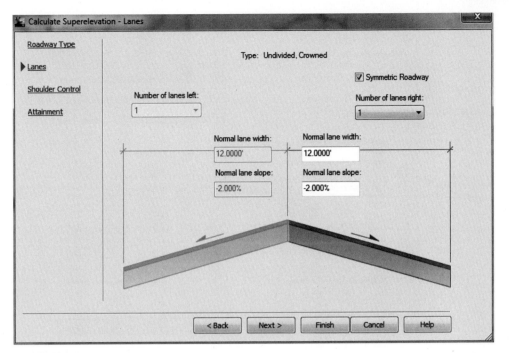

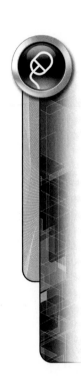

FIGURE 9.19

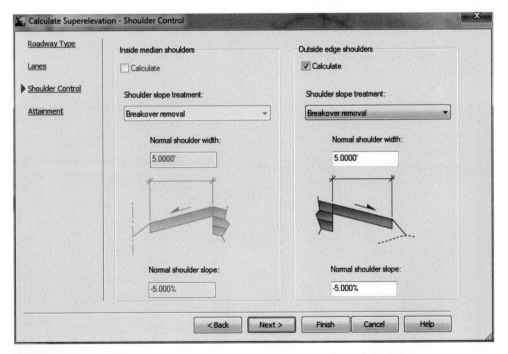

FIGURE 9.20

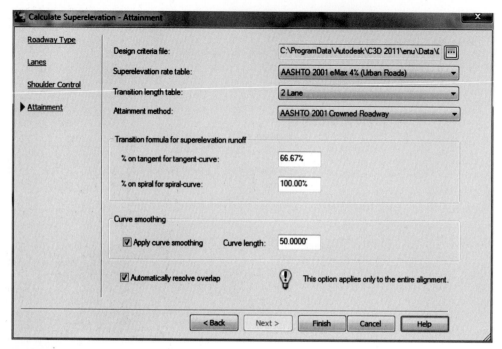

FIGURE 9.21

> 10. When the values are set, click Finish.

The Superelevation Tabular Editor displays. See Figure 9.22.

> 11. Review the superelevation values. When done reviewing the values, close the Tabular Editor.

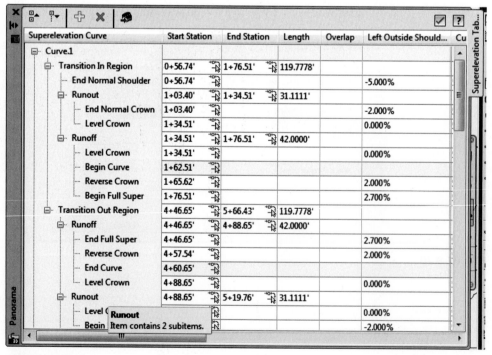

FIGURE 9.22

Review and Edit Superelevation Calculations

After assigning and calculating superelevation values, you can review and set different values for each curve with the Calculate/Edit Superelevation command (see Figure 9.23).

1. On the Ribbon, click the Modify tab.

2. In Modify's Design panel, click the Alignment icon.

3. In the Alignment ribbon, the Modify panel, click the **Superelevation** icon to display a command list. From the command list, select Edit/Calculate Superelevation.

4. The routine prompts you to select an alignment. In the drawing, select the Briarwood alignment to display the Superelevation Curve Manager.

5. Click the Next bottom to review each curve's data. After reviewing the data, close the Manager.

6. At Civil 3D's top left, Quick Access toolbar, click the **Save** icon to save the drawing.

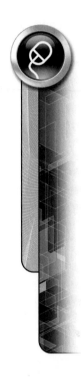

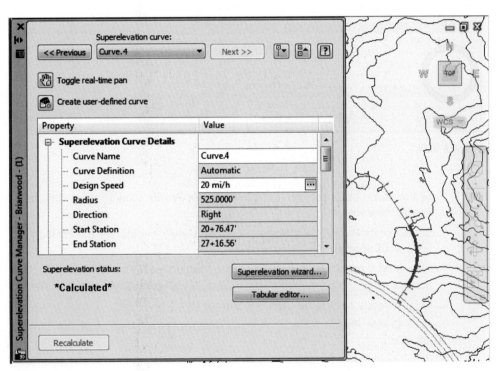

FIGURE 9.23

Create the Briarwood Profiles and View

1. On the Ribbon, click the **View** tab. On the Views panel, select NAMED VIEWS, create a New view using the current display without saving a layer snapshot, and name the view **Briarwood**.

2. Use the PAN command and place the roadway to the right to create open space to the surface's left.

3. Click the Ribbon's **Home** tab. In the Create Design panel, click the **Profile** icon, and from the shortcut menu, select CREATE SURFACE PROFILE.

In Create Profile from Surface, the Briarwood vertical profile should already be in the Profile list area.

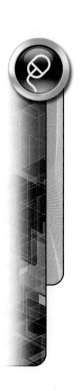

4. In Create Profile from Surface dialog box, set the alignment to **Briarwood – (1)**. At the top right, select the surface **EG**, and in the middle right, click **Add >>** to add the list EG – Surface (1) (see Figure 9.24) to the profile.

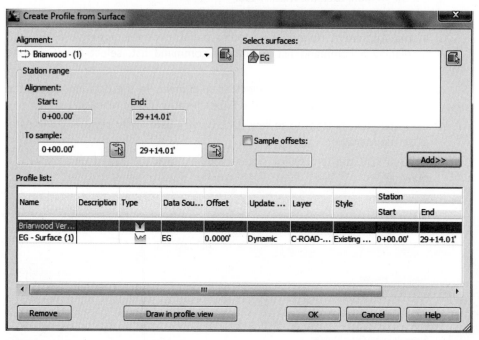

FIGURE 9.24

5. At the dialog box's bottom left, click **Draw in Profile View** to open the Create Profile View wizard.

6. In the Create Profile View wizard – General panel's middle, set the Profile View Style to **Clipped Grid**.

7. Click **Next** three times, and in the Create Profile View – Profile Display Options panel, change the Briarwood Vertical (1) style to **Design Style** and click Next.

8. In Data Bands, set Profile 1 to EG – Surface (1).

9. Click **Create Profile View**, and in the drawing, select a point at the screen's left side.

10. If you need to move the profile, use the MOVE command to relocate the profile.

Your drawing should now look like Figure 9.25.

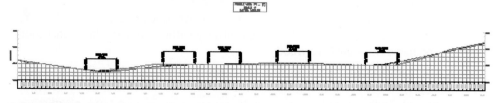

FIGURE 9.25

11. At Civil 3D's top left, Quick Access toolbar, click the **Save** icon to save the drawing.

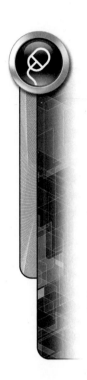

Create the Assembly

Figure 9.26 is the Briarwood assembly.

1. On the Home tab, in the Create Design panel, click the Assembly icon, and from the shortcut menu, select CREATE ASSEMBLY.

2. In the Create Assembly dialog box, name the assembly **Briarwood**, leaving the counter. Click **OK**, and select a point to the left of the Briarwood profile.

Add Subassemblies

1. From the Ribbon's *Home* tab. In the Palettes panel, click the Tool Palettes icon to display the Civil 3D Imperial tool palette.

2. On the tool palette, click the *Lanes* tab.

3. From the Imperial – Lanes palette, select *LaneOutsideSuper*.

4. In Properties, the Parameters area, change the Side to **Right**, change the width to **14** feet, and in the drawing, select the assembly.

5. In Properties, the Parameters area, change the Side to **Left**, create the lane by selecting the assembly, and press ENTER twice to exit the routine.

6. On the tool palette, click the **Shoulders** tab.

7. From the Imperial – Shoulders palette, select *ShoulderExtendSubbase*. In Properties, the Parameters area, set the Side to **Right**, and in the drawing, select the lane's upper-right marker.

8. In Properties, the Parameters area, change the Side to **Left**. In the drawing, select the lane's upper-left outer marker and press the ENTER key twice to exit the routine.

9. At Civil 3D's top left, Quick Access toolbar, click the *Save* icon to save the drawing.

FIGURE 9.26

Create the Corridor

1. On the Ribbon, click the *View* tab. On the Views panel's left, select and restore the named view **Briarwood**.

2. On the Ribbon, click the Home tab. On the Create Design panel, click the Corridor icon, from the shortcut menu, select CREATE SIMPLE CORRIDOR, and in Create Simple Corridor, keeping the counter, enter **Briarwood** for the name, and click **OK** to continue.

3. The command prompts you for an alignment. Press the right mouse button, select **Briarwood – (1)**, and click **OK** to continue.

4. The command prompts you for a profile. Press the right mouse button, and in the Select a Profile dialog box, click the drop-list arrow. From the list, select **Briarwood Vertical (1)**, and click **OK** to continue.

5. The command prompts you for an assembly. Press the right mouse button, and in the Select an Assembly dialog box, click the drop-list arrow. From the list, select **Briarwood – (1),** and click **OK**.

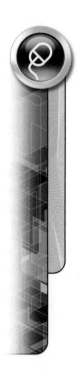

6. The Target Mapping dialog box opens. There are no mappings. Click **OK** to create the corridor.

View Corridor Sections

1. Click the Ribbon's Modify tab. On the Design panel, click the Corridor icon, and in the Modify panel, select CORRIDOR SECTION EDITOR. In the drawing, select the Briarwood corridor to view the sections.

2. After viewing the sections and their superelevation, exit the routine by clicking the Section Editor panel's **Close** icon.

3. At Civil 3D's top left, Quick Access toolbar, click the **Save** icon to save the drawing.

Divided Highway Corridor

A divided highway corridor starts at the site's top left and continues to its middle right. The LandXML file for this highway is in the Chapter 09 folder of the CD that accompanies this textbook.

1. Use the ZOOM EXTENTS command to view the entire drawing.

Import Route 7380 Alignment and Profile

1. On the Ribbon, click the Insert tab. On the Import panel, select LANDXML. Browse to the Chapter 09 folder of the CD that accompanies this textbook, select the file *Route 7380.xml*, and click **Open**.

2. In Import LandXML, if necessary, set the Alignment site to **None**, and then click **OK**.

The alignment appears in the site's northern half.

3. Use the ZOOM and pan the site to the left.

4. If necessary, click the **Prospector** tab.

5. On the Ribbon, click the Home tab. On the Create Design panel, click the Profile icon, and from the shortcut menu, select CREATE SURFACE PROFILE.

6. In Create Profile from Surface, change the alignment to **Route 7380 – (1)**, select the surface **EG**, and click **Add >>** to add the surface to the profile list. At the bottom of the dialog box, click **Draw in Profile View** to display the Create Profile View wizard.

7. In the Create Profile View – General panel, change the style to **Clipped Grid**, and click **Next** three times.

8. In the Create Profile View – Profile Display Options panel, change the Route 7380 Vertical (1) design vertical alignment style to **Design Style**, and click Next.

9. In the Data Bands panel, change Profile 1 to EG – Surface (2), click **Create Profile View**, and in the drawing, to the east of the site, select a point.

10. At Civil 3D's top left, Quick Access toolbar, click the **Save** icon to save the drawing.

Create the Assembly

Figure 9.27 represents the exercise's assembly.

1. From the Ribbon's Home tab, Create Design panel, click the Assembly icon, and from the shortcut menu, select **CREATE ASSEMBLY**. In the Create Assembly dialog box, for the assembly name, enter **Route 7380**, keeping the counter, click **OK**, and place the assembly near the Route 7380 profile.

2. If necessary, from the Ribbon's Home tab, on the Palettes panel, click **Tool Palettes**.

3. Click the **Medians** tab.

4. From the Imperial – Medians palette, select ***MedianDepressedShoulderVert***.

5. In Properties, the Parameters area, change the Hold Ditch Slope to **Hold Ditch at Center, adjust sideslope on high side**. Change the Paved Shoulder Width to **8** feet, and change the Unpaved Shoulder Width to **3** feet. In the drawing, attach the subassembly by selecting the assembly, and press ENTER twice to exit the routine.

The assembly pivots around the median's centerline.

6. Click the **Lanes** tab.

7. From the ***Imperial - Lanes*** palette, select ***LaneOutsideSuper***.

8. In Properties, the Parameters area, change the Side to **Right**, and change the width to **26** feet. In the drawing, select the median's upper-right marker to place the lane.

9. In Properties, in the Parameters area, change the Side to **Left**. In the drawing, select the median's upper-left marker to place the lane.

10. Press ENTER twice to exit the routine.

11. Click the **Shoulders** tab.

12. From the ***Imperial - Shoulders*** palette, select ***ShoulderVerticalSubbase***.

13. In Properties, the Parameters area, set the Side to **Right**, set the Paved Width to **8**, and set the Unpaved Width to **3**. In the drawing, select the lane's upper-right marker to place the subassembly.

14. In Properties, the Parameters area, change the Side to **Left**. In the drawing, select the lane's upper-left outer marker to place the subassembly, and press ENTER twice to exit the routine.

15. Close the Civil 3D - Imperial tool palette.

16. At Civil 3D's top left, Quick Access toolbar, click the **Save** icon to save the drawing.

FIGURE 9.27

Assign Design Speeds

1. Click the **Settings** tab.

2. In Settings, select Alignment, press the right mouse button, and from the shortcut menu, select EDIT FEATURE SETTINGS….

3. Expand the Criteria-Based Design Options, set the Design Speed to **70**, and click **OK** to exit the dialog box.

4. Use the ZOOM and PAN commands to view the Route 7380 alignment.

5. Click the **Prospector** tab.

6. Expand the Alignments branch until you are viewing Route 7380 – (1).

7. Select **Route 7380 – (1)**, press the right mouse button, and from the shortcut menu, select PROPERTIES….

8. Click the **Design Criteria** tab.

9. In Design Criteria, click the **Add Design Speed** icon, set the Start station to **10**, and set the Design speed to **70**.

10. In Design Criteria, click the **Add Design Speed** icon, set the Start station to **10250**, and set the Design speed to **70**.

If warnings appear indicating that speeds are too great, select Manually Change the Speeds to close them. The warning stems from the current superelevation table, max 4%.

11. In the panel's upper right, toggle **ON** Use criteria-based design and Use design criteria file.

12. In Default criteria, change the Minimum Radius Table to **AASHTO 2001 eMax 10%**, change the Transition Length Table to **4 Lane**, and toggle Off Use Design Check set.

13. Click **OK** to set the speeds and default design criteria.

Set Superelevation Properties

The Superelevation editor uses the alignment's design speeds and criteria files to calculate critical superelevation stations.

1. On the Ribbon, click the Modify tab.

2. In Modify's Design panel, click the Alignment icon.

3. In the Alignment ribbon, the Modify panel, click the **Superelevation** icon to display a command list. From the command list, select Edit/Calculate Superelevation.

4. The routine prompts you to select an alignment. In the drawing, select the Route 7380 – (1) alignment.

The alignment has no superelevation data.

5. The Edit Superelevation-No Data dialog box displays. Click the Calculate superelevation now to display the Superelevation Wizard's Roadway type panel.

6. In the Roadway type panel, set the roadway type to Divided Planar with median. When set, click Next.

7. In the Lanes panel, at its top right, make sure Symmetrical Roadway is toggled on and the lane number is set to 1. Next make sure the Normal lane width is set to 26 and the Normal lane slope is set to -2 percent. When set, click Next.

8. In the Shoulder Control panel, toggle on Calculate outside shoulders and set the Shoulder slope treatment to Breakover removal. Set the Normal shoulder width to 8 and the Normal shoulder slope to -6 percent. When set, click Next.

9. In the Attainment panel, use Figure 9.28 to set the values and toggle on the required methods. When entering a curve, the cross slope will be two-thirds of the maximum superelevation value and the transitions will use a 50-foot radius curve. When set, click Finish to display the Superelevation Tabular Editor.

10. Review the critical superelevation values. When done, close the Tabular Editor.

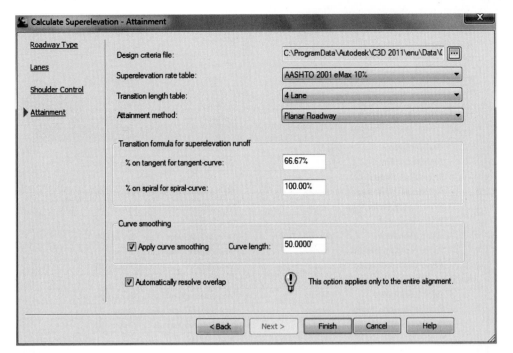

FIGURE 9.28

Create the Corridor

1. If necessary, click the Ribbon's Home tab. On the Create Design panel, click the Corridor icon, and from the shortcut menu, select `CREATE CORRIDOR`.

2. The command line prompts you for an alignment. Press the right mouse button, select **Route 7380 – (1)**, and click **OK** to continue.

3. The command line prompts you for a profile. Press the right mouse button, and in the Select a Profile dialog box, click the drop-list arrow. From the list, select **Route 7380 Vertical (1)**, and click **OK** to continue.

4. The command line prompts you for an assembly. Press the right mouse button, and in the Select an Assembly dialog box, click the drop-list arrow. From the list, select **Route 7380 – (1)**, and click **OK** to continue.

5. In the Create Corridor dialog box, name the corridor **Route 7380**, keeping the counter.

6. At the dialog box's top right, click **Set All Targets**.

7. In Target Mapping, for the Surfaces, Object Name cell, click (<Click here to set all>). In Pick a Surface, select **EG**, and click **OK** (see Figure 9.29).

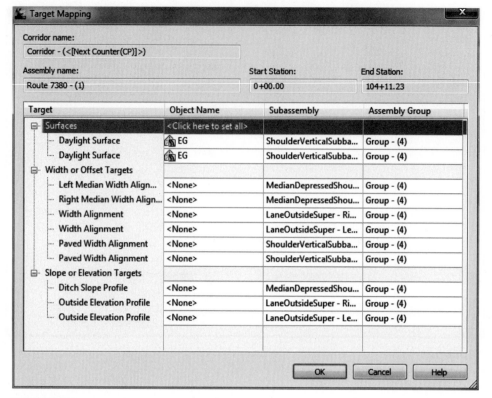

FIGURE 9.29

8. Click **OK** until you have exited the dialog boxes and built the highway.

9. At Civil 3D's top left, Quick Access toolbar, click the **Save** icon to save the drawing.

10. Use the ZOOM and PAN commands to view the new corridor.

The maximum superelevation occurs at stations 32 + 00 to 47 + 00 and 71 + 00 to 80 + 00. Start viewing the first region's sections at station 25 + 00 and the second region at station 54 + 00.

11. Click the Ribbon's Corridor tab. In the Corridor tab's Modify panel, select CORRIDOR SECTION EDITOR, pick any line representing the corridor, and view the assembly behavior through the superelevated curves. When you are finished reviewing the curves, click Close to exit the Section Editor tab.

12. Click Close to exit the Corridor tab.

This completes the transitions and superelevations exercise. The following unit focuses on Civil 3D's Intersection Wizard.

EXERCISE 9-4

After completing this exercise, you will:

- Be able to review Design Criteria values.
- Be able to create an intersection.

Exercise Setup

This exercise uses the *Chapter 09 - Unit 4.dwg* file. To find this file, browse to the CD that accompanies this textbook and open the *Chapter 09 - Unit 4.dwg* file.

1. If you are not in **Civil 3D**, double-click the **Civil 3D** desktop icon to start the application.

2. When you are at the command line, CLOSE the open drawing and **do not save it**.

3. At Civil 3D's top left, Quick Access toolbar, click the Open icon, browse to the Chapter 09 folder of the CD that accompanies this textbook, and open the *Chapter 09 - Unit 4* drawing.

4. At Civil 3D's top left, click Civil 3D's drop-list arrow, from Application Menu, highlight SAVE AS, from the flyout menu, select AUTOCAD DRAWING, browse to the Civil 3D Projects folder, for the drawing name enter **Intersection**, and click **Save** to save the file.

5. On the Ribbon, click the **View** tab. On the Views panel's left, select and restore the named view **Intersection**.

Intersection Wizard

1. On the Ribbon, click the Home tab. In the Create Design panel, click the Intersection icon.

2. The command line prompts for an intersection point. In the drawing, select the intersection between the two roadways.

3. The Create Intersection Wizard displays and reviews its values. The primary road will maintain its crown (Rosewood – 1) and Lorraine will remove its crown. Click Next.

In the intersection, Rosewood – (1) will be 24-feet wide on its left side.

4. Click Offset Parameters, and for the Primary road left side, set the width to 24 feet. When set, click OK to return to the Create Intersection wizard.

5. Click Curb Return Parameters, and for the SE and SW quadrants, set the return radius to 37.5. When done, click OK to return to the Create Intersection wizard.

6. Click Lane Slope Parameters and review the intersection slope settings. When done, click OK to return to the Create Intersection wizard.

7. Click Curb Return Profile Parameters and change the quadrants to view their preview graphics. When done, click OK to return to the Create Intersection wizard.

8. Click Next.

9. At the panel's top left, toggle Create corridors in the intersection area to Create a new corridor and click Create Intersection.

10. At Civil 3D's top left, Quick Access toolbar, click the Save icon to save the drawing.

This ends the exercise on intersection design.

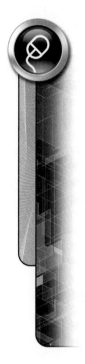

EXERCISE 9-5

After completing this exercise, you will:

- Be able to create a roundabout.
- Be familiar with the Create Roundabout and Add Approach command.

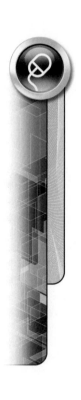

Exercise Setup

This exercise uses the *Chapter 09 - Unit 5.dwg* file. To find this file, browse to the CD that accompanies this textbook and open the *Chapter 09 - Unit 5.dwg* file.

1. If you are not in **Civil 3D**, double-click the **Civil 3D** desktop icon to start the application.

2. When you are at the command line, CLOSE the open drawing and **do not save it**.

3. At Civil 3D's top left, Quick Access toolbar, click the Open icon, browse to the Chapter 09 folder of the CD that accompanies this textbook, and open the *Chapter 09 - Unit 5* drawing.

4. At Civil 3D's top left, click Civil 3D's Menu Browse drop-list arrow, from the Application Menu, highlight SAVE AS, from the flyout menu, select AUTOCAD DRAWING, browse to the Civil 3D Project folder, for the name enter **Round-about**, and click **Save** to save the file.

The drawing contains two alignments: Lorraine – (1) and Rosewood – (1). You will create a roundabout at their intersection. After creating the roundabout, you will add another approach.

Define a Roundabout

1. From the Home ribbon, Create Design panel, click the Intersection icon to view its command list.

2. From the Intersection command list, select **Create Roundabout**. The routine prompts for you to select an intersection.

3. Using the Endpoint object snap, select the Lorraine – (1)'s (green centerline) northern endpoint where it intersects Rosewood – (1) (red centerline).

4. After selecting the intersection, the routine prompts for the approach roads. In the drawing, select Lorraine – (1) and just to the right of the intersection, Rosewood – (1). After selecting the two centerlines, press Enter to display the Create Roundabout wizard's Circulatory panel.

5. Using Figure 9.30 as a guide, match your values to the figure's values. When set, click Next to display the Approach panel.

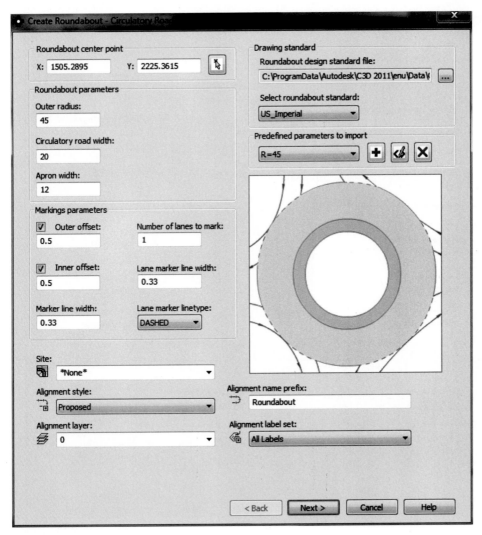

FIGURE 9.30

6. Using Figure 9.31 as a guide, match your values to the figure's values. When set, at the panel's top right, click the button Apply to all.

7. Click the Next button to display the Islands panel.

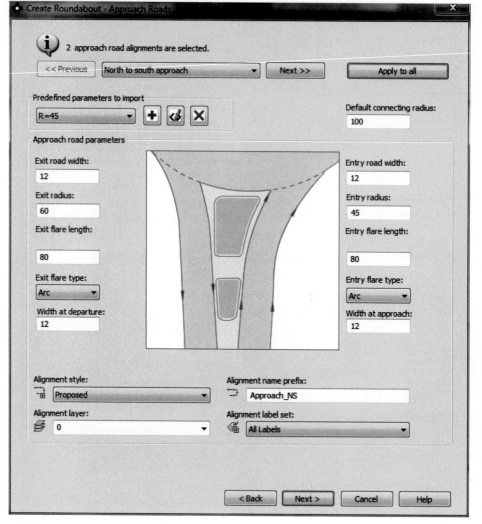

FIGURE 9.31

8. Using Figure 9.32 as a guide, match the values in the figure. When set, at the panel's top right, click the button Apply to all.

9. Click the Next to display the Markings and Signs panel. There are no changes to the panel's value.

10. Click Finish to create the roundabout.

11. From the Quick Access toolbar, click Save.

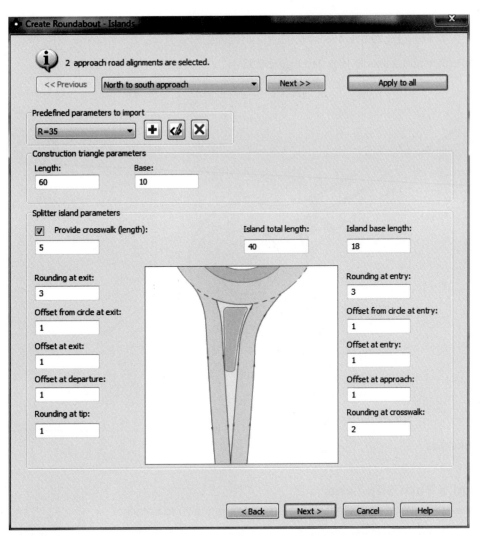

FIGURE 9.32

Add an Approach

The roundabout has two approaches and needs a third for Rosewood – (1)'s west side.

1. From the Home ribbon, Create Design panel, click the Intersection icon to view its command list.

2. From the Intersection command list, select **Add Approach**. The routine prompts for you to select the circular area. In the drawing, select the roundabout outer circle.

3. After selecting the roundabout's circular area, the routine prompts for an approach. In the drawing, select Rosewood – (1)'s red centerline to the west of the roundabout. After selecting the approach, press Enter to display the Create Roundabout – Approach Roads panel.

4. Use Figure 9.31 as a guide and set the values for your panel. When the values are set, click the Next button to display the Islands panel.

5. Use Figure 9.32 as a guide and set the values for your panel. When the values are set, click Next to display the Markings and Signs panel. There are no changes in this panel.

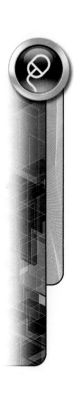

6. Click the Finish button to add the approach.

7. From the Quick Access toolbar, click Save.

Your roundabout should look like Figure 9.33.

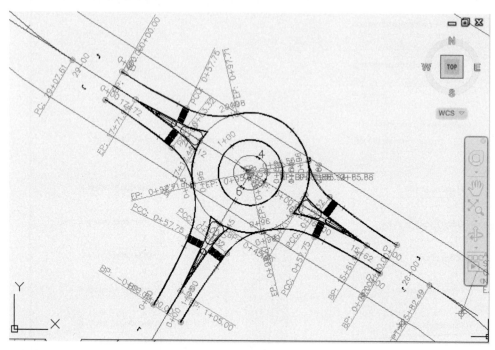

FIGURE 9.33

Add a Turn Slip Lane

Rosewood – (1) needs a slip lane from the west approach to Lorraine.

1. From the Home ribbon, Create Design panel, click the Intersection icon to view its command list.

2. From the Intersection command list, select **Add Turn Slip Lane**. The routine prompts for you to select the entry approach. In the drawing just to the west of the island, select the western approach's southwest line.

3. The routine prompts for an exit approach. In the drawing, just to Lorraine's island's south select its northern outside line.

4. The Draw Slip Lane dialog box displays.

5. Use Figure 9.34 as a guide and set your values to match its.

6. When done, click Finish to draw the turn slip lane.

7. Your roundabout should match Figure 9.35.

8. From the Quick Access toolbar, click Save.

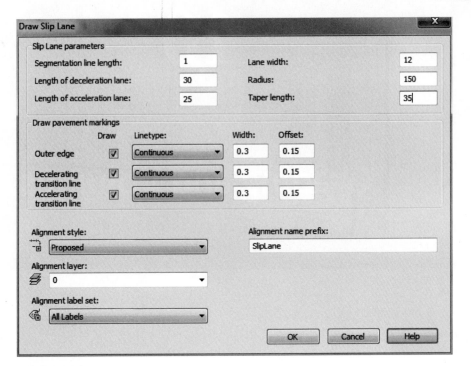

FIGURE 9.34

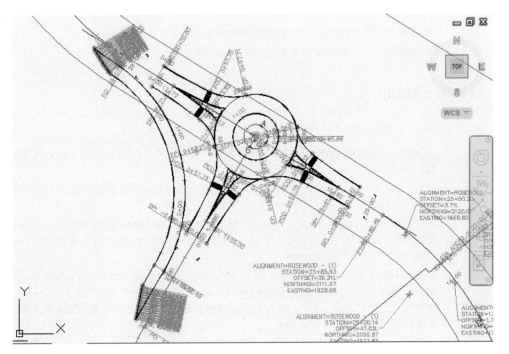

FIGURE 9.35

This ends the exercise on roundabout design.

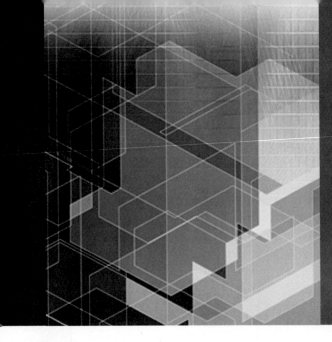

Grading and Volumes

After completing this exercise, you will:

- Be able to set points with elevations from a surface.
- Be familiar with placing points between two existing points.
- Be able to place points on a polyline or contour.
- Be able to place points at a distance or elevation.

Exercise Setup

This exercise uses the *Chapter 10 – Unit 1.dwg* file. To find this file, browse to the CD that accompanies this textbook and open the *Chapter 10 – Unit 1.dwg* file.

1. If you are not in Civil 3D, double-click the **Civil 3D** desktop icon to start the application.

2. When you are at the command prompt, CLOSE the opening drawing and do not save the file.

3. At Civil 3D's top left, Quick Access toolbar, click the **Open** icon, browse to the Chapter 10 folder of the CD that accompanies this textbook, and open the *Chapter 10 – Unit 1* drawing.

4. At Civil 3D's top left icon, click Civil 3D's drop-list arrow, from the Application Menu, highlight SAVE AS, from the flyout menu, select AutoCAD Drawing, browse to the Civil 3D Projects folder, for the drawing name **Surface Points**, and click **Save** to save the file.

5. In Prospector, select the Points heading, press the right mouse button, and from the shortcut menu, select **CREATE...** to display the Create Points toolbar.

6. On the Create Points toolbar's right, click Expand the Create Points dialog (chevron).

7. Expand Points Creation, change Prompt For Elevations to **Manual**, change Prompt For Point Names to **None**, change Prompt For Descriptions to **Automatic**, and set the Default Description to **GP** (see Figure 10.1).

8. At the toolbar's right, click the chevron Collapse the Create Points dialog.

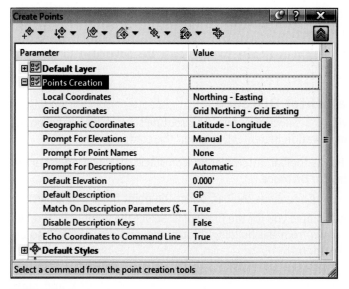

FIGURE 10.1

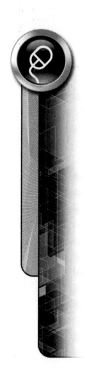

Interpolate — Relative Location

The Interpolate routines use two points between which they calculate elevations for new points. The second point can be an existing or a selected point that you assign an elevation. The new points can have an offset; negative is left and positive is right. You can select two points from the drawing, assign them elevations, and continue using the command. You do not need to select a cogo point for either point.

1. Make sure you can view the Transparent Commands toolbar.

2. From the Miscellaneous icon stack, select **Manual**, and select a point to the west of the building points. For the point's elevation, enter **675.00**, and press ENTER until you have exited the routine.

3. From the Create Points toolbar, Interpolation icon stack, select its drop-list arrow and select **By Relative Location**.

4. To select a cogo point as a starting point, select the **Point Object** filter from the Transparent Commands toolbar.

5. The command line prompts you for the first point. In the drawing along the building's edge, select the northernmost point and press ENTER to accept the elevation of 681.00.

6. The command line prompts you for the second point object. Select the just-created new point and press ENTER to accept its elevation (675.00).

The routine reports the elevation, distance, and grade between the selected points, a direction arrow, and a distance jig from the first point. If necessary, press F2 to view the reported information.

7. The command line prompts you for a distance. Enter **25** and press ENTER.

8. The command line prompts you for an offset. Press ENTER for no offset distance.

A new point appears 25 feet from the first point (building 681.00), whose elevation is interpolated from the distance and elevation between the two points.

9. The command line prompts you for another distance. Enter **45** and press ENTER.

10. The command line prompts you for an offset. Press ENTER for no offset.

11. Press ESC to end the routine.

12. At Civil 3D's top left, Quick Access toolbar, click the **Save** icon to save the drawing.

Interpolate — Incremental Distance

This routine works only with the selected points. The routine's focus is placing points either at a distance or at an elevation increment between the control points.

1. From the Create Points toolbar, Interpolation icon stack, click its drop-list arrow and select **Incremental Distance**.

2. From the Transparent Commands toolbar, select the point filer **Point Object**.

3. The command line prompts you for the first point. Select the point object, select the second point down from the building's north end, and press ENTER to accept the 681.00 elevation.

4. The command line prompts you for the second point. Select the point object, select the point you placed in the drawing (elevation of 675.00), and press ENTER to use its elevation.

5. The command line prompts you for a distance between points. Enter **25** and press ENTER.

6. The command line prompts you for an offset. Press ENTER for no offset.

The routine places points between the control points at 25-foot increments. Their elevations are calculated using a constant slope between the two control points.

7. Press ESC to exit the routine.

8. Use the ERASE command and erase the new points, leaving the points along the building's westerly side.

9. At Civil 3D's top left, Quick Access toolbar, click the **Save** icon to save the drawing.

Slope — Slope/Grade – Elevation

This routine creates points in a direction from an existing or selected point. The elevation is either the point's elevation or one you specify.

1. From the Create Points toolbar, Slope icon stack, click its drop-list arrow and select **Slope/Grade – Elevation**.

2. From the Transparent Commands toolbar, select the **Point Object** override.

3. The command line prompts you for the first point as a point object. Select the northernmost point of the building edge.

The routine prompts you for the next point, to set a direction and default distance.

4. In the drawing, select a point to the west.

5. The command line prompts you for a slope. Enter **8** (for 8:1) and press ENTER.

The command line reports the slope and the grade.

6. The command line prompts you for an ending elevation. Enter **675.00** and press ENTER.

7. The command line prompts you for the number of intermediate points. Enter **4** and press ENTER.

8. The command line prompts you for an offset. Press ENTER for no offset.

9. The command line prompts you about including a point at the end of the distance. Press ENTER for Yes.

The routine creates points that radiate from the control point. The last point's elevation is 675.

10. Press ESC to end the routine.
11. Use the ERASE command to erase the new points from the drawing, leaving those points along the building's side.
12. At Civil 3D's top left, Quick Access toolbar, click the **Save** icon to save the drawing.

Surface — Random Points

Random Points requires a surface. Any surface elevation can be assigned to a point by selecting it, pressing the right mouse button, and from the shortcut menu, selecting Elevations from Surface....

1. In the Layer Properties Manager, thaw the layer **C-TOPO-EXISTING GROUND** and click **X** to exit the palette.
2. From the Create Points toolbar, Surface icon stack, click its drop-list arrow, and from the list, select **Random Points**.
3. The command line prompts you for a surface. In the drawing, select a surface contour to identify the surface.
4. In the drawing, select a few random points, and, after creating them, press ENTER twice to exit the routine.

The routine creates points whose elevations are from the selected surface.

5. ERASE the points you just made and leave the points along the building.
6. At Civil 3D's top left, Quick Access toolbar, click the **Save** icon to save the drawing.

Surface — Along Polyline/Contour

This routine creates points whose elevations are from a selected surface at a measured distance along a polyline's path.

1. If necessary, in the status bar, toggle **OFF** Object Snaps.
2. Start the POLYLINE command, and in the drawing, draw two polylines, each with multiple vertices.
3. From the Create Points toolbar, Surface icon stack, click its drop-list arrow and select **Along Polyline/Contour**.
4. The command line prompts you to select a surface. In the drawing, select a contour to identify the surface.
5. The command line prompts you for a distance. For distance, enter **15** and press ENTER.
6. The command line prompts you to select a polyline or contour. In the drawing, select a polyline and press ENTER.

The routine creates points at 15-foot intervals along the polyline, ignoring the polyline vertices.

Surface — Polyline/Contour Vertices

This routine creates points whose elevations are from a selected surface at each polyline or contour vertex.

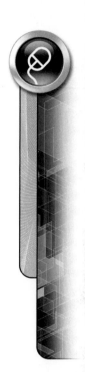

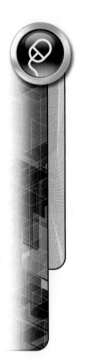

1. From the Create Points toolbar, Surface icon stack, click its drop-list arrow, and from the list, select ***Polyline/Contour Vertices***.

2. The command line prompts you for a surface. In the drawing, select a contour to identify the surface.

3. The command line prompts you to select a polyline or contour. In the drawing, select the remaining polyline.

The routine places a point at each vertex. When you are using a polyline, the point's elevation is the surface's elevation.

4. Press ENTER to exit the routine.

5. Close the Create Points toolbar.

6. At the top right of the drawing display, click the ***Close*** icon, exit the drawing, and do not save the changes.

This ends the exercise on points as grading data. The next unit reviews feature lines and grading objects.

EXERCISE 10-2

After completing this exercise, you will:

- Be familiar with Feature Line Edit Drawing Settings and Edit Feature Settings dialog boxes.
- Be able to adjust a profile PVI and Export Feature Lines from the profile.
- Be able to modify feature line elevations.
- Be able to use a feature line as surface data.
- Be able to convert polylines into feature lines.
- Be able to annotate a feature line.

Exercise Setup

This exercise uses the *Chapter 10 – Unit 2.dwg* file. To find this file, browse to the CD that accompanies this textbook and open the *Chapter 10 – Unit 2.dwg* file.

1. If not in Civil 3D, double-click the ***Civil 3D*** desktop icon to start the application.

2. When you are at the command prompt, CLOSE the opening drawing and do not save the file.

3. At Civil 3D's top left, Quick Access toolbar, click the ***Open*** icon, browse to the Chapter 10 folder of the CD that accompanies this textbook, and open the *Chapter 10 – Unit 2* drawing.

4. At Civil 3D's top left icon, click Civil 3D's drop-list arrow, from the Application Menu, highlight SAVE AS, from the flyout menu, select AutoCAD Drawing, browse to the Civil 3D Projects folder, for the drawing name enter **Feature Lines**, and click ***Save*** to save the file.

Edit Drawing Settings

1. Click the ***Settings*** tab.

2. In Settings, select the drawing name, press the right mouse button, and from the shortcut menu, select EDIT DRAWING SETTINGS….

3. Click the ***Object Layers*** tab.

   Most objects have suffix modifiers that append each object with its name.

4. For **Feature Line, Grading, and Grading-Labeling**, set their modifiers to **Suffix** and their values to -*.

5. Click ***OK*** to exit the dialog box.

Edit Feature Settings

1. In Settings, select the Grading heading, press the right mouse button, and from the shortcut menu, select Edit Feature Settings....

2. Expand the Default Styles and Name Format sections.

The Default section assigns a feature line style and naming templates.

3. Review the settings and then click ***OK*** to exit the dialog box.

Drafting a Feature Line

Drafting a feature line also assigns elevations to it (see Figure 10.2).

1. Click the Ribbon's ***View*** tab. In the Views panel, select and restore the named view **PK-Island**.

2. Click the ***Home*** tab. On the Create Design panel, click the Feature Line icon, and from the shortcut menu, select ***Create Feature Line***.

3. In the Create Feature Lines dialog box, click ***OK*** to accept the defaults.

4. In the drawing, use an *Endpoint* object snap to select the north island's gutter endpoint (see Figure 10.2). The command line prompts you for an elevation. Press ENTER to accept.

5. The command line prompts for the next point. For the next endpoint, in the command line, enter **@40 <270** and press ENTER.

6. The command line prompts you for the new point's elevation. If not grade, in the command line, enter '**G**' and press ENTER. For the grade, enter **-2** and press ENTER.

7. The command line prompts for the next point. For the next endpoint, use the *Perpendicular* object snap, and in the drawing, select a point on the southern island flange line.

8. The command line prompts you for a grade. To enter an elevation in the command line, enter '**E**' and press ENTER. The command line prompts you with the selected point's elevation. Press ENTER to accept the elevation.

9. Press ENTER to exit the command.

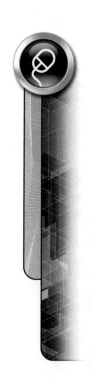

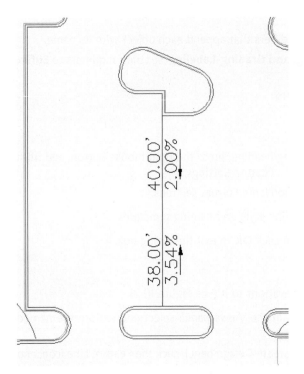

FIGURE 10.2

Label Feature Line

1. Click the Ribbon's Annotate tab, on the Labels & Tables panel, select Add Labels.
2. In the Add Labels dialog box, change the Feature to **Line and Curve**, change the Label type to **Multiple Segment**, and set the Line label style to **Grade over Distance**. Click Add, and in the drawing, select the feature line that was just drawn.
3. Close the Add Labels dialog box.

Elevation Editor...

Elevation Editor displays a feature line's elevations and grades in a vista.

1. PAN the feature line to the right side of the screen.
2. In the drawing, select the feature line. In the Feature Line tab, Modify panel, select the *Edit Elevations* icon to display the Edit Elevations panel. From the Edit Elevations panel, select ELEVATION EDITOR to display the Grading Elevation Editor vista.
3. Press ESC to remove the grips, place the cursor on the vista, and notice the triangles at the feature line control points.
4. In the editor, select a control point. Only its triangle is displayed on the feature line.
5. Select another entry in the vista and notice the redisplayed triangle.
6. Unselect the rows by selecting the *Unselect All Rows* icon at the vista's top center (rightmost icon).
7. Double-click in the Station 0 + 40 elevation cell, change it to **676**, and press ENTER.
8. Click the Panorama's *green checkmark* until it is closed.

Quick Elevation Edit

Quick Elevation Edit adjusts a feature line's elevations and grades.

1. On the Ribbon's **Modify** tab, on the Edit Elevations panel's middle right, select the **Quick Elevation Edit** icon (first row, first icon).

2. In the drawing, slowly drag the cursor up and down the just drawn feature line.

Control points and grade directions and their values are displayed as you move the cursor over the feature line.

3. Place the cursor near the northern segment's midpoint to display its grade. Click the segment with the grade direction pointing south, and, for the new grade, enter **-2**, and press ENTER.

4. Move the cursor to the feature line's middle vertex. The elevation should not be 676.

5. Press ENTER to exit the command.

Edit Elevations

Edit Elevations edits at the feature line's nearest vertex to the selection point and cycles through each feature line vertex.

1. On the Ribbon's **Modify** tab, on the Edit Elevations panel's middle right, select the **Edit Elevations** icon (first row, second icon).

2. In the drawing, select the feature line's northern end and press ENTER several times to cycle through the vertices.

3. Change the middle vertex's elevation to **676.00**, press ENTER, and enter '**X**' to exit the command.

Converting a Feature Line

Converting an object to a feature line, if selected, assigns a style and an elevation.

1. Use the PAN and ZOOM commands to view the building footprint to the left of the parking lot.

2. On the Ribbon, click the **Home** tab. On the Create Design panel, click the **Feature Line** icon, and from the shortcut menu, select CREATE FEATURE LINES FROM OBJECTS.

3. The command line prompts you to select an object. In the drawing, select the building's footprint and press ENTER to continue.

4. The Create Feature Lines dialog box opens. For the feature line name, toggle **ON** Name, enter **470 Willow**, toggle **ON** Style, and from the list, select **Basic Feature Line**, toggle OFF Erase existing entities, toggle **ON** Assign elevations, and click **OK** to continue.

5. In the Assign Elevations dialog box, toggle **OFF** Insert intermediate grade break points, and click **OK**.

6. In the drawing, select the feature line, press the right mouse button, and from the shortcut menu, select ELEVATION EDITOR….

Not inserting intermediate grade break points places elevations only at the feature vertices.

7. Click the vista's **green checkmark** to close it.

8. Use the AutoCAD Erase command to erase the feature line created.

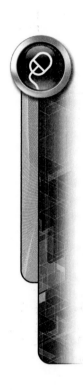

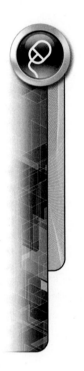

9. On the Ribbon's Home tab, on the Create Design panel, click the **Feature Line** icon, and from the shortcut menu, select the CREATE FEATURE LINES FROM OBJECTS.

10. The command line prompts you to select an object. In the drawing, select the building's footprint, and press Enter to continue.

11. The Create Feature Lines dialog box opens. For the feature line name, toggle **ON** Name, enter **470 Willow**, toggle **ON** Style, and from the list, select **Basic Feature Line**, toggle **ON** Erase existing entities and Assign elevations, and click **OK** to continue.

12. In the Assign Elevation dialog box, toggle **ON** Insert intermediate grade break points and click **OK**.

13. In the drawing, select the feature line to display its grips.

Each round grip represents a TIN intersection with the feature line.

14. In the drawing, select the feature line, press the right mouse button, from the shortcut menu, select ELEVATION EDITOR..., and review the elevation assignments.

15. Click the **green checkmark** to close the Panorama.

Create a Surface from Feature Line

Feature Lines are surface breakline data.

1. Click the **Prospector** tab.

2. In Prospector, select the **Surfaces** heading, press the right mouse button, and from the shortcut menu, select CREATE SURFACE... .

3. In Create Surface, change the style to **Contours and Triangles** and click **OK** to continue.

4. Expand the Surfaces branch until you are viewing the Surface1 Definition tree.

5. In the Definition tree, select **Breaklines**, press the right mouse button, and from the shortcut menu, select ADD... .

6. For the Description enter **470 Willow**, click **OK**. If necessary, in the drawing, select the building footprint.

7. In the drawing, select **Surface1**, right mouse click, and from the shortcut menu, select SURFACE PROPERTIES... .

8. In Surface Properties, review the surface statistics.

9. Click **OK** to exit Surface Properties.

10. In Prospector, select **Surface1**, press the right mouse button, from the shortcut menu, select DELETE..., and click **Yes** to delete the surface.

Create from Alignment

The Create from Alignment icon exports a feature line from an alignment's vertical profile.

1. In Layer Properties Manager, thaw the layer **C-ROAD-CORR-93rd - (1)** and click **X** to exit.

2. Use the ZOOM and PAN commands to better view the 93rd centerline just south of the entrance.

3. If necessary, click the **Prospector** tab.

4. Select the **Sites** heading, press the right mouse button, and from the shortcut menu, select NEW... .

5. In the new Site Properties dialog box, for the name, enter **93rd Alignment** and click **OK**.

6. In Prospector, select the Sites heading, press the right mouse button, and from the shortcut menu, select NEW….

7. In the Site Properties dialog box, for the site name enter Feature Lines from Corridor, and click OK to create the site.

8. On the Ribbon's Home tab, Create Design panel, click the **Feature Line** icon, and from the shortcut menu, select CREATE FEATURE LINES FROM ALIGNMENT.

9. In the drawing, select the 93rd centerline to display the Create Feature Line from Alignment dialog box.

10. Using Figure 10.3 as a guide, at the top of the dialog box, for the name, toggle ON Name, enter **93rd Centerline**, make sure the current profile is 93rd (3), toggle **ON** Style, and from the drop-list, select **Corridor Crown**, and click **OK** to open the Weed Vertices dialog box.

11. Click **OK** to accept the default weeding factors, thus creating the feature line.

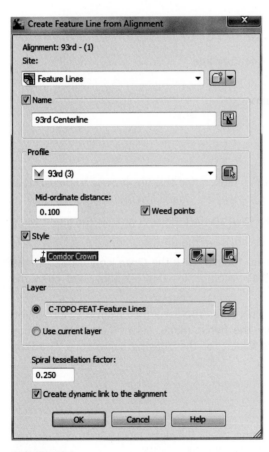

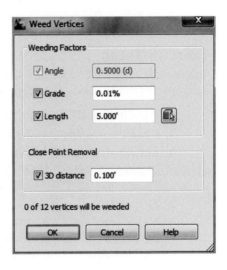

FIGURE 10.3

Feature Lines from Corridor

1. Move your cursor around the corridor until you identify the ETW feature line.

2. On the Ribbon, click the **Modify** tab. On the Design panel, click the Corridor icon, and in the Corridor tab, Launch Pad panel's right side, select CREATE FEATURE LINES FROM CORRIDOR.

The command line prompts you to select a corridor feature line.

3. In the drawing, select the blue corridor ETW feature line, in the Select a Feature Line dialog box, select the feature line **ETW**, and click OK.

The Create Feature Line from Corridor dialog box opens.

4. In the dialog box, change the style to **Corridor Edge of Travel Way** and click **OK** to create the feature line.

5. Export a few more corridor feature lines. Some feature lines may be on top of one another. If this is the case, just export one of the feature lines. Press Enter to end the command.

6. Use the LIST command, select a feature line, and review the AutoCAD object report.

7. At Civil 3D's top left, Quick Access toolbar, click the **Save** icon to save the drawing.

EXERCISE 10-3

After completing this exercise, you will:

- Be able to convert a polyline to a feature line.
- Be able to modify a feature line.
- Be able to create a grading solution.
- Be able to edit a grading.

Exercise Setup

This exercise uses the *Chapter 10 – Unit 3.dwg* file. To find this file, browse to the CD that accompanies this textbook and open the *Chapter 10 – Unit 3.dwg* file.

1. If not in Civil 3D, double-click the **Civil 3D** desktop icon to start the application.

2. Close the opening drawing and do not save the file.

3. At Civil 3D's top left, Quick Access toolbar, click the **Open** icon, browse to the Chapter 10 folder of the CD that accompanies this textbook, and open the *Chapter 10 – Unit 3* drawing.

4. At Civil 3D's top left icon, click Civil 3D's drop-list arrow, from the Application Menu, highlight SAVE AS, from the flyout menu, select AutoCAD Drawing, browse to the Civil 3D Projects folder, for the drawing name enter **Grading**, and click **Save** to save the file.

Edit Drawing Settings

1. Click the **Settings** tab.

2. In Settings, select the drawing name, press the right mouse button, and from the shortcut menu, select EDIT DRAWING SETTINGS....

3. In Drawing Settings, click the **Object Layers** tab, toggle the **Suffix** modifier, and set its value to a dash asterisk (**-***) for the objects **Feature Line, Grading,** and **Grading-Labeling**.

4. Click **OK** to exit and set the modifiers.

5. Click the **Prospector** tab.

6. In Prospector, select **Sites**, press the right mouse button, and from the shortcut menu, select NEW... to open the Site Properties dialog box.

7. In Site Properties, Information panel, for the site name enter **93rd Street** and click **OK**.

Creating the Ditch Feature Line

Grading is ideal for ditch, pond, and surface design.

1. Click Ribbon's **View** tab. On the Views panel's left, select and restore the named view **South Ditch**. You may need to scroll down the list by selecting the down arrow.

2. On the Ribbon, click the **Home** tab. In the Create Design panel, click the **Feature Line** icon, and from the shortcut menu, select CREATE FEATURE LINES FROM OBJECTS, select the ditch outline, and press ENTER to open the Create Feature Lines dialog box.

3. In Create Feature Lines, toggle **ON**, and for the name, enter **South Ditch**, keeping the counter, toggle **ON**, and set the style to **Grading Ditch**, and click **OK** to create the feature line.

4. Click the **Modify** tab. From the Edit Geometry panel, select **Fillet** (second row, second icon), and in the drawing, select the new feature line.

5. The command line prompts you to specify a corner. Enter '**R**', press ENTER to set the radius to **17.5**, and press ENTER to continue.

6. Move the cursor to the three corners to preview the resulting radius.

7. Select the three corners, except the lower left, to create the arc segments.

8. Still in the command, change the radius by entering '**R**', and then press ENTER. For the radius, enter **5.0** and press ENTER.

9. In the lower left of the ditch, select the corner to create the new arc and press ENTER twice to exit the command.

10. At Civil 3D's top left, Quick Access toolbar, click the **Save** icon to save the drawing.

Creating the Ditch

1. In the drawing, select the **South Ditch** feature line, press the right mouse button, and from the shortcut menu, select ELEVATION EDITOR....

2. In the Grading Elevation Editor vista, click the **Raise/Lower** icon, in the elevation cell for the elevation, enter **671.00**, press ENTER to assign the elevation, and click the **green checkmark** until the vistas are closed.

3. Click the Ribbon's Home tab, Create Design panel, click the **Grading** icon, and from the shortcut menu, select CREATE GRADING TOOLS to display the Grading Creation Tools toolbar.

4. On the toolbar's left, click the **Set the Grading Group** icon.

5. In the Site dialog box, select **93rd Street** and click **OK** to open the Create Grading Group dialog box.

6. In Create Grading Group, do not change the grading group Name, toggle **ON Automatic Surface Creation**, set the Surface style to **_No Display**, toggle **ON** and set the Volume base surface to **Existing Ground**, and click **OK** to close the dialog box (see Figure 10.4).

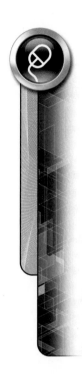

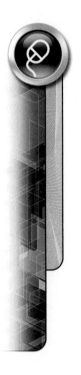

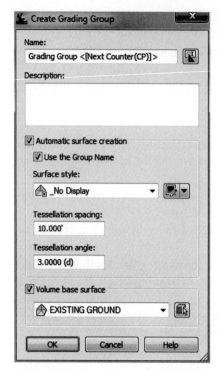

FIGURE 10.4

The Create Surface dialog box opens and has a surface name of Grading Group 1.

7. In Create Surface, for the surface description, enter **Grading Objects**, and click **OK** to exit the dialog box.

8. At the middle of the Grading Creation Tools toolbar, click the criteria drop-list arrow, and from the list, select **Grade to Relative Elevation**.

9. In the toolbar, to the right of criteria (the second icon to its right), click the drop-list arrow, and from the list, select CREATE GRADING.

10. In the drawing, select the southern ditch's feature line.

11. The command line prompts you for a grading side. In the drawing, select a point in the ditch's interior and press ENTER to apply the criteria to the entire feature line length.

12. The command line prompts you for a relative elevation. Enter **-5** and press ENTER.

13. The command line prompts you for the slope format. Press ENTER to accept the slope, for the slope, enter **2**, and press ENTER twice to create the grading and exit the command.

Edit Grading

1. In the grading, select its diamond icon, press the right mouse button, and from the shortcut menu, select GRADING EDITOR….

2. In the Grading Editor vista, change the Relative Elevation to **-7** and press ENTER to change the grading solution.

3. Close the Panorama.

North Ditch

1. Use the ZOOM and PAN commands to view the north ditch polyline.

2. On the Ribbon, click the **Modify** tab. In the Design panel, click the **Feature Line** icon. In the Feature Line tab's Modify panel, click the Edit Geometry panel icon. On the Edit Geometry panel's right, select the **Weed** icon (third row, first icon), select the northern ditch line, and review the weeding parameters.

3. In Weed Vertices, change the 3D distance to **15** and press ENTER.

The red vertices will be weeded from the feature line.

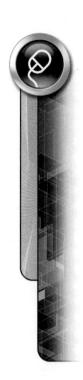

4. Click **OK** to weed the vertices.

5. Click the Ribbon's Home tab. On the Create Design panel, click the **Feature Line** icon, from the shortcut menu, select CREATE FEATURE LINES FROM OBJECTS, select the north ditch polyline, and press ENTER to open the Create Feature Lines dialog box.

6. In the Create Feature Lines dialog box, toggle **ON** name and enter **North Ditch**, toggle **ON** Style and set it to **Grading Ditch**, and click **OK**.

7. From the Grading Creation Tools toolbar, make sure the criteria is **Grade to Relative Elevation**, click **Create Grading**, and in the drawing, select the North Ditch feature line.

8. In the drawing, set the grading side by picking a point in the north ditch's interior, press ENTER to apply the grading to the entire length, for the relative elevation, enter **-6**, press ENTER to accept the Slope format, for the slope enter **2.5**, and press ENTER twice to create the grading and exit the routine.

9. At Civil 3D's top left, Quick Access toolbar, click the **Save** icon to save the drawing.

Grading a Pond

A pond's outline is just to the northwest of the building. The first series of grading commands will use a grading solution to create a second solution. The interior of the pond will have the floor as an infill to put a floor in the pond.

1. Use the PAN and ZOOM commands to view the pond outline to the north of the building.

2. In the Grading Creation Tools toolbar, change the grading criteria to **Grade to Distance** and click **Create Grading**.

3. In the drawing, select the pond's top contour (676), in the Create Feature Lines dialog box, toggle **ON** Style, set the style to **Basic Feature Line**, and click **OK** to continue.

4. The command line prompts you for a grading side. Select a point to the outside of the polyline.

5. The command line prompts you whether to apply the grading to the entire length. Press ENTER to accept.

6. The command line prompts you for a distance. For the distance, enter **6**, and press ENTER to continue.

7. The command line prompts you for a format. Press ENTER for Slope, for the slope value, enter **4**, and press ENTER twice to solve the grading and exit Create Grading.

8. In the Grading Creation Tools toolbar, change the grading criteria to **Grade To Surface** and click **Create Grading**.

9. In the drawing, select the outer grading target line and press ENTER to apply the grading to the entire length.

10. The command line prompts you for a cut format. Press ENTER for **Slope**, for the slope value, enter **4**, and press ENTER to continue.

11. The command line prompts you for a fill slope. Press ENTER for Slope, for the slope value, enter **4**, and press ENTER twice to solve the grading and end the routine.

The grading solution intrudes on the building.

12. In the Grading Creation Tools toolbar, change the criteria to **Grade to Relative Elevation** and click *Create Grading*.

13. In the drawing, select the pond edge feature line, select a point inside the pond for the offset side, and press ENTER to apply the grading to its entire length.

14. The command line prompts you for a relative elevation. Enter **-6** and press ENTER to continue.

15. The command line prompts you for a slope or grade. Press ENTER for slope, for the slope, enter **2.75**, and press ENTER twice to solve the grading and exit Create Grading.

16. In the Grading Creation Tools toolbar, to the right of Create Grading, click the drop-list arrow, from the list, select CREATE INFILL, and in the drawing, click in the pond's center.

17. Press ENTER to exit Create Grading.

18. At Civil 3D's top left, Quick Access toolbar, click the *Save* icon to save the drawing.

Edit Grading

The surface grading slope intrudes on the building and should be changed.

1. In the drawing, select the outer grading icon (diamond), press the right mouse button, and from the shortcut menu, select GRADING EDITOR….

2. In the Grading Editor vista, change the fill slope to **1.25**.

The grading is recomputed.

3. Close the Grading Editor vista.

4. At Civil 3D's top left, Quick Access toolbar, click the *Save* icon to save the drawing.

Viewing the Grading

1. In the drawing, select the pond grading objects, press the right mouse button, and from the shortcut menu, select OBJECT VIEWER….

2. After reviewing the grading, exit the Object Viewer.

Grading Volume

1. In the Grading Creation Tools toolbar, select the *Grading Volume Tools* icon (the fifth icon in from the right) to display the Grading Volume Tools.

The dialog box reports the grading group volume.

2. Click the red **X** to close the toolbars.

3. On the Feature Line tab, click the Close icon.

This completes the grading unit. Grading is a powerful tool when used with other grading design methods.

EXERCISE 10-4

After completing this exercise, you will:

- Be able to build, evaluate, and calculate volumes.
- Be able to create a surface from contour and 3D polyline data.
- Be able to calculate a volume (in a panorama) between two surfaces.
- Be able to create a Triangular Irregular Network (TIN) and grid volume surface.
- Be able to review the volume surface statistics.

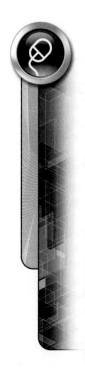

Exercise Setup

This exercise has two parts. The first part involves creating a surface whose data is existing ground contours. This is done for two reasons: to compare the calculated volumes between the existing surface (points and breaklines) and to compare a contour data surface and the design surface.

The second part involves creating a design surface. The second surface sets the stage for the volume calculations.

This exercise uses the *Chapter 10 – Unit 4.dwg* file. To find this file, browse to the CD that accompanies this textbook and open the *Chapter 10 – Unit 4.dwg* file.

1. If you are not in **Civil 3D**, double-click its desktop icon to start the application.
2. CLOSE the current drawing and do not save it.
3. At Civil 3D's top left, Quick Access toolbar, click the **Open** icon, browse to the Chapter 10 folder of the CD that accompanies this textbook, and open the *Chapter 10 – Unit 4* drawing.
4. At Civil 3D's top left icon, click Civil 3D's drop-list arrow, from the Application Menu, highlight SAVE AS, from the flyout menu, select AutoCAD Drawing, browse to the Civil 3D Projects folder, for the drawing name enter **Surface Volumes**, and click **Save** to save the file.

Creating an EGCTR Surface

EGCTR uses the Existing surface contours.

1. If necessary, click the **Prospector** tab.
2. From the Ribbon's Home tab, open the Layer Properties Manager, thaw the layer **C-TOPO-Existing**, and click **X** to exit the palette.
3. On the Ribbon, click the **Modify** tab. On the Ground Data panel, click Surface, and on the Surface Tools panel, select EXTRACT OBJECTS to open the Extract Objects From Surface dialog box.
4. In the dialog box, toggle **OFF** Border and click **OK** to extract the contours.
5. In Prospector, expand the Surfaces branch, select the surface **Existing**, press the right mouse button, and from the shortcut menu, select SURFACE PROPERTIES….
6. Click the **Information** tab, set the style to **_No Display**, and click **OK** to exit.
7. On the Surface tab, click Close.

The drawing contains extracted major and minor contours.

8. On the Ribbon, click the Home tab, open the Layer Properties Manager, create a **New** layer, for the layer's name enter **Existing Contours**, make it the current layer, assign it a color, and click **X** to exit the palette.

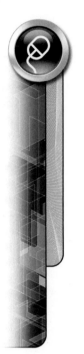

9. In the drawing, select the contours and place them on the Existing Contours layer.

10. In Prospector, select the Surfaces heading, press the right mouse button, and from the shortcut menu, select CREATE SURFACE....

11. In the Create Surface dialog box, for the name, enter **EGCTR**, for the description, enter **From EG Contours**, set the Surface Style to **Contours 1' and 5' (Design)**, and click **OK** to create the surface.

12. In Prospector, expand Surfaces until you are viewing the EGCTR Definition branch.

13. In the EGCTR Definition branch, select **Contours**, press the right mouse button, and from the shortcut menu, select ADD....

14. In Add Contour Data, for the description, enter **FROM EXISTING**, set the Supplemental distance to **35.000**, the Mid-ordinate to **0.100**, click **OK**, and in the drawing, if necessary, select the contours (see Figure 10.5).

15. At Civil 3D's top left, Quick Access toolbar, click the **Save** icon to save the drawing.

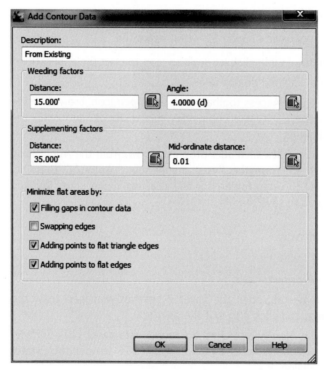

FIGURE 10.5

Base Surface Seed Contours

The Base surface starts with extracted EGCTR contours.

1. On the Ribbon's Home tab, open the Layer Properties Manager, create a **New** layer, for the layer name, enter **Base Contours**, make it the current layer, freeze the Existing Contours layer, and click **X** to exit the palette.

2. On the Ribbon, click the **Modify** tab. On the Ground Data panel, click Surface. On the Surface tab's Surface Tools panel, select EXTRACT OBJECTS, and in the drawing, select the surface to open the Extract Objects From Surface dialog box.

3. In the dialog box, toggle **OFF** Border and click **OK** to extract the contours.

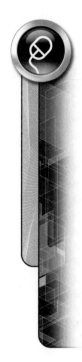

4. In Prospector, the Surfaces branch, select the surface **EGCTR**, press the right mouse button, and from the shortcut menu, select SURFACE PROPERTIES….

5. Click the *Information* tab, set the style to **_No Display**, and click **OK** to exit.

The drawing contains extracted major and minor contours.

6. On the Surface tab, select Close.

7. In the drawing, select the contours and place them on the Base Contours layer.

8. At Civil 3D's top left, Quick Access toolbar, click the **Save** icon to save the drawing.

Modifying Contours

The site's northeast corner is the first editing area. The priority regarding this area is filling in the swale and moving its contours northeasterly to the site's edge. If you are bringing fill into the site and placing it in the swale, the swale elevations rise to move the lower elevations to the northeast.

To create these changes, remove contours from the swale's interior and redraw new contours at the swale's northeastern edge (see Figure 10.6).

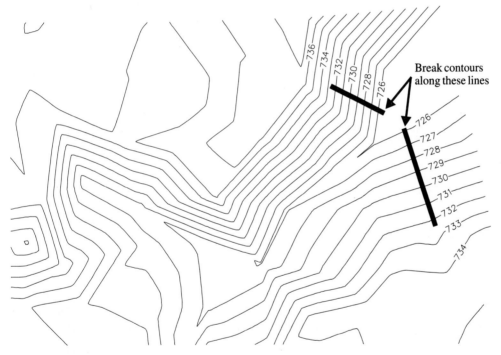

FIGURE 10.6

Northeast Section

1. Use the ZOOM and PAN commands and match your screen to Figure 10.6.

2. Using Figure 10.6 as a guide, use the LINE command and draw the lines as the figure indicates.

3. Using Figure 10.6 as a guide, start the TRIM command. In the drawing, at the cutting edge, select the lines just drawn, press ENTER, and trim the contours to the line's southwest side, removing the unwanted contours.

4. ERASE the two trimming lines.

5. At Civil 3D's top left, Quick Access toolbar, click the **Save** icon to save the drawing.

Draw New Contours

1. In the status bar, toggle **ON** Object Snaps and set *Endpoint*.

Using Figure 10.7, draw a new 726' contour.

2. Start the POLYLINE command, and in the drawing, select the 726' contour's (the innermost contour) north end, toggle **OFF** object snap (F3), select a few points to represent the new 726' contour path, toggle **ON** object snaps (F3), select the 726' contour south end, and press ENTER to exit the polyline command.

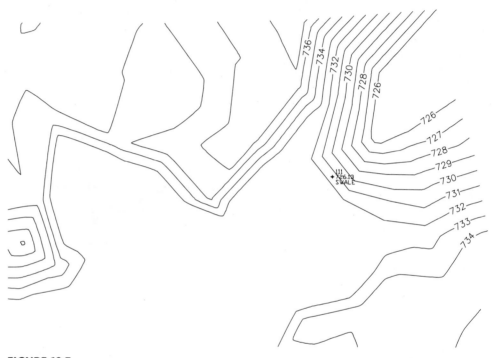

FIGURE 10.7

3. Check the new contour's elevation by selecting it, pressing the right mouse button, from the shortcut menu, selecting PROPERTIES..., and reading the elevation value.

The polyline's Elevation value should match the connected exiting contour's elevation.

4. If the Elevation value does not match the 726' elevation, in Properties, change its elevation to **726**.

The engineer wants an 8:1 slope for the new contours. This means you must offset the contour 8' horizontally and raise its elevation 1'. You use Feature Line's Stepped Offset routine for this task. Stepped Offset's multiple mode creates several new contours from the one just drawn.

5. On the Ribbon, click the **Modify** tab. In the Edit Geometry panel, select STEPPED OFFSET (third row, second icon). Set the Offset Distance to **8**, press ENTER, select the just-drawn polyline, in the command line, enter '**M**' for multiple, press ENTER, select a point to the southwest of the polyline, set the

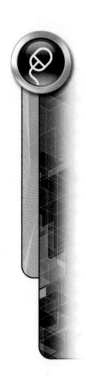

Elevation Difference to **1**, press ENTER, continue picking points to each new polyline's southwest, thus creating six additional contours with the same Elevation Difference, and press ENTER to exit the command.

6. To connect the polylines to the existing contours, activate the new polylines' grips, toggle **ON** object snaps (F3), and stretch the polylines until they are connected to their corresponding existing contours.

7. At Civil 3D's top left, Quick Access toolbar, click the **Save** icon to save the drawing.

Northwest Section

Next, you modify the site's northwest side. This expands the central flat area by erasing the hill and redrafting contours (see Figures 10.8 and 10.9).

1. Use the ZOOM and PAN commands to make your view match Figure 10.8.

2. Use the ERASE command, and in the drawing, erase the hill contours around point 122.

3. Using Figure 10.8 as a guide, use the BREAK command to break the two contours at the indicated locations.

4. Press F3 to toggle **ON** Object Snaps.

5. Using Figure 10.9 as a guide, use the POLYLINE command and the **Endpoint** object snap, and redraw the contours connecting the southern endpoints to the northern endpoints.

6. In the drawing, select one of the new contours, press the right mouse button, and from the shortcut menu, select PROPERTIES….

7. If necessary, change the contour to the correct elevation.

8. Repeat the previous two steps and, if necessary, edit the remaining contour elevations.

9. At Civil 3D's top left, Quick Access toolbar, click the **Save** icon to save the drawing.

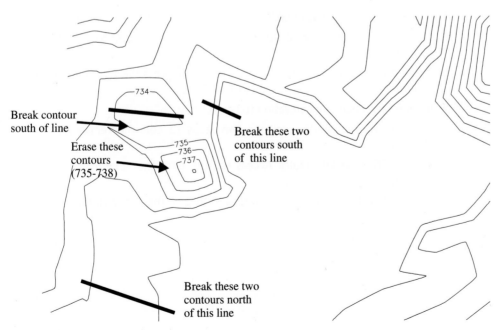

FIGURE 10.8

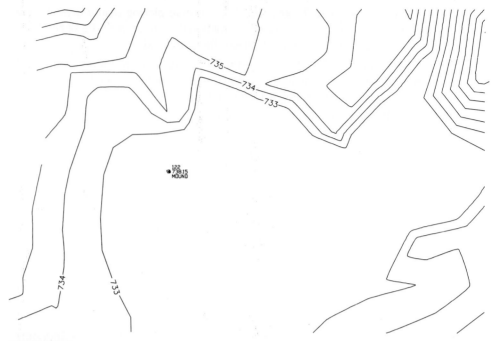

FIGURE 10.9

Southeast Section

Next, you edit contours in the site's southeast side. This expands the flat interior area to the southeast (see Figures 10.10 and 10.11). You also erase two closed contours.

1. Use the ZOOM and PAN commands to match your view to Figure 10.10.
2. Using Figure 10.10 as a guide, use the ERASE command to erase the two closed contours.
3. Using Figure 10.10 as a guide, use the BREAK command to break the 733 and 734 contours.
4. Toggle **ON** Object Snaps (F3).
5. Using Figure 10.11 as a guide, use the POLYLINE command to redraw the two contours.
6. In the drawing, select one of the new contours, press the right mouse button, and from the shortcut menu, select PROPERTIES….
7. If necessary, change its elevation to correct it.
8. If necessary, repeat the previous two steps to edit the remaining contour elevations.
9. At Civil 3D's top left, Quick Access toolbar, click the *Save* icon to save the drawing.
10. Use the ZOOM EXTENTS (ZE) command to view the entire surface.

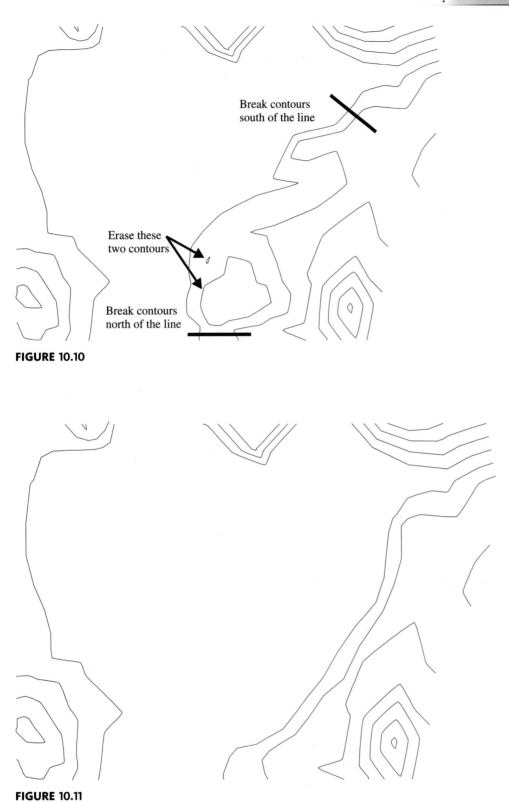

Break contours
south of the line

Erase these
two contours

Break contours
north of the line

FIGURE 10.10

FIGURE 10.11

The drawing should be similar to Figure 10.12.

FIGURE 10.12

Designing Site Drainage

The current design's problem is that the site's center is flat and water will not drain. One strategy to solve this problem is to use swales and a small berm to move the water to the northeast. The shallow swale uses points from the flat area's west side toward the northeast. Figure Lines define a second swale and a berm.

Swale by Points

1. Click Ribbon's *View* tab. From the Views panel's left, select and restore the named view **Central**.

2. Toggle OFF Object Snaps (F3).

3. Click the Ribbon's Home tab, open the Layer Properties Manager, thaw the V-NODE layer, and click **X** to exit to close the palette.

4. In Prospector, select the **Points** heading, press the right mouse button, and from the shortcut menu, select CREATE....

5. At the toolbar's right side, click the Expand the Create Points dialog chevron.

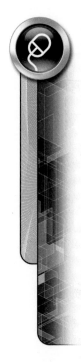

6. Expand the Point Identity section and set the Next Point Number to **500**.

7. Expand the Points Creation section and set the following values:

   Prompt For Elevations: Manual

   Prompt For Descriptions: Automatic

   Default Description: SWCNTRL

8. At the toolbar's right side, click the Collapse the Create Points dialog chevron.

9. From the Create Points toolbar, click the Miscellaneous icon stack's drop-list arrow (the first icon in from the left), and from the list, select MANUAL.

10. Using Figure 10.13 as a guide, place three points and use Table 10.1 to assign the listed elevations.

11. Press ENTER to exit the routine and return to the command line.

TABLE 10.1

Point Number	Elevation
500	732.90
501	732.40
502	732.10

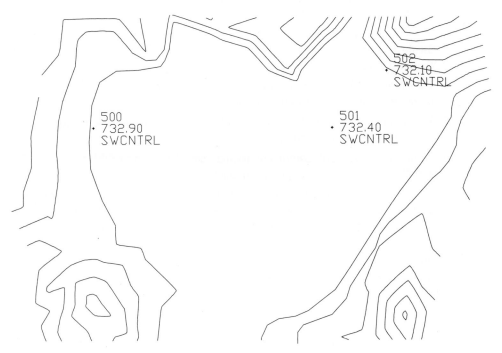

FIGURE 10.13

These three points represent the upper, mid, and end swale points. To complete the swale, place points between the swale points, so the surface correctly triangulates the swale. The Interpolate routine is used to create these points.

12. In the Create Points toolbar, the Interpolate icon stack (the fifth icon in from the left), select its drop-list arrow, and from the list, select INTERPOLATE.

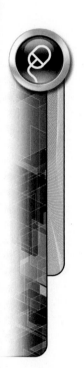

13. The command line prompts you to select a point object. In the drawing, select point **500**.

14. The command line prompts you to select a second point object. In the drawing, select point **501**.

15. The command line prompts you for the number of points between the two selected points. Enter **5** and press ENTER.

16. The command line prompts you for an offset value. Press ENTER for 0 (zero) to create the interpolated points.

17. Press ESC to end the routine and return to the command line.

18. Repeat Steps 11–16 and create interpolated points between points 501 and 502. For the number of interpolated points, enter **3**, for the offset press ENTER for 0 (zero), and press ESC to end the routine.

19. At Civil 3D's top left, Quick Access toolbar, click the **Save** icon to save the drawing.

Creating a Feature Line Swale

The second swale is a feature line with an elevation of 732.9 at its southern end and its northern end elevation (732.4) is from point 501.

1. Make sure the Transparent Commands toolbar is displayed and use Figure 10.13 as a guide for drawing the feature line.

2. On the Ribbon's **Home** tab. On the Create Design panel, click the **Feature Line** icon, and from the shortcut menu, select the **Create Feature Line** to open the Create Feature Lines dialog box.

3. In the Create Feature Lines dialog box, toggle **ON** and name the feature line **Swale 1**, toggle **ON** Style, click the drop-list arrow, from the styles list, select **Grading Ditch**, and click **OK** to continue.

4. The command line prompts you for a starting point. In the drawing, select a point near the site's southern entrance.

5. The command line prompts you for an elevation. For the elevation, enter **732.90** and press ENTER to continue.

6. From the Transparent Commands toolbar, select the **Point Object** filter ('PO), and click anywhere on the point 501 label.

7. The command line may prompt you for a grade of <0.00>. If so, for elevation, enter '**E**' and press ENTER. The command line echoes point 501's elevation (732.40). Press ENTER to accept this elevation (see Figure 10.14).

8. The command line prompts you for a point object. To end the override and exit, press ESC twice.

9. At Civil 3D's top left, Quick Access toolbar, click the **Save** icon to save the drawing.

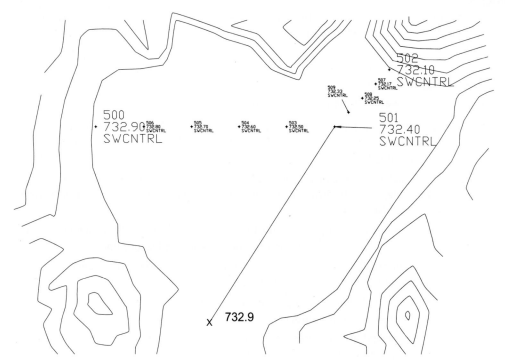

FIGURE 10.14

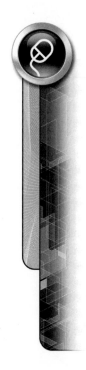

The following is the feature line command sequence:

```
Command:
Specify start point: <Select a point in the southern entrance
area>
Specify elevation or [Surface] <0.000>: 732.9 <Press ENTER>
Specify the next point or [Arc]: '_PO
>>
Select point object: <Select point 501>
Resuming DRAWFEATURELINE command.
Specify the next point or [Arc]: (5292.54 5351.15 732.4)
Distance 220.546', Grade -0.23, Slope -441.09:1,Elevation
732.400'
Specify grade or [SLope/Elevation/Difference/SUrface]
<0.00>: e
Specify elevation or [Grade/SLope/Difference/SUrface]
<732.400>: <Press ENTER>
Specify the next point or [Arc/Length/Undo]:
>>
Select point object: *Cancel* <Press ESC>
>>
Specify the next point or [Arc/Length/Undo]:
Resuming DRAWFEATURELINE command.
Specify the next point or [Arc/Length/Undo]: <Press ESC>
Command:
```

The feature line has only two vertices: its northern and southern ends. These two are the only swale data points. When you define the swale as a breakline, it can supplement the feature line's data.

Creating a Berm

The berm sheds water to the point and feature line swale.

1. On the Home tab, Create Design panel, click the ***Feature Line*** icon, and from the shortcut menu, select the ***Create Feature Line*** to open the Create Feature Lines dialog box.

2. In Create Feature Lines, toggle **ON** Name, for the name, enter **Berm 1**, toggle **ON** Style, click its drop-list arrow, from the styles list, select **Grading Ditch**, and click ***OK*** to continue.

3. The command line prompts you for a starting point. In the drawing, select a point in the southwest between the points and the feature line.

4. The command line prompts you for an elevation. For the elevation, enter **732.90**, and press ENTER to continue.

5. The command line prompts you for a second point. In the drawing, select a point between point number 507 and the feature line.

6. The command line prompts you for an elevation. For the elevation, enter **732.75** and press ENTER twice to set the elevation and exit the routine.

Adding a High/Low Point

The berm needs a middle high point. Positive grades from each end vertex create a high point near its middle.

1. Click Ribbon's Modify tab. On the Edit Elevations panel's middle right, select ***Insert High/Low Elevation Point*** icon (first row, fourth icon).

2. The command line prompts you to select an object (feature line, survey figure, parcel line, or 3D polyline). In the drawing, select the feature line that was just drawn.

3. The command line prompts you for the start point. In the drawing, select the feature line's southern end.

4. The command line prompts you for the end point. In the drawing, select the feature line's northern end.

The routine echoes Start Elevation 732.900', End Elevation 732.750', and Distance 206.176'.

5. The command line prompts you for the ahead slope or grade. Enter '**S**' and press ENTER, for the slope enter **25**, and press ENTER.

6. The command line prompts you for the back slope or grade. For the slope, enter **25** and press ENTER twice to create the high point and exit the routine.

7. Select the feature line, press the right mouse button, and from the shortcut menu, select ELEVATION EDITOR....

The feature line has an elevation point that is higher than either end.

8. Close the Grading Elevation Editor vista.

9. At Civil 3D's top left, Quick Access toolbar, click the ***Save*** icon to save the drawing.

Create the Base Surface

The base surface has points, breaklines, and contour data.

Create Point Group

To assign points to a surface, they must belong to a point group.

1. If necessary, click the **Prospector** tab.
2. In Prospector, select the **Point Groups** heading, press the right mouse button, and from the shortcut menu, select NEW….
3. In the **Information** tab, for the point group Name, enter **Base**.
4. Click the Include tab, toggle ON With Numbers Matching, and for the range, enter **500–520**.
5. Click the **Point List** tab to review the selected points.
6. Click **OK** to exit the dialog box.

Create Surface

1. In Prospector, select the Surfaces heading, press the right mouse button, and from the shortcut menu, select CREATE SURFACE….
2. In the Create Surface dialog box, set the surface name to **Base**, for the description, enter **Prelim Design**, set the Surface Style to **Border & Triangles & Points**, and click **OK** to exit.
3. In Prospector, expand the Surfaces branch until you are viewing the Base's Definition data type list.
4. From the data type list, select **Point Groups**, press the right mouse button, and from the shortcut menu, select ADD….
5. In Point Groups, select **Base**, and click **OK**.
6. In the data type list, select **Breaklines**, press the right mouse button, and from the shortcut menu, select ADD….
7. In the Add Breaklines dialog box, for the description, enter **Breaklines**, set the type to **Standard**, toggle **ON** Supplementing factor's Distance, set it to **10**, set the Mid-Ordinate distance to **0.01**, click **OK**, and in the drawing, select the swale and berm feature lines.
8. From the Home tab's, Layers panel, select ISOLATE. If necessary, set layers not isolated to Off and paperspace viewports to VPfreeze, and select a contour.
9. In the Definition tree type list, select Contours, press the right mouse button, and from the shortcut menu, select ADD….
10. In Add Contour Data, for the description, enter **EG AND NEW**, set the remaining values as shown in Figure 10.15, and when they match, click **OK**.
11. In the drawing, select all contours and press ENTER to exit.
12. If the Event Viewer displays, close it by clicking the green check mark in the panorama's upper-right corner.
13. From the Home tab's, Layers panel, select UNISOLATE.
14. In Layer Properties Manager, for the current layer, set it to 0 (zero), freeze the layer **Base Contours**, **C-TOPO-Base**, and **C-TOPO-FEAT**, and click **OK** to exit.

This completes building the new surface.

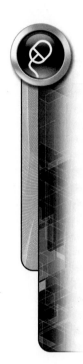

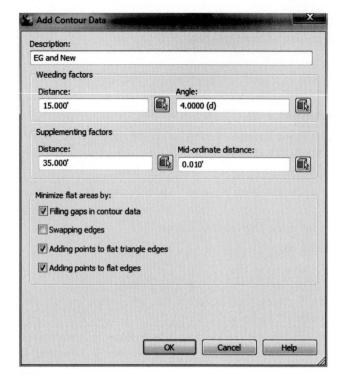

FIGURE 10.15

Calculating a Volume Summary

1. On the Ribbon, click the Analyze tab. On the Volumes and Materials panel, click the Volumes icon, and from the shortcut menu, select VOLUMES to display the Composite Volumes vista.

2. In the vista, at its top left, click the **Create new volume entry** icon to create a calculation entry.

3. In the vista, click in the Index 1 Base Surface cell (<select surface>), click the drop-list arrow, and from the list, select **Existing.**

4. In the vista, click in the Index 1 Comparison Surface cell (<select surface>), click the drop-list arrow, and from the list, select **Base** (see Figure 10.16).

5. Repeat Steps 2 through 4 to calculate a second volume using **EGCTR** and **Base**.

6. Note the Cut, Fill, and Net volumes for each comparison.

7. In the Panorama, click its **X** to close it.

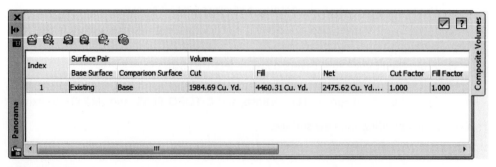

FIGURE 10.16

Volume Surface Style

A volume surface uses style like any other surface object. For this exercise, you will define a style that shows a 2D view of contour ranges for cut and fill and in 3D as a model.

1. Click the **Settings** tab.
2. Expand the Surface branch until you are viewing the Surface Styles list.
3. From the styles list, select **Contours 1' and 5' (Design)**, press the right mouse button, and from the shortcut menu, select COPY....
4. In the Surface Style dialog box, click the **Information** tab, and for the name, enter **Volume Contours**.
5. Click the **Contours** tab, expand the Contour Ranges section, and set the following values (see Figure 10.17):

   Major Color Scheme: Rainbow

   Minor Color Scheme: Land

   Group Values by: Quantile

   Number of Ranges: 12

   Use Color Scheme: True

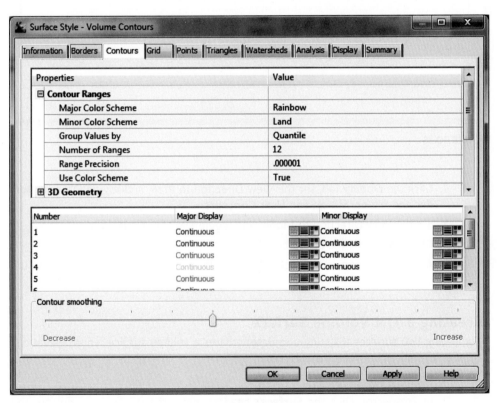

FIGURE 10.17

6. Click the **Analysis** tab, expand Elevations, and set the following values (see Figure 10.18):

   Scheme: Rainbow

   Group by: Quantile

   Number of Ranges: 12

   Display Type: 3D Faces

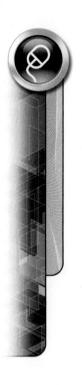

Elevations Display Mode: Exaggerate Elevation

Exaggerate Elevations by Scale Factor: 4

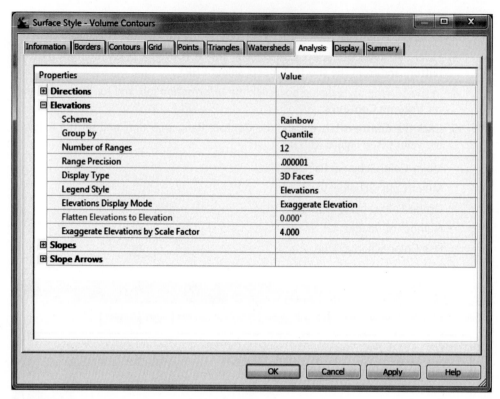

FIGURE 10.18

7. Click the *Display* tab, set the View Direction to **Plan**, and toggle ON only **Border** and **Major**, **Minor**, and **User Contours**.

8. Set the View Direction to **Model** and toggle ON only **Elevations**.

9. Click *OK* to exit the dialog box.

10. At Civil 3D's top left, Quick Access toolbar, click the *Save* icon to save the drawing.

Creating a TIN Volume Surface

Comparing two surfaces (Existing and Base) creates a volume surface whose elevations are the differences between the surfaces.

1. Click the *Prospector* tab.

2. In Prospector, click the **Surfaces** heading, press the right mouse button, and from the shortcut menu, select CREATE SURFACE... .

3. In the Create Surface dialog box, set the surface Type to **TIN Volume Surface**, for the name, enter **EXISTING – BASE TIN VOL**, and enter the remainder of the information found in Figure 10.19.

4. At the bottom of the dialog box, click in the Base Surface value cell, click its ellipsis, in Select Base Surface, select **Existing**, and click *OK*.

5. Repeat the previous step, but for the Comparison Surface, select the surface **Base**.

6. Click *OK* to exit the dialog box and create the EXISTING – BASE TIN VOL surface.

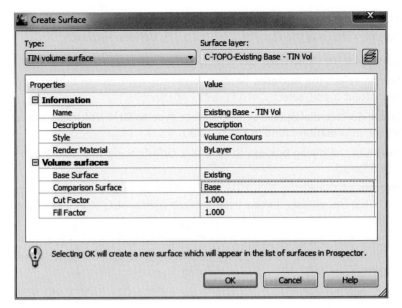

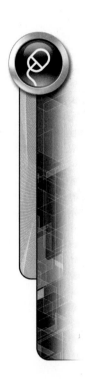

FIGURE 10.19

Surface Properties

A volume surface's property dialog box reports the amount of cut and fill. The surface contours display the amounts of surface cut and fill.

1. From Prospector's Surfaces, select the EXISTING – BASE TIN VOL, press the right mouse button, and from the shortcut menu, select SURFACE PROPERTIES….

2. Click the **Analysis** tab, set the Analysis Type to **User-Defined Contours**, set the Number of Ranges to **12**, and click the **Run Analysis** icon.

3. In the Range Details section, change contour 6's elevation to **0.00** and click **Apply**.

4. Change the Analysis Type to **Elevations**, set the Number of Ranges to **12**, and click the **Run Analysis** icon.

5. Select the **Statistics** tab and expand the Volume section to list the earthwork calculation.

6. Compare this volume to the volumes calculated previously.

7. Click **OK** to exit the Surface Properties dialog box.

The contours represent cut and fill amounts. The style displays a 3D view of the cut and fill amounts. The cut areas are the deepest holes, and the fill areas are the highest hills.

8. In the drawing, select any surface component, press the right mouse button, from the shortcut menu, select OBJECT VIEWER…, and view the surface from different angles.

9. Exit the Object Viewer.

10. In the drawing, select any surface component, press the right mouse button, and from the shortcut menu, select SURFACE PROPERTIES….

11. In Surface Properties, click the **Information** tab, set the Surface Style to **_No Display**, and click **OK** to exit the dialog box.

12. At Civil 3D's top left, Quick Access toolbar, click the **Save** icon to save the drawing.

Creating a Grid Volume Surface

Creating a Grid Volume surface is the same as creating a TIN Volume surface.

1. If necessary, click the **Prospector** tab.

2. In Prospector, select the **Surfaces** heading, press the right mouse button, and from the shortcut menu, select CREATE SURFACE….

3. In the Create Surface dialog box, set the surface type to **Grid Volume Surface** and enter the information found in Figure 10.20.

4. In the middle of the dialog box, set the X and Y Grid Spacing to **5**, and set the rotation to **0** (zero).

5. At the bottom of the dialog box, click in the Base Surface value cell. At its right side, click the ellipsis, in the Base Surface dialog box, select **Existing**, and click **OK**.

6. At the bottom of the dialog box, click in the Comparison Surface value cell, at its right side, click the ellipsis, in the Comparison Surface dialog box, select **Base**, and click **OK**.

7. Click **OK** to exit the dialog box and create the EXISTING – BASE GRID VOL surface.

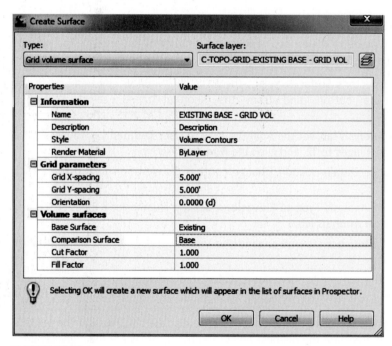

FIGURE 10.20

Create User Contours

1. From Prospector's Surfaces list, select EXISTING – BASE GRID VOL, press the right mouse button, and from the shortcut menu, select SURFACE PROPERTIES….

2. Click the **Analysis** tab, set the Analysis type to **User-Defined Contours**, set the Number of Ranges to **12**, and click the **Run Analysis** icon.

3. Set the Analysis to **Elevations**, set the ranges to **12**, and click the **Run Analysis** icon.

4. Click the **Statistics** tab and expand the Volume section to list the cut and fill volumes.

5. Note the volume values and compare them to the previously calculated values.

6. Click **OK** to exit the Surface Properties dialog box.

7. In the drawing, select any surface component, press the right mouse button, from the shortcut menu, select OBJECT VIEWER..., and review the surface from different angles.

8. Exit Object viewer.

9. Select any surface component, press the right mouse button, and from the shortcut menu, select SURFACE PROPERTIES....

10. Click the Information tab, change the Surface Style to **Border & Triangles & Points**, and click **OK** to display the surface volume grid triangulation.

11. At Civil 3D's top left, Quick Access toolbar, click the **Save** icon to save the drawing.

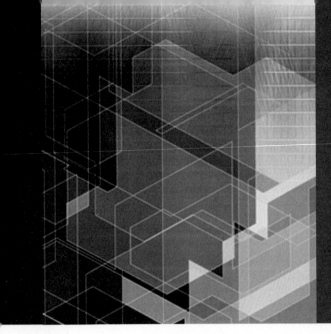

CHAPTER
11

Pipe Networks

After completing this exercise, you will:

- Be familiar with the pipe network's Edit Drawing Settings.
- Be familiar with the pipe network's Edit Feature Settings.
- Be able to draft a pipe network.
- Be able to assign materials to pipes and structures.
- Be able to assign pipe rules.
- Be able to add structures to a parts list.
- Be able to assign pay item codes to part list items.

Exercise Setup

This exercise uses an existing drawing. To find this file, browse to the Chapter 11 folder of the CD that accompanies this textbook and open the *Chapter 11 - Unit 1.dwg* file.

1. If you are not in *Civil 3D*, double-click the Civil 3D desktop icon to start the application.

2. When you are at the command prompt, close the opening drawing and do not save it.

3. At Civil 3D's top left, Quick Access toolbar, click the *Open* icon, browse to the Chapter 11 folder of the CD that accompanies this textbook, and open the *Chapter 11 - Unit 1* drawing.

4. At Civil 3D's top left, click Civil 3D's drop-list arrow, and from the Application Menu, highlight SAVE AS, from the flyout menu, select AUTOCAD DRAWING, browse to the Civil 3D Projects folder, for the drawing name enter **Pipe Network-1**, and click *Save* to save the file.

Edit Drawing Settings

Edit Drawing Settings set several styles and values that affect pipe networks.

1. Click the **Settings** tab.

2. At Settings' top, select the drawing's name, press the right mouse button, and from the shortcut menu, select EDIT DRAWING SETTINGS....

3. Click the **Object Layers** tab and review its settings. If necessary, change your settings to match those in Figure 11.1. All Pipe and Structure layers except Structure and Pipe with a modifier (**suffix**) have a value of (**-***).

4. Click **OK** to exit the Edit Drawing Settings dialog box.

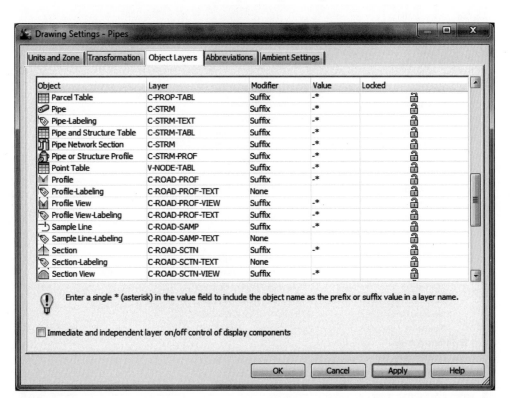

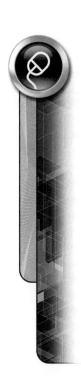

FIGURE 11.1

Edit Feature Settings

Pipe Network's Edit Feature Settings values affect both pipes and structures. These values assign pipe segments, structure their initial styles, and render materials (see Figures 11.2 and 11.3).

1. In Settings, select the **Pipe Network** heading, press the right mouse button, and from the shortcut menu, select EDIT FEATURE SETTINGS....

2. Expand and review the values for the Default Styles and Default Name Format sections.

3. In the Default Styles section, click in the Render Material's value cell to display an ellipsis. Click the ellipsis to open the Render Material dialog box.

4. In the Render Material dialog box, click the drop-list arrow, select **Concrete.Cast.In.Place.Flat.Grey.1**, and click **OK** to return to Edit Feature Settings.

5. Expand the section Storm Sewers Migration Default. Change the Parts List Used for Migration to **Storm Sewers**.

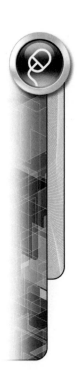

6. Expand the Default Rules, Default Profile Label Placement, and Default Section Label Placement sections and review their values.

Default Rules sets the pipes and structures default rule. The Default Profile Label Placement section sets initial label anchor points. The Default Section Label Placement section sets the initial labels of pipes and structures in a section view.

7. Click **OK** to exit the Edit Feature Settings dialog box.
8. At Civil 3D's top left, Quick Access toolbar, click the **Save** icon to save the drawing.

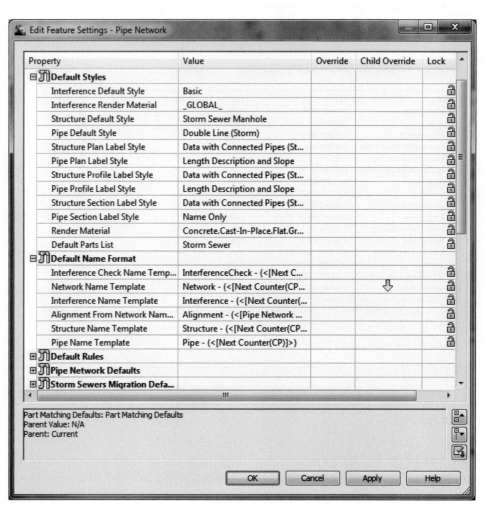

FIGURE 11.2

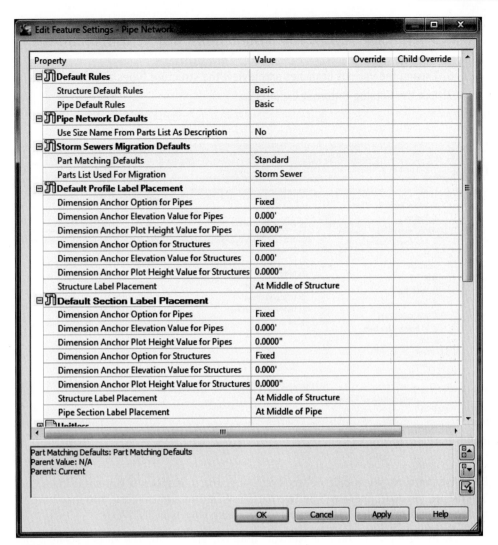

Property	Value	Override	Child Override
⊟ 🗊 **Default Rules**			
Structure Default Rules	Basic		
Pipe Default Rules	Basic		
⊟ 🗊 **Pipe Network Defaults**			
Use Size Name From Parts List As Description	No		
⊟ 🗊 **Storm Sewers Migration Defaults**			
Part Matching Defaults	Standard		
Parts List Used For Migration	Storm Sewer		
⊟ 🗊 **Default Profile Label Placement**			
Dimension Anchor Option for Pipes	Fixed		
Dimension Anchor Elevation Value for Pipes	0.000'		
Dimension Anchor Plot Height Value for Pipes	0.0000"		
Dimension Anchor Option for Structures	Fixed		
Dimension Anchor Elevation Value for Structures	0.000'		
Dimension Anchor Plot Height Value for Structures	0.0000"		
Structure Label Placement	At Middle of Structure		
⊟ 🗊 **Default Section Label Placement**			
Dimension Anchor Option for Pipes	Fixed		
Dimension Anchor Elevation Value for Pipes	0.000'		
Dimension Anchor Plot Height Value for Pipes	0.0000"		
Dimension Anchor Option for Structures	Fixed		
Dimension Anchor Elevation Value for Structures	0.000'		
Dimension Anchor Plot Height Value for Structures	0.0000"		
Structure Label Placement	At Middle of Structure		
Pipe Section Label Placement	At Middle of Pipe		
⊟ 🗊 **Unitless**			

Part Matching Defaults: Part Matching Defaults
Parent Value: N/A
Parent: Current

FIGURE 11.3

Pipe and Structure Catalog

The Pipe and Structure Catalog contains basic pipe and structure shapes and parameters. There is a simple, but poorly documented, interface to add or modify the entries of this file: Partbuilder. The catalog is in the Documents and Settings, All Users path (XP), and for Windows 7, the path is ProgramData\Autodesk\C3D 2011\enu\Pipes Catalog\US Imperial Pipes. The Catalog is an HTML file located in the US Imperial Pipes and Structures folders: US Imperial Pipes.htm and US Imperial Structures.htm. Double-click the HTML file (see Figure 11.4) to review their values. You may get a message about scripts or ActiveX controls, click the message to allow blocked content, and click Yes in the security Warning dialog box.

1. Start Windows Explorer and make US Imperial Pipes the current folder.

2. Double-click the *US Imperial Pipes.htm* file, displaying Internet Explorer with the US Imperial Pipes Catalog.

3. Expand the various branches and review the pipe types and their values.

4. Exit the Catalog (Internet Explorer).

5. Change to the US Imperial Structures folder and double-click the *US Imperial Structures.htm* file displaying Internet Explorer with the US Imperial Structure Catalog.

6. Expand the various branches and review the structure types and their values.

7. Exit the Catalog (Internet Explorer).

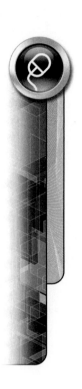

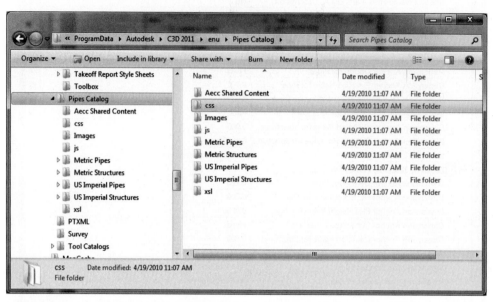

FIGURE 11.4

Parts list contains typical pipe network pipe and structure specs. If you are using a Civil 3D content template, there will be two parts lists: storm and sanitary. You may have to add to the lists or change their part values to match your typicals.

1. In Settings, expand Pipe Network until you are viewing the Parts Lists list.

2. From the list, select **Storm Sewer**, press the right mouse button, and from the shortcut menu, select EDIT....

3. Click the Pipes tab to view and, if necessary, expand Storm Sewer and the Concrete Pipe list.

This panel lists more concrete pipes sizes than you would use in a typical project. But over several projects, the list may be complete. You could define a rule for each size and type of pipe. If you are doing so, this is where rules are assigned.

4. Click the **Structures** tab.

5. If necessary, expand Storm Sewer and each part family to view its contents.

Each structure has a descriptive name and has an appropriate object style (see Figure 11.5).

6. Click **OK** to exit the Storm Sewer Parts List.

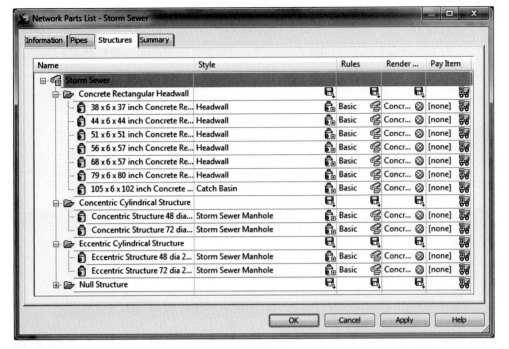

FIGURE 11.5

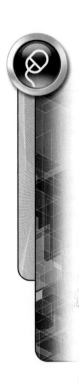

Parts List — Sanitary Parts List

1. In the Pipe Network's Parts Lists, select **Sanitary Sewer**, press the right mouse button, and from the shortcut menu, select EDIT....

2. Click the ***Pipes*** tab and, if necessary, expand the Sanitary Sewer's PVC Pipe list.

The panel lists more PVC pipe sizes than would be typical for one project. But over several projects, the list may be complete. If you are defining a rule for each pipe size, this is the place to assign them.

3. Click the ***Structures*** tab and, if necessary, expand each Sanitary Sewer cylindrical parts family and review its contents.

4. Click **OK** to exit the Sanitary Sewer Parts List.

Pipe Styles

1. In Settings, expand the Pipe branch until you are viewing the Pipe Styles list.

2. From the styles list, select **Single Line (Sanitary)**, press the right mouse button, and from the shortcut menu, select EDIT....

3. Click each tab to review its contents.

4. Click the ***Display*** tab to review component visibility settings.

This style draws only the pipe's centerline.

5. Click **OK** to exit the dialog box.

6. From the list, select **Double Line (Storm)**, press the right mouse button, and from the shortcut menu, select EDIT....

7. Click each tab to review its contents.

8. Click the ***Display*** tab to review component visibility settings.

The pipe's inside diameter draws the two lines.

9. Click **OK** to exit the dialog box.

Pipe Rules

1. In Settings, expand the Pipe branch until you are viewing the Pipe Rule Set list.
2. From the list, select **Basic**, press the right mouse button, and from the shortcut menu, select EDIT....
3. Click the ***Rules*** tab.
4. If necessary, expand each rule section to review its values.

Cover and Slope set a pipe's minimum and maximum slope and cover. Pipe Length Check defines a pipe's shortest and longest length (see Figure 11.6).

5. Click **OK** to exit the dialog box.

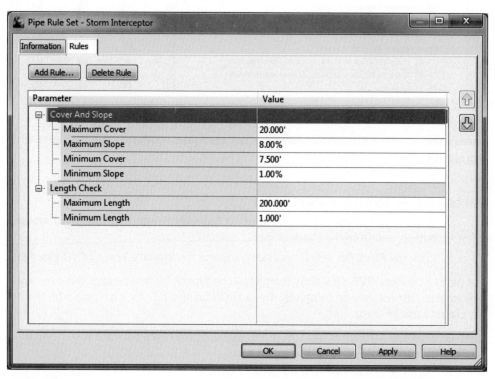

FIGURE 11.6

Pipe Label Styles

Pipe labels occur while you are drawing or after you have designed the network. Unit 4 of this chapter reviews these styles and how they are used.

Structure Styles

Structure Styles define how storm and sanitary sewer systems display manholes, flared end sections, catch basins, and so on, in Plan, Profile, and Section Views.

1. In Settings, expand the Structure branch until you are viewing the Structure Styles list.
2. From the styles list, select **Storm Sewer Manhole**, press the right mouse button, and from the shortcut menu, select EDIT....
3. Click the ***Model*** tab.

The catalog defines the structure's model shape, or you can choose a simpler shape.

4. Click the ***Plan*** tab.

This panel sets a structure's symbol. These settings also affect the symbol's size. A symbol can be an actual size, exaggerated, or resized each time the display area changes.

     5. Click the **Profile** tab.

This panel sets how a structure is displayed in Profile View.

     6. Click the **Section** tab to view its contents.

This panel sets how a structure is displayed in Section View.

     7. Click the **Display** tab to review the component visibility settings.

This panel controls what components are visible for this structure in Plan, Model, Profile, or Section Views.

     8. Click **OK** to exit the dialog box.

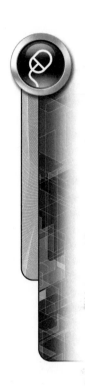

Structure Rules

1. Expand Settings' Structure branch until you are viewing the Structure Rule Set list.
2. Select **Basic**, press the right mouse button, and from the shortcut menu, select EDIT....
3. Click the **Rules** tab.
4. If necessary, expand each section to view its contents.
5. Click **OK** to exit the dialog box.
6. At Civil 3D's top left, Quick Access toolbar, click the **Save** icon to save the drawing file.

Structure Label Styles

Label structures as you draw them or after you have created the network. Unit 4 of this chapter discusses these styles and their uses.

This completes the Pipe Networks settings review exercise. Pipe networks use an extensive styles library, rules, and catalog values. Next, you learn how to create pipe networks.

EXERCISE 11-2

After completing this exercise, you will:

- Be able to draft a pipe network.
- Be able to set structures and add pipes.

Exercise Setup

This exercise continues with the previous exercise's drawing, Pipe Network-1. If you did not complete the previous exercise, browse to the Chapter 11 folder of the CD that accompanies this textbook and open the *Chapter 11 - Unit 2.dwg* file.

1. If not open, open the previous exercise's drawing or browse to the Chapter 11 folder of the CD that accompanies this textbook and open the *Chapter 11 - Unit 2* drawing.

Existing Storm (93rd Street)

Along the centerline of 93rd Street is an existing storm sewer line (see Figure 11.7). Table 11.1 contains structure names and types for the network. The network starts at 93rd Street's eastern end and is drawn upslope.

TABLE 11.1

Manhole #	Type	Material
EX-Stm-01	Concentric 72"	Reinforced Concrete
EX-Stm-02	Concentric 72"	Reinforced Concrete
EX-Stm-03	Concentric 72"	Reinforced Concrete
EX-Stm-04	Concentric 48"	Reinforced Concrete

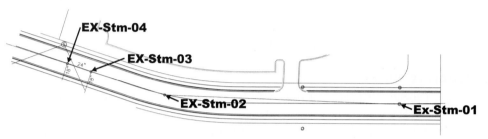

FIGURE 11.7

This pipe network has several different pipe sizes. The pipe size from EX-Stm-03 to EX-Stm-01 is 36". The pipe size between EX-Stm-04 and EX-Stm-03 is 24". Both EX-Stm-04 and EX-Stm-03 have 18" laterals feeding their barrels. Drawing this pipe network requires changing drafting modes and pipe sizes.

1. Use the ZOOM and PAN commands to view 93rd Street's easterly end and EX-Stm's first manhole.

The 93rd Street storm sewer uses a special rule, Storm Interceptor. This rule allows the network to be deeper than other storm systems (i.e., parking lot catch basins). You must set the rule before you define the next network. Otherwise, you have to apply a rule to each existing pipe.

2. If necessary, click the **Settings** tab.

3. In Settings, select the **Pipe Network** heading, press the right mouse button, and from the shortcut menu, select EDIT FEATURE SETTINGS....

4. If necessary, expand the Default Rules section. For Pipe Default Rules, click in the Value cell to display an ellipsis, and click the ellipsis to open the Pipe Default Rules dialog box.

5. In the Pipe Default Rules dialog box, click the drop-list arrow, from the list, select **Storm Interceptor**, and click **OK** until you have returned to the command line.

6. Click the **Prospector** tab.

7. On the Home tab, Create Design panel, click the **Pipe Network** icon, and from the shortcut menu, select PIPE NETWORK CREATION TOOLS to open the Create Pipe Network dialog box.

8. Using Figure 11.8 as a guide, for the pipe network name, enter **EX-STM**, keeping the counter. For the Network parts list, select the **Storm Sewer**, set the Surface name to **Proposed**, set the Alignment name to **93rd - (1)**, and click the **OK** button to exit the dialog box and display the Network Layout Tools toolbar.

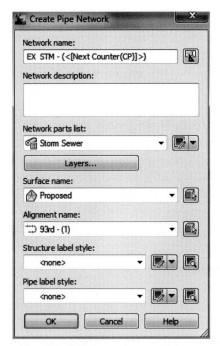

FIGURE 11.8

Along the toolbar's bottom, Parts List should be Storm Sewer, the Surface should be Proposed, and the Alignment should be 93rd - (1). Use Figure 11.1 as a guide for drafting this network.

9. At the toolbar's center, click the Structure drop-list arrow, and from the Concentric Cylindrical Structure parts list, select the **Concentric Structure 72" dia**.

10. At the toolbar's right of center, click the Pipe Size drop-list arrow, and from the list, select **36 inch RCP**.

11. In the toolbar, click the *Upslope/Downslope* icon (the second to the right of the pipe size) and set the design to **Upslope**.

12. In the toolbar, from the Drafting Mode icon stack, click the drop-list arrow, and from the list, select PIPES AND STRUCTURES.

13. Referencing Figure 11.1 and using a **Center** object snap, in the drawing, place the easternmost manhole (EX-Stm-01).

14. In the drawing, also using the Center object snap, locate the structures EX-Stm-02 and EX-Stm-03.

15. In the toolbar's center, click the Structure drop-list arrow, and from the Concentric Cylindrical Structure parts list, select **Concentric Structure 48" dia**.

16. In the toolbar, click the Pipe Sizes drop-list arrow, and from the pipe size list, select **24 inch RCP**.

Changing pipe sizes does not interrupt the drafting process.

17. Using the **Center** object snap, in the drawing, locate EX-Stm-04.

18. In the toolbar, click the Pipe Sizes drop-list arrow, and from the list, select **18 inch RCP**.

19. In the toolbar, click the Drafting Mode icon stack drop-list arrow, and from the list, select PIPES ONLY.

This breaks the cursor's connection to EX-Stm-04. To correctly draft the pipe, you need to reconnect the 18" pipe to EX-Stm-04 before you draft the pipe.

20. In the drawing, place the cursor near EX-Stm-04. A structure connection icon appears. Click near the EX-Stm-04 structure to start drafting the pipe.

21. In the drawing, click a point to the west of the structure as the pipe end.

22. To indicate that you are drafting a new pipe segment connecting to a different structure, click the **Pipe Only** drafting mode icon.

23. In the drawing, place the cursor near EX-Stm-04. The attach structure icon appears. Select a point near the structure and draw an 18" pipe EX-Stm-04 toward the south past the sidewalk polyline line.

24. To start a new pipe segment, click the **Pipe Only** drafting mode icon to place a new 18" pipe segment from EX-Stm-03 toward the south.

25. In the drawing, place the cursor at EX-Stm-03. A structure connection icon appears. Click near the structure to start drawing the pipe and select a second point south of the structure as the opposite end of the pipe.

26. Use the PAN command to view the EX-Stm-01 structure.

27. In the toolbar, click the Pipe Size drop-list arrow, and from the list, select **36 inch RCP**.

28. In the toolbar, click the **Upslope/Downslope** icon and set the mode to **Downslope**.

29. To draft a new pipe segment, in the toolbar, click the **Pipe Only** drafting mode icon to place a new 36" pipe segment from EX-Stm-01 toward the east.

30. In the drawing, place the cursor at EX-Stm-01. A structure connection icon appears. Click near the structure to start drawing the pipe, and select a second point east of the structure (near the end of the corridor's centerline) as the opposite end of the pipe.

31. Click the toolbar's red **X** to close it.

32. At Civil 3D's top left, Quick Access toolbar, click the **Save** icon to save the drawing.

Editing a Pipe Network

This simplest way to change structures and pipes names is by editing their properties, rather than trying the vista.

1. In the drawing, select each structure and edit its properties, changing each structure's name using the entries in Table 11.1.

2. In the drawing, select each pipe and edit its properties, changing each pipe's name using the entries in Table 11.2.

3. At Civil 3D's top left, Quick Access toolbar, click the **Save** icon to save the drawing.

TABLE 11.2

Pipe Name	From/to Manholes	Size
EX-Stm-P01	East end of 93rd to EX-Stm-01	36
EX-Stm-P02	EX-Stm-01 to EX-Stm-02	36
EX-Stm-P03	EX-Stm-02 to EX-Stm-03	36
EX-Stm-P04	EX-Stm-03 to EX-Stm-04	24
EX-Stm-P05	EX-Stm-04 to West end of 93rd	18
EX-Stm-P06	EX-Stm-03 to south of sidewalk	18
EX-Stm-P07	EX-Stm-04 to south of sidewalk	18

Catch Basin Network

The last network is a series of catch basins and pipes with an outfall FES to a pond at the site's southeast corner. The drafting strategy for this network is to first place a 48″ × 48″ rectangular structure at the parking lot's southeast corner next to the detention pond. After placing the structure, you next locate the three catch basins in the western part of the parking lot. After you locate the three western catch basins, you next locate a catch basin in the parking lot's northeast. All catch basins connect to the southeast manhole. Finally, you connect a pipe to the manhole and locate the FES in the detention pond. Use Figure 11.9 and Table 11.3 as guides for placing and naming the various network elements.

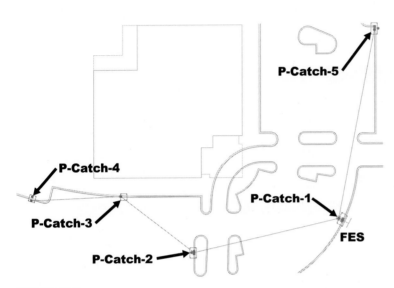

FIGURE 11.9

TABLE 11.3

Structure	Type	Pipe Size	Pipe Name	Material
P-Catch-1	48″ × 48″ Rectangular	12″ Concrete	P-Catch-P1	Reinforced Concrete
P-Catch-2	24″ × 24″ Rectangular	12″ Concrete	P-Catch-P2	Reinforced Concrete
P-Catch-3	18″ × 18″ Rectangular	12″ Concrete	P-Catch-P3	Reinforced Concrete
P-Catch-4	15″ × 15″ Rectangular	12″ Concrete	P-Catch-P4	Reinforced Concrete
P-Catch-5	24″ × 24″ Rectangular	12″ Concrete	P-Catch-P5	Reinforced Concrete
FES	44″ × 44″ Rectangular Headwall	24″ Concrete	P-Catch-P6	Reinforced Concrete

Adding a Part Family and Part to the Storm Sewer Parts List

The storm sewer parts list does not contain all the required structures. The parts need to be added to the storm sewer parts list.

1. Click the **Settings** tab.
2. In Settings, expand the Pipe Network branch until you are viewing the Parts Lists list.
3. Select the **Storm Sewer** parts list, press the right mouse button, and from the shortcut menu, select EDIT....
4. In the Network Parts List - Storm Sewer dialog box, click the **Structures** tab.

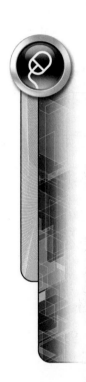

5. In the Structures tab, at the top left, select the **Storm Sewer** heading, press the right mouse button, and from the shortcut menu, select ADD PART FAMILY….

6. In the Part Catalog dialog box, toggle ON **Rectangular Structure Slab Top Rectangular Frame**, and click **OK** to return to the Network Parts List dialog box.

7. From the list of parts, select **Rectangular Structure Slab Top Rectangular Frame**, press the right mouse button, and from the shortcut menu, select ADD PART SIZE….

8. If necessary, change the Inner Structure Width and Length to **15"** and click **OK** to add the part. Use Figure 11.10 as a guide.

9. Repeat the previous two steps and add the following sizes: **18 × 18**, **24 × 24**, and **48 × 48**.

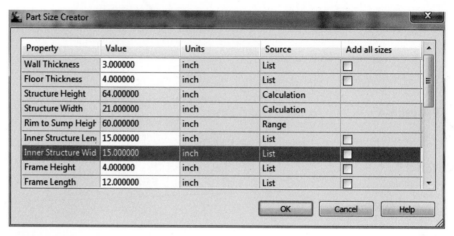

Property	Value	Units	Source	Add all sizes
Wall Thickness	3.000000	inch	List	☐
Floor Thickness	4.000000	inch	List	☐
Structure Height	64.000000	inch	Calculation	
Structure Width	21.000000	inch	Calculation	
Rim to Sump Heigh	60.000000	inch	Range	
Inner Structure Len	15.000000	inch	List	☐
Inner Structure Wid	15.000000	inch	List	☐
Frame Height	4.000000	inch	List	☐
Frame Length	12.000000	inch	List	☐

FIGURE 11.10

The style for the new structures is storm sewer manhole.

10. In the Part List dialog box, click the **Style** icon to the right of Rectangular Structure Slab Top Rectangular Frame to open the Structure Style dialog box. Click the drop-list arrow, from the style list, select **Catch Basin**, and click **OK** to return to the Network Parts List dialog box.

11. Click **OK** to exit the Network Parts List - Storm Sewer dialog box.

Drafting the Catch Basin Network

1. Use the ZOOM and PAN commands to view the entire parking lot.

2. In the Layer Properties Manager, toggle **ON** the layer **PROP-STM-LIN** and click **X** to exit.

3. From the Ribbon's Home tab, Create Design panel, click the Pipe Network icon, and from the shortcut menu, select PIPE NETWORK CREATION TOOLS.

4. In the Create Pipe Network dialog box, for the Network name, enter **P-Catch**, leaving the counter, set the Network parts list to **Storm Sewer**, and set the Surface name to **Proposed**. Use Figure 11.11 as a guide.

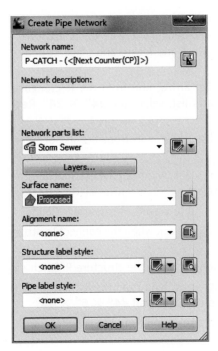

FIGURE 11.11

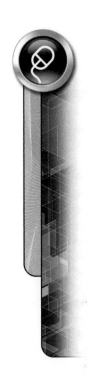

5. Click **OK** to exit the dialog box and display the pipe Network Layout Tools toolbar.

6. In the toolbar's center, click the Structure drop-list arrow, and from the Rectangular Structure Slab Top Rectangular Frame section, select the **48″ × 48″ Rect Structure**.

7. In the toolbar, set **Upslope/Downslope** to **Upslope** (the second icon to the right of the pipe size).

8. In the toolbar, click the **Drafting Mode** icon stack drop-list arrow (to the right of the pipe size), and from the list, select STRUCTURES ONLY.

9. Using Figure 11.9 as a guide, in the drawing at the southeast parking lot's corner, place P-Catch - 1.

10. In the toolbar, click the Structure drop-list arrow, and from the Rectangular Structure Slab Top Rectangular Frame parts list, select the **24″ × 24″ Rect Structure**.

11. In the toolbar, click the Pipe Size drop-list arrow, and from the list, select **18 inch RCP**.

12. In the toolbar, click the **Drafting Mode** icon stack drop-list arrow (to the right of the pipe size), and from the list, select PIPES AND STRUCTURES.

13. In the drawing, place the cursor near the structure you just placed. The connect to structure icon appears. Select the point to attach the pipe.

14. Use the PAN command toward the west and locate and place the P-Catch - 2 structure.

15. In the toolbar, click the **Structure** drop-list arrow, and from the Rectangular Structure Slab Top Rectangular Frame parts list, select the **18″ × 18″ Rect Structure**.

16. In the toolbar, click the **Pipe Size** drop-list arrow, and from the list of sizes, select **15 inch RCP**.

17. In the drawing, PAN west and locate and place structure P-Catch - 3.

18. After placing the structure, in the toolbar, click the **Structure** drop-list arrow, from the Rectangular Structure Slab Top Rectangular Frame parts list, select the **15″ × 15″ Rect Structure**, and PAN west and locate and place structure P-Catch - 4.

19. In the toolbar, click the **Structure** drop-list arrow, and from the Rectangular Structure Slab Top Rectangular Frame parts list, select the **24″ × 24″ Rect Structure**.

20. In the toolbar, click the **Pipe Size** drop-list arrow, and from the list of sizes, select **18 inch RCP**.

21. In the toolbar, click the **Drafting Mode** icon stack, and from the list, indicate the start of a new branch to the network by selecting PIPES AND STRUCTURES.

22. If necessary, use the PAN and ZOOM command to view the structure P-Catch - 1.

23. In the drawing, place the cursor near P-Catch - 1, and when the connect to structure icon appears, select the point.

24. Use the PAN command to pan north, locate and place the structure P-Catch - 5.

25. In the toolbar, click the **Structure** drop-list arrow, and from the list in the Concrete Rectangular Headwall part family, select **Headwall 44″ × 6″ × 44″**.

26. In the toolbar, click the **Pipe Size** drop-list arrow, and from the list of sizes, select **30 inch RCP**.

27. In the toolbar, click the **Downslope/Upslope** toggle to set it to **Downslope**.

28. In the toolbar, click the **Drafting Mode** icon.

29. In the drawing, place the cursor at P-Catch - 1. When the connect to structure icon appears, select the point and select a second point to the southeast in the detention pond to place the FES.

30. Click the Network Layout Tools toolbar's red **X** to close it.

31. At Civil 3D's top left, Quick Access toolbar, click the **Save** icon to save the drawing.

This completes the pipe networks drafting exercise.

EXERCISE 11-3

After completing this exercise, you will:

- Be able to review and edit pipe network properties.
- Be able to graphically edit pipe network segments.
- Be familiar with pipe or structure properties and be able to edit them.
- Be familiar with pipes or structures and be able to swap them out.

Exercise Setup

This exercise continues with the previous exercise's drawing. If you did not complete the previous exercise, browse to the Chapter 11 folder of the CD that accompanies this textbook and open the *Chapter 11 - Unit 3.dwg* file.

1. If not open, open the previous exercise's drawing or browse to the Chapter 11 folder of the CD that accompanies this textbook and open the *Chapter 11 - Unit 3* drawing.

Pipe Network Properties

Pipe Network Properties lists assigned values when the network is created.

1. Use the ZOOM and PAN commands until you are viewing the EX-Stm - (1) pipe network.
2. In the drawing, select a pipe or structure, press the right mouse button, and from the shortcut menu, select NETWORK PROPERTIES....
3. Click each tab and review its settings and values.
4. Click **OK** to exit the Pipe Network Properties dialog box.

Structure Properties

Structure Properties displays structure information. You can adjust structure parameters, the structure's name, and review connected pipes.

1. If necessary, click the ***Prospector*** tab.
2. In Prospector, expand the Pipe Networks branch until you are viewing the EX-STM – (1) pipe network's Pipes and Structures.
3. Click **Structures** to list the structures in preview.
4. In Preview, select **EX-Stm-01**, press the right mouse button, and from the shortcut menu, select STRUCTURE PROPERTIES....
5. Click each tab to review the structure's settings and values, and click **OK** to exit the dialog box.

Swap Parts

Swap pipes or structures to larger or smaller sizes.

1. In Preview, select **EX-Stm-01**, press the right mouse button, and from the shortcut menu, select ZOOM TO.
2. In the drawing, at the eastern side of EX-Stm-01, select the pipe (EX-Stm-P01), press the right mouse button, and from the shortcut menu, select SWAP PART... to display a pipe size list.
3. From the list, select a **48 inch RCP**, and click **OK** to swap the part.
4. At Civil 3D's top left, Quick Access toolbar, click the **Save** icon to save the drawing.

Profile Review of EX-STM – (1)

Civil 3D drafts some or all of a pipe network in Profile View. The 93rd Street storm system needs to be reviewed in Profile View.

1. Click Ribbon's ***Modify*** tab. In the Design panel, click the Pipe Network icon. From the Pipe Networks tab, on the Network Tools panel's right side, select DRAW PARTS IN PROFILE.
2. The command line prompts you for a network. In the drawing, select a pipe from the EX-STM – (1) network and press ENTER. The command line prompts you to select a Profile View. Use the Pan command until you are viewing the 93rd Street Profile View, and then select it.

The EX-STM – (1) pipe network is drawn in the Profile View.

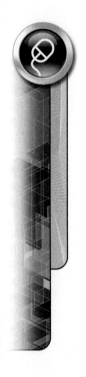

Edit Pipes in Profile View

Using Table 11.4 as a guide, edit each pipe's start and end inverts in each pipe's Pipe Properties dialog box.

TABLE 11.4

Pipe	Start Invert	End Invert
EX-Stm-P01	669.00	668.00
EX-Stm-P02	669.50	675.00
EX-Stm-P03	675.50	679.00
EX-Stm-P04	680.00	681.00
EX-Stm-P05	681.25	682.00
EX-Stm-P06	681.50	684.00
EX-Stm-P07	681.25	682.00

1. In the 93rd Street Profile View, select the pipe EX-Stm-P01, press the right mouse button, and from the shortcut menu, select PIPE PROPERTIES....

2. In the Pipe Properties dialog box, click the **Part Properties** tab, scroll down to the Start and End Invert Elevation entries, and using the values in Table 11.4, change the inverts for pipe EX-Stm-P01. Click **OK** to exit the Pipe Properties dialog box.

3. Repeat the previous two steps and edit the inverts for the EX-STM – (1) pipe network (EX-Stm-P02...P5).

4. In the drawing, select any entity that represents EX-STM – (1), press the right mouse button, and select from the shortcut menu, EDIT NETWORK....

5. In the Network Layout Tools toolbar, click the **PIPE NETWORK VISTAS** icon (right side).

6. Click the **Pipes** tab and edit the Start and End Inverts for EX-Stm-P06 and EX-Stm-P07 using the values in Table 11.4.

7. After editing the inverts, click the **X** to close the vistas and Network Layout Tools toolbar.

8. At Civil 3D's top left, Quick Access toolbar, click the **Save** icon to save the drawing.

Intersection Between EX-STM – (1) and EX-San – (1)

The EX-San – (1) pipe network crosses the EX-STM – (1) pipe network. The intersection needs to be checked for interferences.

1. Use the PAN and ZOOM commands to view the intersection of EX-San – (1) and EX-STM – (1).

2. If necessary, click the **Prospector** tab.

3. Expand the Pipe Networks branch until you view the Interference Checks heading.

4. Select **Interference Checks**, press the right mouse button, and from the shortcut menu, select CREATE INTERFERENCE CHECK....

5. The command line prompts you to select a part from the first network. In the drawing, select a pipe from the **EX-San – (1)** network.

6. The command line prompts you to select a part from the second network. In the drawing, select a pipe from the **EX-STM – (1)** network.

The Create Interference Check dialog box opens.

7. In the Create Interference Check dialog box, click **3D Proximity Check Criteria...**.

8. In the Criteria dialog box, toggle **ON** Apply 3D Proximity Check, set the distance to **5.0**, and click **OK** to return to the Create Interference Check dialog box.

9. Click **OK** to exit Create Interference Check.

There are three interferences.

10. Click the **OK** button to close the interference report.

Viewing a Pipe Network Model

Viewing a network with 3D Orbit or Object Viewer produces a 3D view of selected components.

1. In the drawing, select the interference symbols, pipes, and structures of the two pipe networks, press the right mouse button, and from the shortcut menu, select OBJECT VIEWER....

2. Rotate the model until you are viewing the network objects and interferences.

3. After reviewing the network and interferences, close the Object Viewer.

4. Use the ZOOM and PAN commands to view the P-Catch network in the parking lot.

5. In the drawing, select all of the P-Catch elements, press the right mouse button, and from the shortcut menu, select OBJECT VIEWER....

6. After reviewing the network, close the Object Viewer.

7. At Civil 3D's top left, Quick Access toolbar, click the **Save** icon to save the current drawing.

Graphical Editing

Graphically editing a pipe network relocates pipes or structures. Pipes and structures can be disconnected and reconnected by clicking the network part, and from the shortcut menu, selecting Reconnect or Disconnect.

1. Use the ZOOM and PAN commands to view the western end of the 93rd Street existing STM – (1) and San – (1) pipe networks.

2. In the drawing, select the **EX-SAN–1** structure and, using its grips, relocate some of its structure and its pipes.

3. Clear the grips by pressing ESC.

4. In the drawing, select one of the pipes connected to the EX-SAN–1 structure, click the grip nearest the structure, and move the pipe end to a new location.

The pipe disconnects from the structure.

5. In the drawing, select the pipe that was just disconnected, activate its northern grip, move it to the structure, and when the connect to structure icon appears, select it, connecting the pipe to the structure.

6. Close the drawing and **do not save** the changes.

This ends the exercise on editing a pipe network. The next unit reviews annotating a network's pipes and structures.

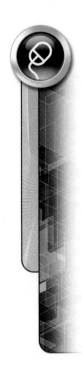

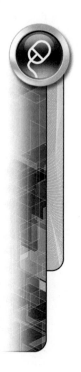

EXERCISE 11-4

After completing this exercise, you will:

- Be able to annotate individual pipes and structures.
- Be able to annotate entire networks in plan and profile.

Exercise Setup

This exercise continues with the previous exercise's drawing. If you did not complete the previous exercise, browse to the Chapter 11 folder of the CD that accompanies this text-book and open the *Chapter 11 - Unit 4.dwg* file.

1. Open the previous exercise's drawing or browse to the Chapter 11 folder of the CD that accompanies this textbook and open the *Chapter 11 - Unit 4* drawing.

Annotating an Entire Pipe Network in Plan View

Add Labels annotates an entire pipe network in Plan View.

1. Use the ZOOM and PAN commands to view the entire 93rd Street storm network in Plan View.
2. Click the Ribbon's Annotate tab. In the Labels & Tables panel, click Add Labels to display the Add Labels dialog box.
3. In Add Labels, set the Feature to Pipe Network, set the Label Type to **Entire Network Plan**, set the Pipe Label Style to **Length Description and Slope**, and set the Structure Label Style to **Data with Connected Pipes (Storm)**.
4. Click **Add** and select a pipe or structure from the 93rd Street storm network.
5. Use the ZOOM and PAN commands to review the labels.

To make pipe labels respond to rotated views, a user must toggle ON plan readability.

6. Drag some labels to view their dragged state.

Annotating an Entire Pipe Network in Profile View

Annotate an entire Pipe Network in Profile View by setting the correct Add Labels parameters.

1. Use the ZOOM and PAN commands to view the entire 93rd Street storm network in Profile View.
2. In Add Labels, set the Label Type to **Entire Network Profile**, set the Pipe Label Style to **Length Description and Slope**, set the Structure Label Style to **Data with Connected Pipes (Storm)**, click **Add**, and in Profile View, select a pipe or structure from the 93rd Street storm network.
3. Use the ZOOM and PAN commands to review the labels.
4. Drag some labels to view their dragged state.
5. Click **Close** to exit Add Labels.
6. Close the drawing and **do not save** it.
7. Reopen the drawing file.

Annotating Individual Parts in Plan View

Add Labels annotates individual pipes and structures in Plan View.

1. Use the ZOOM and PAN commands to view the EX-SAN-1 structure and its pipes at the 93rd Street storm sewer network's western end.

2. Click the Annotate tab. In the Labels & Tables panel, click Add Labels to display the Add Labels dialog box.

3. In the Add Labels dialog box, set the Feature to Pipe Network, set the Label Type to **Single Part Plan**, set the Pipe Label Style to **Length Description and Slope**, set the Structure Label Style to **Data with Connected Pipes (Storm)**, click **Add**, and in the drawing, select a structure or a pipe.

Only the selected part receives a label.

4. Label additional pipes and structures by selecting a network pipe or structure.

Annotating Single Parts in Profile View

Add Labels annotates individual pipes and structures in Profile View.

1. Use the ZOOM and PAN commands to view the 93rd Street Storm sewer network in Profile View.

2. In Add Labels, set the Label Type to **Single Part Profile**, set the Pipe Label Style to **Length Description and Slope**, set the Structure Label Style to **Data with Connected Pipes (Storm)**, click Add, and select a structure or pipe.

Only selected structures or pipes are labeled.

3. Label additional pipes and structures by selecting a part.

4. Click **Close** to exit Add Labels.

5. Close the drawing and **do not save** it.

6. Reopen the drawing file.

Drawing Intersecting Pipes

1. In the drawing, select **Profile View**, press the right mouse button, and from the shortcut menu, select PROFILE VIEW PROPERTIES....

2. Click the *Pipe Networks* tab.

3. If necessary expand the EX-San – (1) section.

4. Toggle **ON** EX-San-Pipe-(6) and (7), for each pipe toggle **ON** Style Override, and in Pick Pipe Style, click the drop-list arrow, from the list, select **Pipe Crossing Pipe (Sanitary)**, and click **OK**.

5. Click **OK** to exit Profile View Properties, and the crossing pipes are drawn in Profile View.

Pipes in Road Section Views

1. In Prospector, expand the Corridors branch, if the **93rd - (1)** Corridor is out-of-date, rebuild it.

2. Use the AutoCAD ZOOM and PAN commands to view the entire 93rd Street storm network in Plan View.

3. Click the Home tab, on the Profile & Section Views panel, click the *Sample Lines* icon.

4. The command line prompts you for an Alignment. Press the right mouse button, from the list, select **93rd - (1)**, and click **OK**.

5. In the Create Sample Line Group dialog box, uncheck the P-Catch - (1) pipe network, and click **OK** to accept the defaults.

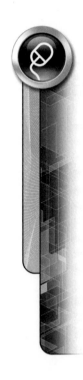

The routine opens in select At a Station mode. To create a section centered on a structure, select its station using an AutoCAD Center object snap.

6. Using a **Center** object snap, select EX-Stm-01 to define the first section.
7. After selecting the structure, the command line prompts you for a swath width. Enter the right and left swath width as **60'**.
8. Repeat this three times for each EX-Stm manhole and press ENTER.

Note the sample line stations.

9. Use the PAN and ZOOM commands to view an empty area to the right of the site.
10. In the Home tab's Profile & Section Views panel, click the **Section Views** icon, and from the shortcut menu, select CREATE MULTIPLE VIEWS.
11. In Create Multiple Section Views' General panel, click Next.
12. In the section Placement panel, Placement Options, click the browse button to display the Select Layout as Sheet Template dialog box. In the dialog box, select ARCH D Section 40 Scale and click OK to return to the Create Multiple Section Views dialog box.
13. In dialog box's middle left, click the Plot Group Style drop-list arrow, and from the list, select **Plot All**.
14. Click **Next** three times until you are viewing the Section Display Options panel.
15. In the Section Display Options panel, set Change Labels to **_No Labels**, set style for Proposed to **Finished Ground**, click **Create Section Views**, and in the drawing, select a point to place the sections in the drawing (see Figure 11.12).

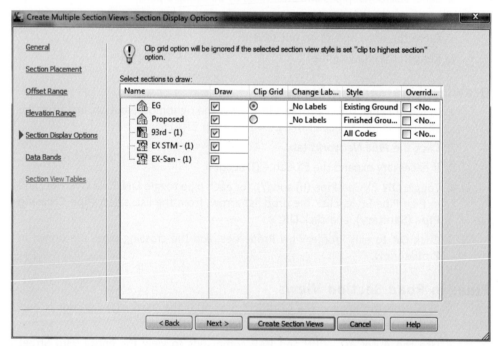

FIGURE 11.12

16. Use the ZOOM and PAN commands to better view the sections.
17. At Civil 3D's top left, Quick Access toolbar, click the **Save** icon to save the current drawing.

Pipe networks have two labels types: pipe and structure. You can place labels while you are drafting a network or after you have completed its design.

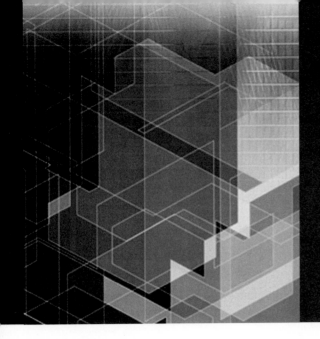

Civil 3D Shortcuts

After completing this exercise, you will:

- Be able to set a working folder.
- Be able to create a new shortcut folder.
- Be able to switch between shortcut folders.
- Be familiar with creating Civil 3D data shortcuts.
- Be familiar with referencing data shortcuts.
- Be familiar with synchronizing drawing and data shortcut values.
- Be familiar with promoting references to drawing objects.

Exercise Setup

This exercise uses the *Data Shortcuts.dwg* file. To find this file, browse to the Chapter 12 folder of the CD that accompanies this textbook and open the *Data Shortcuts.dwg* file. This drawing has several Civil 3D objects. After you create the shortcuts, a new drawing references these shortcuts.

1. If you are not in **Civil 3D**, start the application by double-clicking its desktop icon.

2. When you are at the command prompt, close the opening drawing and do not save it.

3. At Civil 3D's top left, Quick Access toolbar, click the **Open** icon. In the Open dialog box, browse to the Chapter 12 folder of the CD that accompanies this textbook, select the *Data Shortcuts* drawing, and click **Open**.

4. At Civil 3D's top left, click Civil 3D's drop-list arrow, and from the Application Menu, highlight Save As, from the flyout menu, select AutoCAD Drawing, browse to the Civil 3D Projects folder, for the drawing name enter **Data Shortcuts – work**, and click **Save** to save the file.

Set a Working Folder

1. If necessary, click the **Prospector** tab.

2. If necessary, at the top of Prospector, click the View drop-list arrow, and from the list of views, select Master View.

3. If necessary, expand Data Shortcuts [ ].

4. Select **Data Shortcuts [ ]**, press the right mouse button, and from the shortcut menu, select SET WORKING FOLDER ... to open the Browse For Folder dialog box.

5. In the Browse For Folder dialog box, browse to and select the **C:\Civil 3D Projects** folder, and click **OK**.

Create a Shortcut folder

1. Select **Data Shortcuts [ ]**, press the right mouse button, and from the shortcut menu, select NEW DATA SHORTCUTS FOLDER ... to open the New Data Short-cut Folder dialog box.

2. In the New Data Shortcut Folder dialog box, for the shortcut folder name, enter **HC3D**; for the description, enter **Exercise folders for shortcuts**, and click **OK**.

3. In Prospector, the Data Shortcuts heading now lists the working folder and the current shortcut folder: Data Shortcuts [C:\Civil 3D Projects\HC3D].

An HC3D folder now contains a _Shortcuts folder with five subfolders: Alignments, Pipe Networks, Profiles, Surfaces, and ViewFrameGroups.

Create Data Shortcuts

The current drawing contains several Civil 3D objects: surfaces, alignment and its profiles, and pipe networks.

1. In Prospector, select the Data Shortcuts heading, press the right mouse button, and from the shortcut menu, select CREATE DATA SHORTCUTS....

2. In Create Data Shortcuts - Share Data panel, toggle **ON** all of the object types to share, **Surfaces, Alignments** and **Pipe Networks**, and click **OK**.

3. At Civil 3D's top left, Quick Access toolbar, click the **Save** icon to save the drawing.

4. Close the drawing.

Create a Data Shortcut Reference

1. If necessary, click **Prospector**.

2. Expand Drawing Templates and AutoCAD branches, from the template list, select _AutoCAD Civil 3D (Imperial) NCS, press the right mouse button, and from the shortcut menu, select CREATE NEW DRAWING.

3. Click the Civil 3D's icon drop-list arrow at the top left, from the Application Menu, highlight Save As, from the flyout menu, select AutoCAD Drawing, browse to the Civil 3D Projects folder, for the drawing name enter **Data Short-cuts Destination**, and click **Save** to save the file.

4. In Prospector, expand Data Shortcuts' Surfaces branch. From the list of surfaces, select **EG**, press the right mouse button, and from the shortcut menu, select CREATE REFERENCE ... to open the Create Surface Reference dialog box.

5. In the Create Surface Reference dialog box, review the settings, and click **OK** to create the surface reference.

6. In Prospector, expand the Data Shortcuts Destination's Surfaces branch and note the reference icon to the left of EG.

7. Repeat Steps 4 and 5 to create references for the surface **Proposed**, select **_No Display** for the surface style, alignment **93rd – (1)**, profile **93rd (1)**, and Pipe Network **EX-Stm – (1)** (make sure to use the Storm Sewer parts list). When you create the references, the dialog boxes will be different for each object type.

Create a Profile View from Reference Data

1. Use the PAN and ZOOM commands to move the site to the screen's left side.

2. On the Ribbon's Home tab, Create Design panel, click the ***Profile*** icon, and from the shortcut menu, select CREATE SURFACE PROFILE.

3. In the Create Profile from Surface dialog box, select the EG surface, and at the middle right, click ***Add*** >>.

At the bottom of the dialog box, the profile list already includes the 93rd (1) vertical design profile.

4. Click ***Draw in Profile View***, set the Profile View Style to **Major Grids**, and click Next three times.

5. In the Profile Display Options panel, change the style for 93rd (1) to Design Profile. Also, assign 93rd (1) the Complete Label Set. When set, click Next.

6. In the Pipe Network Display panel, toggle on EX-Stm – (1) and then click NEXT.

7. In the Data Bands Panel, set Profile 1 to EG – Surface 1. When set, click ***Create Profile View***, and in the drawing, select a point to place the profile in the drawing.

8. At Civil 3D's top left, Quick Access toolbar, click the ***Save*** icon to save the drawing.

Edit the Data Shortcut Source Drawing

1. In Prospector, Data Shortcuts, select the *EG* surface, press the right mouse button, and from the shortcut menu, select OPEN SOURCE DRAWING....

2. In Prospector, under Data Shortcuts – works' Surfaces, expand the EG branch until you view the Definition heading's data list.

3. In the data list, select Edits, press the right mouse button, and from the shortcut menu, select RAISE/LOWER SURFACE.

4. The command line prompts you for the amount to add. Enter **1** and press ENTER to raise the surface 1 foot.

5. If necessary, REBUILD the EG surface.

6. At Civil 3D's top left, Quick Access toolbar, click the ***Save*** icon to save the drawing.

7. Close the Data Shortcuts – work drawing.

At the Data Shortcuts Destination drawing's lower right, a change notification balloon should be displayed.

8. In Prospector, expand the Data Shortcuts Destination's Surfaces and the Alignments branches to view their out-of-date icons.

9. In the change balloon, click SYNCHRONIZE. A Panorama is displayed and notifies you that all shortcuts were updated.

10. Close the Panorama.

11. At Civil 3D's top left, Quick Access toolbar, click the ***Save*** icon to save the drawing.

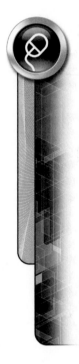

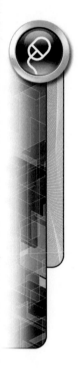

Promote a Reference

1. In Prospector, select the **EG** surface reference, press the right mouse button, and in the shortcut menu, select PROMOTE.

This creates an object instance that breaks the shortcut link. The surface starts its definition with a snapshot. This allows the drawing to move to new locations and not go out-of-date.

2. Close the Data Shortcuts Destination drawing and do not save the changes.

Multiple Shortcut Folders

A working folder can contain multiple shortcut folders. If this is the case, the shortcut folder needs to be set before making a reference.

1. In Prospector, Data Shortcuts, select the **EG** surface, right mouse click, and from the shortcut menu, select OPEN SOURCE DRAWING ....

2. Select Data Shortcuts, press the right mouse button, and from the shortcut menu, select NEW DATA SHORTCUTS FOLDER ... to open the New Data Short-cut Folder dialog box.

3. In the New Data Shortcut Folder dialog box, for the shortcut folder name, enter **HC3D-2**, for the description, enter **Exercise folders for shortcuts-2**, and click **OK**.

4. In Prospector, the Data Shortcuts heading now lists the working folder and the current shortcut folder: Data Shortcuts [C:\Civil 3D Projects\HC3D-2].

5. Select **Data Shortcuts**, press the right mouse button, and from the shortcut menu, select **Set Data Shortcuts Folder ...** to open the Set Data Shortcuts Folder dialog box.

The dialog box displays both shortcut folders; the current folder has a green checkmark.

6. In the dialog box, select the name **HC3D**, and click **OK** to change the current shortcut folder.

7. The Data Shortcuts heading updates accordingly.

8. Close all of the drawings.

Edit Shortcut Paths

1. Minimize Civil 3D.

2. Click Start; from All Programs, Autodesk, the AutoCAD Civil 3D 2010 flyout menu, select DATA SHORTCUTS EDITOR.

3. In the Data Shortcuts Editor, click the **Open** icon and the Browse to Folder dialog box opens.

4. In the dialog box, browse to the Civil 3D Projects folder, and click **OK**.

The Editor displays the shortcuts, and their values are available for editing.

5. Close the Data Shortcut Editor and maximize AutoCAD Civil 3D.

This ends the data shortcuts exercise.

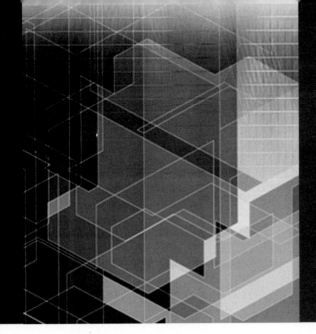

Hydraulics and Pipe Design

EXERCISE 13-1

After completing this exercise, you will:

- Be able to define an IDF chart.
- Be able to create an SCS chart.
- Be able to calculate culverts.
- Be able to calculate a weir.
- Be able to calculate a channel.

Create an IDF Chart

1. If you are not in Civil 3D, start the application by double-clicking its desktop icon.

2. On the Ribbon, click the **Analyze** tab. On the Design panel, select LAUNCH EXPRESS.

3. In the Hydraflow Express application, at the top left, click the IDF icon to open the Rainfall IDF Curve dialog box.

4. In the dialog box's top left, click the **Coefficients** tab, and make sure that **FHA** is the active tab.

5. Click the **IDF Table** tab, and at its bottom, click **Clear**.

6. Using Table 13.1 as a guide, enter the following values for the DuPage county IDF table:

TABLE 13.1

Year	5-Minute	15-Minute	30-Minute	60-Minute
2	0.44	0.84	1.12	1.38
5	0.49	0.94	1.29	1.62
10	0.58	1.11	1.54	1.96
25	0.67	1.26	1.78	2.31
50	0.76	1.43	2.04	2.89
100	0.84	1.58	2.28	3.05

7. Click SAVE, browse to the Civil 3D Projects folder, for the IDF name, enter **DuPage**, click **Save**, and click **Exit** to close the Rainfall IDF Curve.

SCS Chart

1. In the Hydraflow Express application, at the top left, click the **Precip** icon to open the SCS Precipitation Data dialog box.
2. In SCS Precipitation Data, at its top left, click **Clear**.
3. Using Table 13.2 as a guide, enter the following values for the SCS Precipitation Data for DuPage county:

TABLE 13.2

Hr	1	2	5	10	25	50	100
24	2.44	2.96	3.79	4.50	5.62	6.63	7.82
6	1.79	2.06	2.48	3.10	3.79	4.56	5.33

4. Click **Apply**.
5. In the top left of SCS Precipitation Data, click SAVE, for the SCS name, enter **DuPage**, click **Save**, and click **Exit** to close the SCS Precipitation Data dialog box.

Culvert — Area-B 10-Inch Culvert

This exercise section calculates four culvert values: two for a 10-inch and two for an 8-inch culvert. The resulting values will be a part of the Unit 2 flow values. The target discharge from parking lot catch basins into a pond needs to be around 3 fps per catch basin. This target fps is for erosion control and for using the catch basins as delay structures in the event of heavy rains.

1. At the top left of Hydraflow Express, click **Culverts** to open the Culvert calculator.
2. Using Figure 13.1 as a guide, set the values in the Input grid. When you enter the culvert length, do not include the foot mark.

Section	Item	Input
Pipe	Inv Elev Dn =	706.50
	Length (ft) =	18.00
	Slope (%) =	2.00
	Inv Elev Up =	706.86
	Rise (in) =	10.0
	Shape =	Cir
	Span (in) =	10.0
	No. Barrels =	1
	n-value =	0.024
	Inlet Edge =	Projecting
Embank	Top Elev =	709.33
	Top Width (ft) =	10.00
	Crest Len (ft) =	30.00
Calcs	Q Min (cfs) =	0.90
	Q Max (cfs) =	1.50
	Q Incr (cfs) =	0.10
	Tailwater (ft) =	(dc+D)/2

Clear	Run

FIGURE 13.1

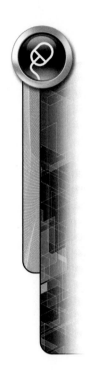

3. After entering the values, click **Run** to compute the culvert values.
4. At the top center, for the name, enter **Area-B-10-culvert**.
5. From the File menu, select SAVE AS..., save the computations as **Area B-10-culvert**, click **Save**, and click **OK** to continue.
6. Review the culvert values in the results grid, select different results, and view the changes to the culvert diagram.
7. Click **P-Curve** to view the chart.
8. While in P-Curve mode, click **Diag** to view the P-Curve diagram.
9. Click **Diag** (now Plot) and then click **Plot** to return to the initial view.

The fps at the flow rate of 1cfs (the runoff rate Q) is different for the up and down velocity. The target is 3 fps for the up and down rate.

Culvert — Area-B 8-Inch Culvert

1. Using Figure 13.2 as a guide, change the pipe rise to 8 (for an 8-inch pipe) and click **Run** to calculate the new flow rates for an 8-inch culvert.

Section	Item	Input
Pipe	Inv Elev Dn =	706.50
	Length (ft) =	18.00
	Slope (%) =	2.00
	Inv Elev Up =	706.86
	Rise (in) =	8.0
	Shape =	Cir
	Span (in) =	8.0
	No. Barrels =	1
	n-value =	0.024
	Inlet Edge =	Projecting
Embank	Top Elev =	709.33
	Top Width (ft) =	10.00
	Crest Len (ft) =	30.00
Calcs	Q Min (cfs) =	0.90
	Q Max (cfs) =	1.50
	Q Incr (cfs) =	0.10
	Tailwater (ft) =	(dc+D)/2

Clear	Run

FIGURE 13.2

2. At the top center, for the name, enter **Area-B-8-culvert**.

3. From the File menu, select SAVE AS…, save the computations as **Area B-8-culvert**, click **Save**, and click **OK** to continue.

4. Review the culvert values in the results grid and select different results to view the changes to the culvert diagram.

5. Click **P-Curve** to view the chart.

6. While in P-Curve mode, click **Diag** to view the P-Curve diagram.

7. Click **Diag** (now Plot) and then click **Plot** to return to the initial view.

The fps at the flow rate of 1cfs (the runoff rate Q) is approximately 3 fps, the target velocity. The Area-B catchbasin will also use an 8-inch outfall.

Culvert — Area-C 10-Inch Culvert

1. Using Figure 13.3 as a guide, set the following values in the Culvert Design Input grid. When you enter the culvert length, do not include the foot mark.

Section	Item	Input
Pipe	Inv Elev Dn =	706.00
	Length (ft) =	18.00
	Slope (%) =	2.00
	Inv Elev Up =	706.36
	Rise (in) =	10.0
	Shape =	Cir
	Span (in) =	10.0
	No. Barrels =	1
	n-value =	0.024
	Inlet Edge =	Projecting
Embank	Top Elev =	708.26
	Top Width (ft) =	10.00
	Crest Len (ft) =	30.00
Calcs	Q Min (cfs) =	0.90
	Q Max (cfs) =	1.50
	Q Incr (cfs) =	0.10
	Tailwater (ft) =	(dc+D)/2

| Clear | | Run |

FIGURE 13.3

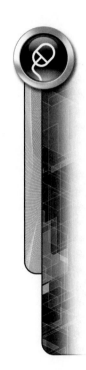

2. After entering the values, click **Run** to compute the culvert values.
3. At the top center, for the name, enter **Area-C-10-culvert**.
4. From the File menu, select SAVE AS..., save the computations as **Area C-10-culvert**, click **Save**, and click **OK** to continue.
5. Review the culvert values in the results grid, and select different results to view the changes to the culvert diagram.
6. Click **P-Curve** to view the chart.
7. While in P-Curve mode, click **Diag** to view the P-Curve diagram.
8. Click **Diag** (now Plot) and then click **Plot** to return to the initial view.

The fps at the flow rate of 1cfs (the runoff rate Q) is different for the up and down velocity. The target is 3 fps for the up and down rate.

Culvert — Area-C 8-Inch Culvert

1. Using Figure 13.4 as a guide, change the pipe rise to 8 (for an 8-inch pipe).

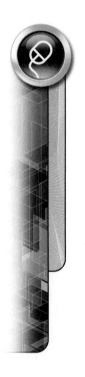

Section	Item	Input
Pipe	Inv Elev Dn =	706.00
	Length (ft) =	18.00
	Slope (%) =	2.00
	Inv Elev Up =	706.36
	Rise (in) =	8.0
	Shape =	Cir
	Span (in) =	8.0
	No. Barrels =	1
	n-value =	0.024
	Inlet Edge =	Projecting
Embank	Top Elev =	708.26
	Top Width (ft) =	10.00
	Crest Len (ft) =	30.00
Calcs	Q Min (cfs) =	0.90
	Q Max (cfs) =	1.50
	Q Incr (cfs) =	0.10
	Tailwater (ft) =	(dc+D)/2

Clear	Run

FIGURE 13.4

2. After entering the values, click **Run** to compute the culvert values.

3. At the top center, for the name, enter **Area-C-8-culvert**.

4. From the File menu, select SAVE AS..., save the computations as **Area C-8-culvert**, click **Save**, and click **OK** to continue.

5. Review the culvert values in the results grid and select different results to view the changes to the culvert diagram.

6. Click **P-Curve** to view the chart.

7. While in P-Curve mode, click **Diag** to view the P-Curve diagram.

8. Click **Diag** (now Plot) and then click **Plot** to return to the initial view.

The fps at the flow rate of 1cfs (the runoff rate Q) is approximately 3 fps, the target velocity. The Area-C catchbasin will also use an 8-inch outfall.

Culvert — Area-C-8-Culvert Report

1. From the File menu, Print flyout, select REPORT....

2. In the Print dialog box, select a printer and click PRINT.

3. If you are printing to a file, enter a report name.

The report is for the selected Results Grid item.

4. From the File menu, Print flyout, select RESULTS GRID....

5. In the Print dialog box, select a printer and click **OK**.

6. If you are printing to a file, enter a report name.

The report is the current structure's results grid with no other annotation.

Channel

The next exercise has a channel that empties Area-A into a pond. This section calculates the channel's flow into the pond.

1. Click **Channels** and, if necessary, click **Clear** to reset the values to zeros.
2. At the panel's top left, click the **Trapezoidal** icon.
3. Using Figure 13.5 as a guide, enter the channel's values.

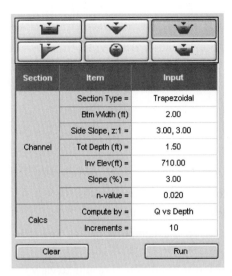

Section	Item	Input
	Section Type =	Trapezoidal
	Btm Width (ft)	2.00
	Side Slope, z:1 =	3.00, 3.00
Channel	Tot Depth (ft) =	1.50
	Inv Elev(ft) =	710.00
	Slope (%) =	3.00
	n-value =	0.020
Calcs	Compute by =	Q vs Depth
	Increments =	10

FIGURE 13.5

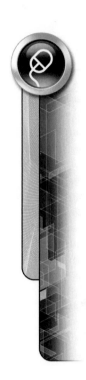

4. After entering the values, click **Run** to compute the channel values.
5. At the top center, for the name, enter **Area-A-channel**.
6. From the File menu, select SAVE AS…, save the computations as **Area A-channel**, click **Save**, and click **OK** to continue.
7. Review the channel values in the results grid and select different results to view the changes to the culvert diagram.
8. Click **P-Curve** to view the chart.
9. While in P-Curve mode, click **Diag** to view the P-Curve diagram.
10. Click **Diag** (now Plot) and then click **Plot** to return to the initial view.

Weir

1. At Express's top, click **Weirs** to open the Weir calculator.
2. At the panel's top left, click the **V-Notch** icon.
3. Using Figure 13.6 as a guide, enter the values into the calculator.

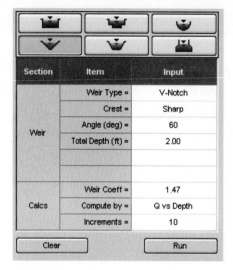

Section	Item	Input
Weir	Weir Type =	V-Notch
	Crest =	Sharp
	Angle (deg) =	60
	Total Depth (ft) =	2.00
Calcs	Weir Coeff =	1.47
	Compute by =	Q vs Depth
	Increments =	10

Clear   Run

FIGURE 13.6

4. After entering the values, click **Run** to compute the channel values.

5. At the top center, for the name, enter **Seward's-Pond-Weir**.

6. From the File menu, select SAVE AS…, save the computations as **Sewards-Pond-Weir**, click **Save**, and click **OK**.

7. Review the weir values in the results grid and select different results to view the changes to the culvert diagram.

Select some results and notice that the plot changes to show the current values in schematic form.

8. Click **P-Curve** to view the chart.

9. While in P-Curve mode, click **Diag** to view the P-Curve diagram.

10. Click **Diag** (now Plot) and then click **Plot** to return to the initial view.

11. Close Hydraflow Express and return to Civil 3D.

This ends the Hydraflow Express extension exercise. Express's focus is hydrological structures and their performance. The next unit's focus is the Hydrograph extension.

EXERCISE 13-2

After completing this exercise, you will:

- Be able to calculate runoff.
- Be able to define a pond.
- Be able to build a hydrograph model.
- Be able to review and modify values.
- Print a report.

Exercise Setup — SCS and IDF Charts

If you have not entered the values for the SCS and IDF charts for DuPage, you can complete the Create IDF and Create SCS charts of the previous exercise's sections. Or, the Chapter 13 folder of the CD that accompanies this textbook contains these files if you did not create them.

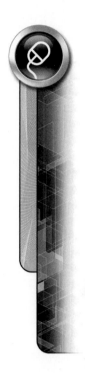

1. If you are not in Civil 3D, start the application by double-clicking its desktop icon.

2. On the Ribbon, click the **Analyze** tab. On the Design panel, select LAUNCH HYDROGRAPHS.

3. At Hydrograph's top, click **IDF** to open the Rainfall IDF Curve dialog box.

4. Click **Open**, browse to and select *DuPage.idf*, click **Open**, and at the top left, click **Exit**.

5. At the top of Hydrograph, click **Precip** to open the Event Manager dialog box.

6. Click **Open**, browse to and select *DuPage.pcp* (or DuPage-Express from the Chapter 13 folder of the CD that accompanies this textbook), click **Open**. In the Event Manager, for Return Periods, toggle off the 1 year storm, click Apply, and at the bottom right, click **Exit**.

Runoff Calculations — SCS — Area A

You first calculate runoff values for three zones. Area A is the entrance to the site and has the following statistics:

Name: Area A

Area: 0.14 Acres

Basin Data

Type	Coefficient	Area 1
Asphalt	98	0.07 Acres
Heavy Soil	85	0.07 Acres

Time of Concentration: TR-55

Distance: 115 Ft.

Slope: 1

1. Select the **Hydrographs** tab, and change the rain period to **2-yr**.

2. For Hydrograph 1, select the Hydrograph Type cell, press the right mouse button, and from the Runoff Hyd. flyout, select SCS METHOD... to open the SCS Runoff Hydrograph dialog box.

3. For Description, enter **Area A** and check that the current SCS is DuPage by clicking **Event Manager** at the bottom right.

4. If necessary, change to DuPage.pcp (or DuPage-Express from the Chapter 13 folder of the CD that accompanies this textbook) by clicking the folder icon in the upper left, browse to and select *DuPage.pcp*, click **Open**, and click **Exit**.

5. For the Drainage Area (Ac) enter **0.14**.

The area's coefficient is a combination coefficient; it is the effect of the two types of zone materials.

6. In SCS Runoff Hydrograph, to the right of Curve Number (CN), click % (the percentage icon) to open the Composite CN dialog box.

7. Using Figure 13.7 as a guide and the Area A values from earlier in this unit, enter them, and when you are done, click **OK** to return to the SCS Runoff Hydrograph dialog box.

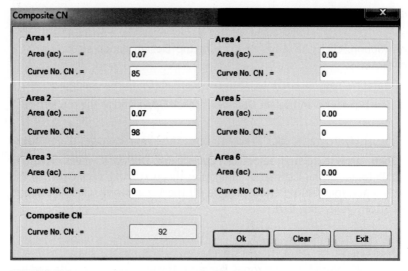

FIGURE 13.7

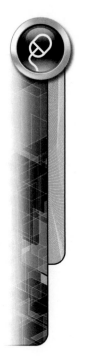

8. Change the Time of Concentration (Tc) Method to TR-55 by clicking the option button under the percent icon.

9. To the right of Time of Concentration, toggle TR55, and click *TR-55* to open the TR-55 Tc Worksheet dialog box.

10. Using Figure 13.8 as a guide, and the Area A values from earlier in this exercise section, enter **115** for the flow length and **1** for the slope. When you click in the two-year rainfall cell, the rainfall amount is automatically entered from the current SCS values.

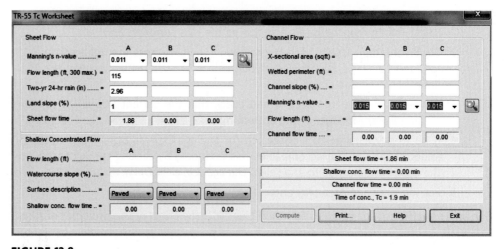

FIGURE 13.8

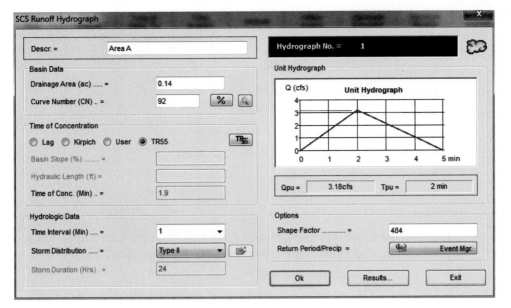

FIGURE 13.9

11. When you are done, click **Compute** to view the calculated values. Finally, click **Exit** to return to the SCS calculator.

12. In the dialog box's lower left, set the Time Interval (Min) to **1** and the Storm Distribution to **Type II**. Your calculator should look like Figure 13.9.

13. Review the calculated Hydrograph by clicking **OK** and then clicking **Results...** (see Figure 13.10).

14. Click **Exit** until you have returned to the Hydrographs page.

15. From the File menu, select SAVE PROJECT AS... and name the file **DuPage Lakewood**.

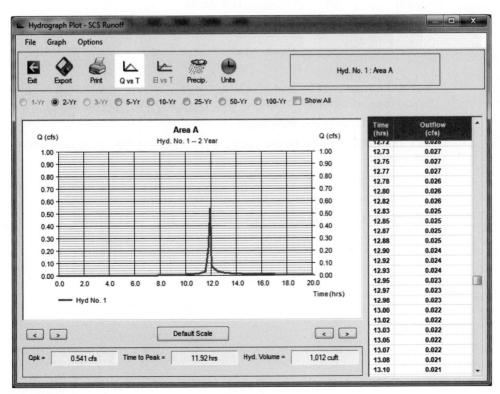

FIGURE 13.10

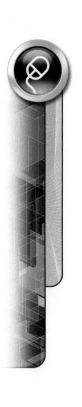

Runoff Calculations — Area B

Area B is the eastern half of the entrance road and half of the parking lot and has the following statistics:

> Name: Area B
>
> Area: 0.17 Acres
>
> Basin Data

Type	Coefficient	Area 1
Asphalt	98	0.17 Acres

> Time of Concentration: TR-55
>
> Distance: 150 Ft.
>
> Slope: 2

1. Repeat the steps used to create Area A for Area B. Set the time interval to 1 and click in the Hydrograph Type cell for Hyd. No. 2 and set it to Type II.
2. Click OK to produce the hydrograph.
3. Click Results to see the hourly values. After reviewing the values, click Exit.
4. Click OK to exit the Area B calculator.
5. Click Exit to return to the Hydrograph application.
6. Save the calculations after you have established Area B.

Runoff Calculations — Area C

Area C is the parking lot's northern half and has the following statistics:

> Name: Area C
>
> Area: 0.16 Acres
>
> Basin Data

Type	Coefficient	Area 1
Asphalt	98	0.16 Acres

> Time of Concentration: TR-55
>
> Distance: 115 Ft.
>
> Slope: 1

1. Repeat the steps used to create Area A for Area C. Set the time interval to 1 and click in the Hydrograph Type cell for Hyd. No. 3 and set it to Type II.
2. Click OK to compute the hydrograph.
3. Click Results to view the hourly values. After reviewing the hydrograph values, click Exit.
4. Click OK to exit the Area C calculator.
5. Save the calculations after you have established Area C.
6. Your Hydrograph editor should look like Figure 13.11.

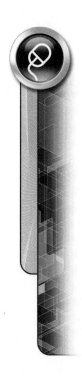

Hyd. No.	Hydrograph type	Peak flow	Time interval	Time of conc. Tc	Time to peak	Volume
	(origin)	(cfs)	(min)	(min)	(min)	(cuft)
1	SCS Runoff	0.541	1	1.90	715.00	1,012
2	SCS Runoff	0.755	1	1.70	715.00	1,578
3	SCS Runoff	0.711	1	1.90	715.00	1,486
4						
5						

FIGURE 13.11

Move Area A's Runoff to a Channel

Area A's Runoff drains to the detention pond through a channel. Channels are reaches; that is, they are concentrators and conveyors of runoff with no storage capacity. A reach can be defined in the Model or Hydrographs panel.

1. Select the **Model** tab; it should look like Figure 13.12. If the labels don't match, from the Options menu, Model flyout, select Show Hydrograph Numbers, and deselect Show Hydrograph Descriptions.

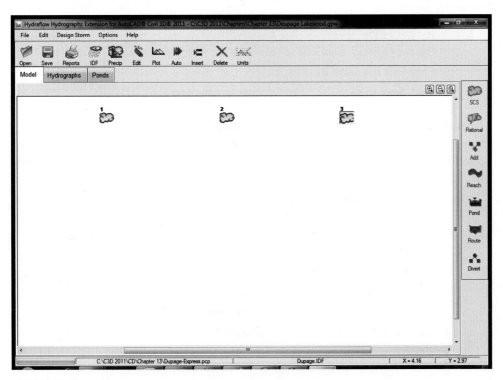

FIGURE 13.12

2. On the right side of the panel, click **Reach** to open the Reach dialog box. If Reach is grayed out, you will need to click in an open area of the panel to select Reach.

3. In the Reach dialog box, enter the values found in Figure 13.13. At the top left, name the reach **Channel A**.

4. In the Reach data section, click the drop list arrow Inflow Hyd. No. and from the list select Area A's hydrograph (1 – SCS Runoff).

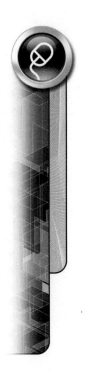

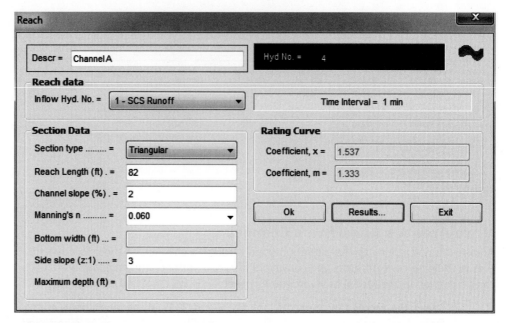

FIGURE 13.13

5. Use Figure 13.13 to enter the channel values. When you are done entering the data, click **OK**, and then click **Results...** to view the hydrograph. After reviewing the hydrograph, click **Exit** to return to the Reach dialog box.

6. Click **Exit** to create the Reach.

7. Click **Save** to save the current model and hydrographs.

Move Area B's Runoff to Catchbasin B

Runoff from Areas B and C flows through catch basins that discharge into a pond. The catch basins act like mini-culverts and take the water in from the parking lot and pass the flow to the pond. The catch basins are reservoirs with little or no detention capacity.

In the model, create a reservoir (catch basin) with little or no capacity. This, however, is not quite correct. If the rainfall intensity is sufficient, water will back up in the catch basin because of its restrictive outfall (8-inch pipe).

1. Click the **Ponds** tab.

2. Double-click the Pond1 Pond Name cell (<New Pond>) to open the Stage / Storage / Discharge Setup dialog box.

3. If necessary, at the top left, click **Contours** to set the Pond definition type.

4. For the pond's name, enter **Catchbasin B**.

5. Using Figure 13.14 as a guide, create the pond Catchbasin B.

6. At the top left, for Bottom Elevation, enter **705.9** and click **Apply**.

7. Click in Row 1's elevation cell, and for the elevation enter **706.00**.

8. Click in Row 1's area cell, for the area enter **13**, and press ENTER.

9. Click in Row 2's elevation cell, and for the elevation enter **709.50**.

10. Click in Row 2's area cell, for the area enter **13**, and press ENTER.

11. In the bottom right, click **Done**.

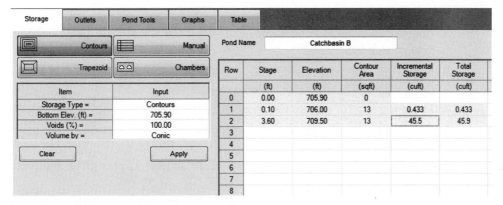

FIGURE 13.14

Catchbasin B has an 8-inch outfall.

12. Click the **Outlets** tab.
13. At the top left, Culv/Orifice, in column A, for the Rise enter **8**.
14. Click in column A's Span and it automatically sets to 8.
15. For the number of Barrels cell, set it to **1**.
16. Click in the Invert Elev. cell to fill in the value.
17. For Length enter **18**, and for slope enter **1**.
18. In the middle of the panel, click **Compute** to calculate the orifice's behavior.

Your Outlets panel should look similar to Figure 13.15.

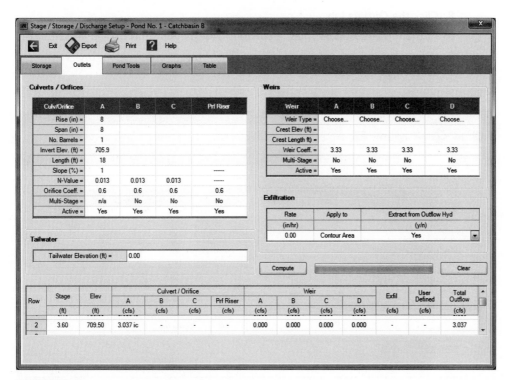

FIGURE 13.15

19. Click **Exit** to return to Hydrographs.
20. At the top left, click **Save** to save the file.

Catchbasin C

1. Double-click Pond2's Pond Name cell to open the Stage / Storage / Discharge Setup dialog box.

2. Repeat the steps from the previous section (Catchbasin B), but for Catchbasin C.

3. Using Figure 13.16 as a guide, enter the values for the Catchbasin C.

4. At the top left, for Bottom Elevation, enter **706.5** and click Apply.

5. Click in Row 1's elevation cell, enter **706.60** and set its area of **13**, and press ENTER.

6. Click in Row 2's elevation cell, enter **709.00** and set its area of **13**, and press ENTER.

7. After entering the values, in the dialog box's lower right, click **Done**.

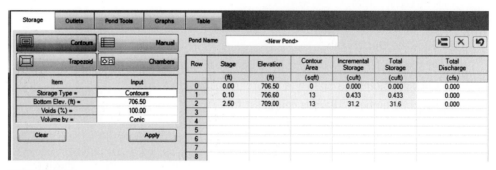

FIGURE 13.16

8. Click the **Outlets** tab.

9. At the top left, Culv/Orifice, in column A, for the Rise enter **8** and press ENTER.

10. Using the values shown in Figure 13.17, finish entering in the values for Catchbasin C – Outlets.

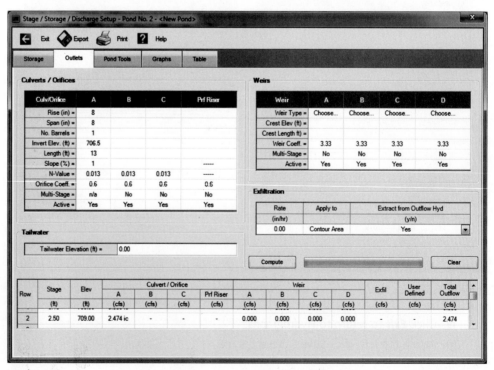

FIGURE 13.17

11. Click **Exit** to return to Hydrographs.
12. At the top left, click **Save** to save the file.
13. Click the **Model** Tab.

Connect Area B to a Reservoir (Catchbasin B)

1. On Model's right side, select **Route**. If Route is grayed out, you will need to click in an open area of the panel to select Route.
2. Using Figure 13.18, enter the necessary values. For the description, enter **Catchbasin B**, set the Inflow Hydrograph to 2 – SCS Runoff - Area B, and set the pond name to 1. Catchbasin B.

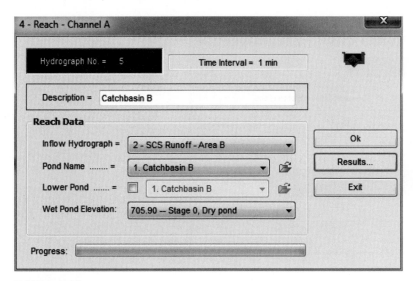

FIGURE 13.18

3. When the settings are correct, click **OK**.
4. Click **Results...** to review the hydrograph.
5. In the Hydrograph Plot - Reservoir dialog box, at its top left, click **Exit**.
6. Click **Exit** to close the Reservoir Route dialog box and return to the Model panel.

Connect Area C to a Reservoir (Catchbasin C)

1. Repeat the previous Steps 1–6, and using Figure 13.19, connect Area C with Reservoir (Catchbasin C).

FIGURE 13.19

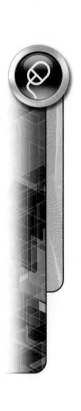

2. Click the *Save* icon to save the calculations.

Combine the Inflow Hydrographs

All three areas discharge into the pond. The Model panel's Add icon merges hydrographs to a single model icon.

1. From the Model panel's right side, select *Add*. If Add is grayed out, you will need to click in an open area of the panel to select Add.

2. Using Figure 13.20, enter the following values. For the description, enter **Combined Inflow Detention Pond**, and select from list with the CNTRL key held down **Reach – Channel A, Reservoir - Catchbasin B**, and **Reservoir - Catchbasin C**.

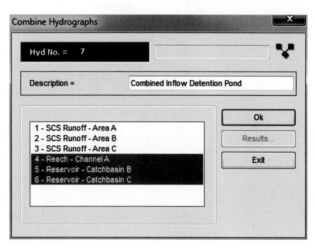

FIGURE 13.20

3. When the settings are correct, click *OK*.
4. Click *Results...* to review the hydrograph.
5. In the Hydrograph Plot - Combine dialog box, at its top left, click *Exit*.
6. Click *Exit* to close the Combine Hydrographs dialog box.
7. Click the *Save* icon to save the calculations.

Your model should look like Figure 13.21.

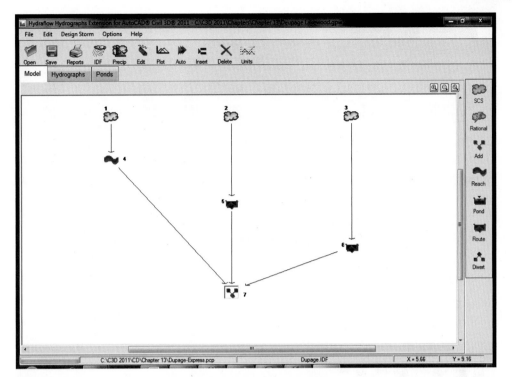

FIGURE 13.21

Define the Detention Pond

1. Click the ***Ponds*** tab.

2. Double-click the Pond3 Pond name cell, to open the Stage / Storage / Discharge Setup dialog box.

3. If necessary, click ***Contours*** at the top left to set the Pond definition type.

4. For the pond name, enter **Detention Pond**.

5. At the top left, for Bottom Elevation, enter **704.9** and click ***Apply***.

6. Using Figure 13.22 as a guide to create the Detention Pond.

7. Use Table 13.3 to enter the elevation and area values. Fill in only the necessary cells.

TABLE 13.3

Elevation	Area
705	3468
706	4810
707	6257
708	7889

8. After carefully entering the elevation and area values, at the bottom right, click ***Done***.

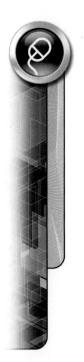

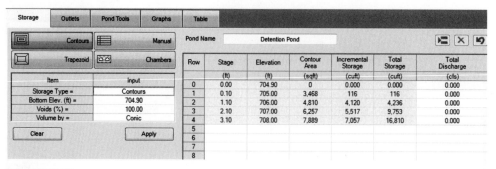

Row	Stage	Elevation	Contour Area	Incremental Storage	Total Storage	Total Discharge
	(ft)	(ft)	(sqft)	(cuft)	(cuft)	(cfs)
0	0.00	704.90	0	0.000	0.000	0.000
1	0.10	705.00	3,468	116	116	0.000
2	1.10	706.00	4,810	4,120	4,236	0.000
3	2.10	707.00	6,257	5,517	9,753	0.000
4	3.10	708.00	7,889	7,057	16,810	0.000

FIGURE 13.22

9. Click the **Outlets** tab.
10. At the top left, Culv/Orifices, in column A, for the Rise, enter **4**.
11. Click in column A's Span and No. Barrels cells and set their values to **4** and **1**.
12. For length enter **82**, and for slope, enter **1**.

Your Outlet should look similar to Figure 13.23.

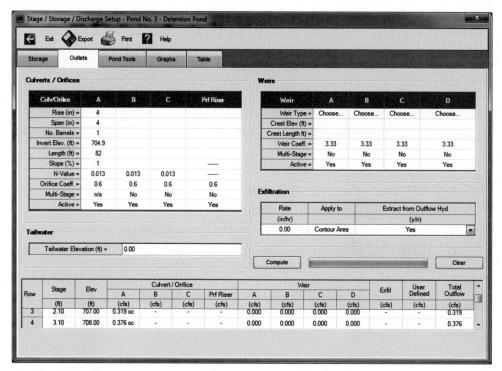

FIGURE 13.23

13. In the middle, click **Compute** to display the outfall values at the bottom.
14. Click **Exit** to return to Hydrographs.
15. At the top left, click the **Save** icon to save the file.

Add the Inflow to a Detention Pond

1. At the top left, click the **Model** tab.
2. From the panel's right side, select **Route**. If Route is grayed out, you will need to click in an open area of the panel to select Route.

3. Using Figure 13.24, enter the following values. For the description, enter **Detention Pond** and for the Inflow Hydrograph select 7 -COMBINE - COMBINED INFLOW DENTION POND.

FIGURE 13.24

4. Set the Pond Name to **3 - Detention Pond**.

5. When the settings are correct, click **OK**.

6. Click **Results...** to review the hydrograph.

7. In the Hydrograph Plot – Reservoir – Detention Pond dialog box, at its top left, click **Exit**.

8. Click **Exit** to close the Reservoir Route dialog box.

9. Click the **Save** icon to save the calculations.

Design Report

1. At the Hydraflow Hydrographs' top, click the Reports icon.

2. In the Reports Print Menu dialog box set its values to match those in Figure 13.25. When set, click Preview and review the report.

3. When done previewing the report, click Exit.

4. Close Hydraflow Hydrographs and return to Civil 3D.

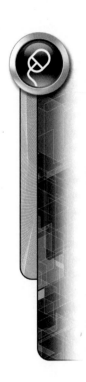

Reports Print Menu ✕

Report Options

☑ Watershed Schematic
  ☑ Legend
☑ Return Period Recap
☑ Summary Page
☑ Hydrograph Reports

  ◉ Graphic   ○ Numeric

From: [1] to: [99] [✔ All]

Q's >= [1] % of peak

Print interval = [1] nth

○ Minutes ◉ Hours
☑ Tc Worksheets
☐ Storm Hyetographs
☑ Pond Report
  ☑ Numeric
  ☐ Expanded
☐ Rainfall Report
☑ Page No's. Begin with: [1]
☐ Table of Contents

Frequencies

☐ 1-Yr
☑ 2-Yr
☐ 3-Yr
☑ 5-Yr
☑ 10-Yr
☑ 25-Yr
☑ 50-Yr
☑ 100-Yr

Output Units

◉ English
○ SI

[Preview...]
[Print...]
[Cancel]

FIGURE 13.25

EXERCISE 13-3

After completing this exercise, you will:

- Be able to create an IDF.
- Import and Export a pipe network to Storm Sewer.
- Be able to set Code Standards.
- Be able to add runoff to inlets.
- Be able to modify a design.

Exercise Setup

This exercise uses the *Chapter 13 – Unit 3.dwg* file. To find this file browse to the Chapter 13 folder of CD that accompanies this textbook, and open the *Chapter 13 – Unit 3.dwg* file.

1. If it is not running, start **Civil 3D** by double-clicking its desktop icon.

2. When you are at the command prompt, close the open drawing and do not save it.

3. At Civil 3D's top left, Quick Access Toolbar, click the OPEN icon, browse to the Chapter 13 folder of the CD that accompanies this textbook, select the *Chapter 13 – Unit 3.dwg* file, and click **Open**.

4. At Civil 3D's top left, click Civil 3D's drop-list arrow, in the Application Menu highlight SAVE AS, from the flyout select AutoCAD Drawing, browse to the Civil 3D Projects folder, for the drawing name enter **Tulsa Parking Lot – Work** and click **SAVE** to save the file.

Edit Pipe Migration Settings

To export and import Civil 3D pipe networks to Storm Sewers, you must set up an equivalency chart. Civil 3D and Storm Sewers do not use the same terms for structures. The current pipe network uses eccentric cylindrical structures with a rectangular grate.

1. Click the Settings tab.

2. From the Settings list, select Pipe Network, right mouse click, and from the shortcut menu, select EDIT FEATURE SETTINGS....

3. Expand the Storm Sewers Migration Defaults section.

4. Click in the Part Matching Defaults value cell to display an ellipsis. Click the ellipsis and click it to display the Part Matchup Settings dialog box.

5. In the Part Matchup Setting dialog box, click the Import tab.

6. In the Importing Rectangular Structures section locate **Grate Inlet Rectangular**. Click in the Civil 3D Part Type column for Grate Inlet Circular to display an ellipsis. Click the ellipsis to display the Part Catalog dialog box.

7. In the Part Catalog dialog box, toggle on **Eccentric Cylindrical Structure** and click **OK** to return to the Part Matchup Settings dialog box.

8. Click the Export tab.

9. In the Exporting Structures section, locate the Civil 3D Part Type, **Eccentric Cylindrical Structure**. For the Eccentric Cylindrical Structure, click in the Storm Sewers Part Type column, click the drop-list arrow, and from the part's type list select **Grate Inlet Rectangular**.

10. Click OK to return to Edit Feature Settings.

11. In the Storm Sewers Migration Defaults section, Parts List Used for Migration, click in the value cell to display an ellipsis.

12. Click the ellipsis and in the Parts List Used For Migration, select **Storm Sewer** and click **OK** to return to Edit Feature Settings.

13. Click **OK** to exit Edit Feature Settings.

14. From the Quick Access toolbar, select Save.

Export to Storm Sewers

1. In the Ribbon, click the *Modify* tab. On the Design panel, click the *Pipe Network* icon displaying the Pipe Networks tab.

2. On the Pipe Networks tab, Analyze panel, click the *Storm Sewers* icon, and from the shortcut menu, select EDIT IN STORM SEWERS to display the Export to Storm Sewers dialog box.

3. From the pipe networks list, select PKLOT - PSTM – (1) and click *OK* to display Export Storm Sewers to File dialog box.

4. In Export Storm Sewers to File dialog box, browse to the folder C:\Civil 3D Projects, name the file *PKLOT - PSTM – (1)*, and click *Save* to display Hydraflow Storm Sewers application.

5. After exporting the file, at the Ribbon's right, click the Close icon.

Hydraflow Storm Sewers Setup — Set the Project to the Exported File

Even though the pipe network shows in Storm Sewers, if you were to save the project, it would not save to the file just exported. You need to open the exported file and all subsequent saves will be to the correct file.

1. On the ribbon, click the Analyze tab.

2. In the Analyze tab, the Design panel, click the Launch Storm Sewers icon.

3. At Storm Sewer's top left, from the File menu, select OPEN PROJECT....

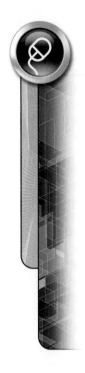

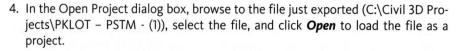

4. In the Open Project dialog box, browse to the file just exported (C:\Civil 3D Projects\PKLOT – PSTM - (1)), select the file, and click **Open** to load the file as a project.

This sets the file as the project file and any subsequent saves will update this file. This file will in turn update your Civil 3D pipe network.

Hydraflow Storm Sewers Setup — Tulsa IDF

The *Tulsa.IDF* and Codes must be set. The Chapter 13 folder of the CD that accompanies this textbook contains the Tulsa IDF.

1. At Storm Sewer's top, click the **IDF** icon to open the Rainfall IDF Curve dialog box.

2. Click the **Open** icon, browse to the Chapter 13 folder of the CD that accompanies this textbook, select the Tulsa.idf file and click Open. When the Tulsa.idf file has been displayed, click the **Exit** icon.

Hydraflow Storm Sewers Setup — Codes

1. From the application's top, click the **Codes** icon to open the Design Codes dialog box.

2. If necessary, click the **Pipes** tab.

3. Adjust your settings to match those in Figure 13.26.

4. When the settings are correct, click the **Inlets** tab.

5. Adjust your settings to match those in Figure 13.27.

6. When the settings are correct, click the **Calculations** tab.

7. Adjust your settings to match those in Figure 13.28.

8. Click **OK** to close the dialog box.

9. From the Edit menu, select COST CODES....

10. Review the per-unit costs.

11. After reviewing the chart's values, click **Exit**.

12. At Storm Sewers top left click **Save**, and click OK to save the project.

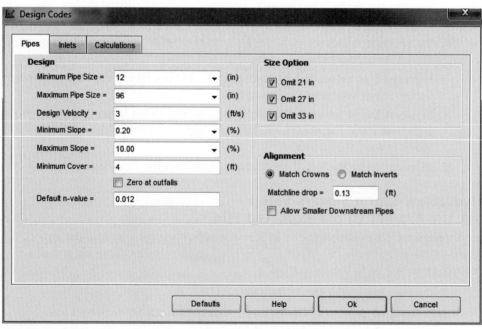

FIGURE 13.26

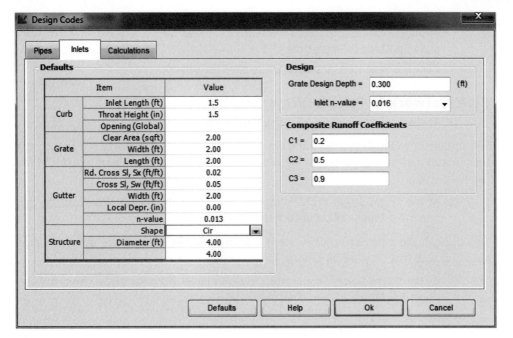

FIGURE 13.27

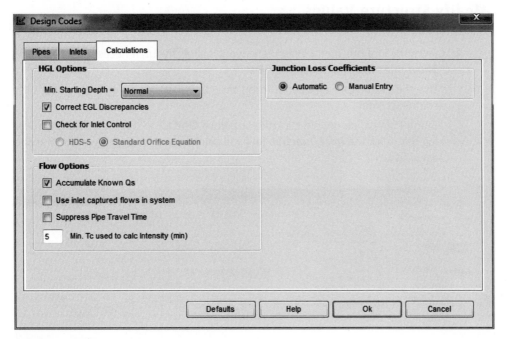

FIGURE 13.28

Import a Background DXF

In the Chapter 13 folder of the CD that accompanies this textbook is a file ACAD_CONTOURS.dxf. It contains contours that provide a background to the Hydraflow pipe network.

1. At Hydraflow Storm Sewers' top right, click the ***Import DXF*** icon, and from its shortcut menu, select BACKGROUND IMAGE FROM DXF….

2. In the Add DXF Background Image dialog box, browse to the Chapter 13 folder of the CD that accompanies this textbook, select the file *ACAD-CONTOURS.dxf*, and click **Open** to import the DXF.

3. Click the Save icon, and click OK to save the project.

Pipe Network in Profile View

1. At the application's top, click the **Profile** icon.

This displays the Outfall structure and pipe. Storm Sewer can display only one branch at a time. A user must adjust the beginning and ending pipe segments at the top center of the panel.

2. At the Storm Sewer Profile's top, click To Line's drop-list arrow, from the list, select **4 – Pipe – W1**, and click the Update icon.

This displays the pipe network's western branch in Profile View. Write down the pipe sizes and note that they are all 12-inch.

3. At the Storm Sewer Profile's top, click To Line's drop-list arrow, from the list, select **8 – Pipe – E1**, and click the Update icon.

This displays the pipe network's eastern branch in Profile View. The pipes are also 12-inch pipes.

4. At the top left, click the **Exit** icon to return to plan view.

Modify Structure Values

A structure must be an inlet to accept runoff.

1. From the Options menu, Plan View flyout, Labels flyout, if necessary, toggle **ON Show Inlet IDs** and **Show Line IDs**.

2. Use the Zoom tool and zoom in on the Outfall structure at the parking lot's southeast side.

3. In the Plan panel, double-click the structure **Outfall** to open the Add/Edit Dialog box. Click the **Inlet/Junction** tab, and using Figure 13.29 as a guide, enter the listed values.

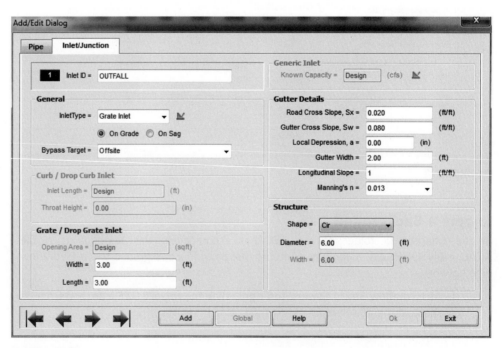

FIGURE 13.29

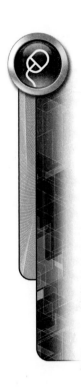

```
General
      Inlet Type = Grate Inlet
      On Grade
Grate / Drop Grate Inlet
      See Table 13.4
Gutter Details
      Road Cross Slope, Sx = 0.020
      Gutter Cross Slope, Sw = 0.080
      Local Depression, a = 0.000
      Gutter Width = 2.00
      Longitudinal Slope = 1
      Manning's n = 0.013
Structure
      Shape = Cir
      Diameter = 6.00
```

4. Click **OK** to set the values.
5. At the panel's lower left, click the **right-facing blue arrow** until you have reached structure **West 3**. You may have to click OK to reactivate the arrows.
6. Using Figure 13.30 as a guide, enter the listed values for West 3.

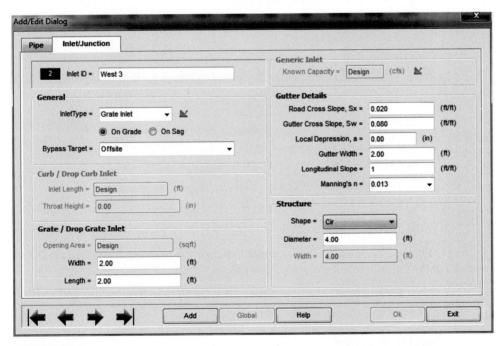

FIGURE 13.30

7. Click **OK** to set the values.
8. At the lower left of the Inlet/Junction panel, click the **RIGHT** arrow. You may have to click OK to reactivate the arrows. Use the same Gutter Details for all Inlet/Junction structures. However, the grate size changes for some of the

structures (see Table 13.4). You will cycle from West 3 to West 1 and then East 4 to East 1. The Gutter details are the same for all structures.

9. After setting each structure's values, click **Exit** to return to the plan panel.

10. At Storm Sewer's top left, click the **Save** icon, and click OK to save the project.

TABLE 13.4

Structure Name	Structure Size	Grate Type	Rectangular Grate
Outfall	6 ft Circular	Grate Inlet	3 ft × 3 ft
West 1	4 ft Circular	Grate Inlet	2 ft × 2 ft
West 2	4 ft Circular	Grate Inlet	2 ft × 2 ft
West 3	4 ft Circular	Grate Inlet	3 ft × 3 ft
East 1	4 ft Circular	Grate Inlet	1.5 ft × 1.5 ft
East 2	4 ft Circular	Grate Inlet	1.5 ft × 1.5 ft
East 3	4 ft Circular	Grate Inlet	2 ft × 2 ft
East 4	4 ft Circular	Grate Inlet	2 ft × 2 ft

Setting Pipe Flows

After setting the structure values, next is setting all the flow values for each structure's pipe. This is done in each structure's Pipe panel.

1. In the Storm Sewers plan panel, double-click the **Outfall** structure to display the Inlet/Junction dialog box.

2. Click the Pipe tab.

3. In the Pipe tab's lower left enter the following values for flows (also in Table 13.5):

```
Flows
     Known Q: 1.594
     Drainage Area: 0.22
     Runoff Coefficient: 0.86
     TC Method: User: Inlet Time: 1.26
Physical
     Manning's n = 0.013
```

In the Physical section for each Pipe you **must** change the Manning's n value to 0.013 or you will get errors in the pipe network's analysis.

4. When done entering the flow values, click **OK** to save the changes and activate the arrows.

5. At the panel's bottom left, click the **RIGHT** arrow **three** times to view the pipe panel for Pipe – W1.

6. Using the values in Table 13.5, enter the flow values for **Pipe – W1**. Set Manning's n to 0.13.

7. When done entering the values for Pipe – W1, click OK.

8. To move to Pipe – W2, at the panel's lower left, click the **LEFT** arrow.

9. Using the entries in Table 13.5, enter the flow values for **Pipe – W2**. Set Manning's n to 0.13.

10. When done entering the values for Pipe – W2, click OK.

11. To move to Pipe – W3, at the panel's lower left, click the **LEFT** arrow.

12. Using the values in Table 13.5, enter the flow values for **Pipe – W3**. Set Manning's n to 0.13.

13. When done entering the values for Pipe – W3, click OK.

14. After entering the values for Pipe – W3, click the **RIGHT** arrow **three** times to move to **Pipe – E4**.

15. If necessary, click **OK** to activate the panel's lower left arrows and click the **RIGHT** arrow **three** times to move to **Pipe – E1**.

16. Using the values in Table 13.5, enter the flow values for **Pipe – E1**. Set Manning's n to 0.13.

17. To move to Pipe – E2, at the panel's lower left, click the **LEFT** arrow.

18. Using the entries in Table 13.5, enter the flow values for **Pipe – E2**. Set Manning's n to 0.13.

19. Repeat steps 16 and 17 until arriving at Pipe – W1. Set Manning's n to 0.13.

20. Click OK and then click **Exit**.

21. At Storm Sewer's top left, click the **Save**, and click OK to save the project.

TABLE 13.5

Structure	Pipe Name	Known Q	Drainage Area	Runoff Coefficient	Inlet Time
Outfall	Pipe - Outfall	1.594	0.22	0.86	1.26
West 1	Pipe - W1	3.564	0.45	0.96	1.60
West 2	Pipe – W2	4.515	0.57	0.96	2.19
West 3	Pipe – W3	1.549	0.21	0.92	1.67
East 1	Pipe – E1	1.561	0.22	0.86	1.97
East 2	Pipe – E2	1.535	0.20	0.90	1.55
East 3	Pipe – E3	1.594	0.20	0.92	1.46
East 4	Pipe – E4	1.077	0.15	0.87	1.65

Run the Analysis and Reset Pipe Sizes

Next, you run an analysis using the current pipe sizes. The pipe sizes are too small to handle the runoff amounts. When you run the analysis, you will allow the routine to resize them.

1. At the Plan panel's top, click the **Run** icon to open the Compute System dialog box.

2. In the Hydrology section, make sure it is set to **10 year**.

3. In the Design Options section, toggle **ON Reset Pipes Sizes**.

4. At Compute System's bottom, Starting HGLs, set the Starting HGL to **Normal**.

5. Click **OK** to evaluate and resize the pipes.

The Storm Sewer Design dialog box opens and displays the outfall structure.

6. At the lower right, click **Up** to review each structure and pipe in the system.

7. After reviewing the values, click **Finish**.

8. At the top left, click the **Save** icon, and click **OK** to save the project.

Review Pipe Results

1. At the top left, click the **Pipes** tab.

This opens the pipes panel that is displaying the adjusted results. A user can change any values in this worksheet.

2. Review the values for the pipe analysis.

Review Inlet Results

1. At the top left, click the **Inlets** tab.

This opens the Inlets panel that is displaying the adjusted results. A user can change any values in this worksheet.

2. Review the values for the Inlet analysis.

Review Results

1. At the top left, click the **Results** tab.

This opens the Results panel that is displaying the analysis in several different formats. One of the Results tabs is a cost estimate for the network.

2. Review the values for the various Results tabs.

3. Click the Save icon to save the storm sewer results.

4. Close the Storm Sewer application.

Import the Revised Pipe Network

The last step is updating the PKLOT – PSTM – (1) pipe network in Civil 3D.

1. If necessary, restart **Civil 3D** and open the *Tulsa Parking Lot – Work* drawing.

2. If necessary, in the Ribbon, Click the **Modify** tab. On the Design panel, click the **Pipe Network** icon displaying the Pipe Networks tab.

3. In the Pipe Networks tab, the Analysis panel, click the **Storm Sewers** icon and from the shortcut menu, select IMPORT FILE.

4. In the Import Storm Sewers Files dialog box, browse to and select *PKLOT - PSTM – (1)* and click **Open** to read the file.

5. The Existing Networks Found dialog box displays. Select **Update** the existing pipe network to modify the network to the new values from Storm Sewers.

6. An Events panorama is displayed. Review the event and then click the **green checkmark** to close the panorama.

7. In the Pipe Networks tab, click Close.

8. In the drawing, select a few of the pipes, right click, and from the shortcut menu select Pipe Properties. Review the new pipe diameters.

9. In the drawing select a few pipes to view their new sizes.

10. At Civil 3D's top left, Quick Access Toolbar, click the **Save** icon to save the drawing.

11. Close the drawing.

You will have to reset the structure types because they follow the Pipe Network defaults.

EXERCISE 13-4

After completing this exercise, you will:

- Be able to create a View Frame Group.
- Be able to create a Plan and Profile sheet set.

Exercise Setup

This exercise uses the *Chapter 13 – Unit 4.dwg* file. To find this file browse to the Chapter 13 folder of CD that accompanies this textbook, and open the *Chapter 13 – Unit 4.dwg* file.

1. If you are not in Civil 3D, start the program by double-clicking on **Civil 3D's** desktop icon.
2. At Civil 3D's top left, Quick Access Toolbar, click the **Open** icon. In the Open dialog box, browse to the Chapter 13 folder of the CD that accompanies this textbook, select the file *Chapter 13 – Unit 4*, and click **Open**.
3. At Civil 3D's top left, click the **Civil 3D** drop-list arrow, and from the Application Menu, highlight SAVE AS, from the flyout select AUTOCAD DRAWING, browse to the Civil 3D Projects folder, for the drawing name enter **Plan and Profile Sheets,** and click **Save** to save the file.

Create View Frames

1. On the Ribbon, click the **Output** tab. On the Plan Production panel, click the CREATE VIEW FRAMES icon displaying the Create View Frames wizard's Alignment panel.
2. Using Figure 13.31 as a guide, set the alignment to **Rosewood – (1)** and click **Next** to open the Sheets panel.
3. Using Figure 13.32 as a guide, if necessary, set the sheet type to **Plan and Profile** and, at the bottom, toggle the sheet orientation to **Along alignment**.
4. In the middle right of the Sheets panel, click the ellipsis to open the Select Layout as Sheet Template dialog box.
5. In the Select Layout as Sheet Template dialog box, select **Arch D Plan and Profile 20 scale** and click **OK** to return to the Sheets panel.

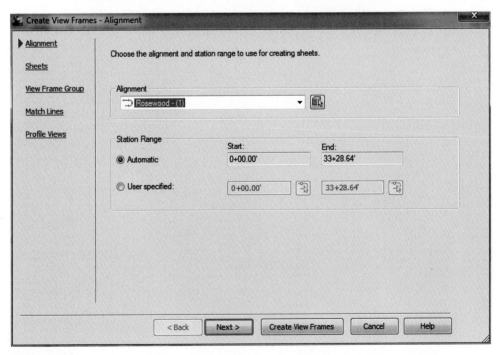

FIGURE 13.31

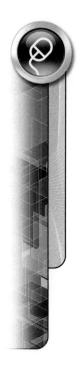

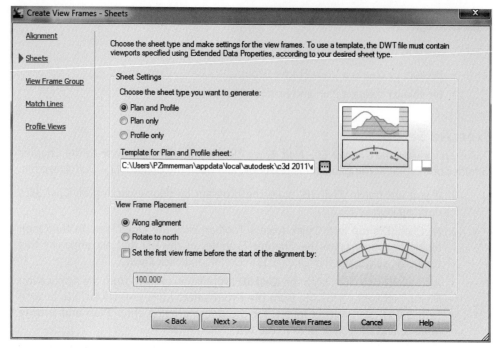

FIGURE 13.32

If no templates appear in the Select Layout as Sheet Template dialog box, browse to the Chapter 13 folder of the CD that accompanies this textbook and in the Plan Production folder, select from the template list.

6. Click **Next** to open the View Frame Group panel (see Figure 13.33).

7. There are no changes, so click **Next** to open the Match Lines panel (see Figure 13.34).

8. There are no changes, so click **Next** to open the Profile Views panel (see Figure 13.35).

9. Click the Select band set style drop-list arrow and, from the list, select **Plan Profile Sheets – Elevations and Stations**.

10. Click **Create View Frames** to create the view frames.

11. At Civil 3D's top left, Quick Access Toolbar, click the **Save** icon to save the drawing.

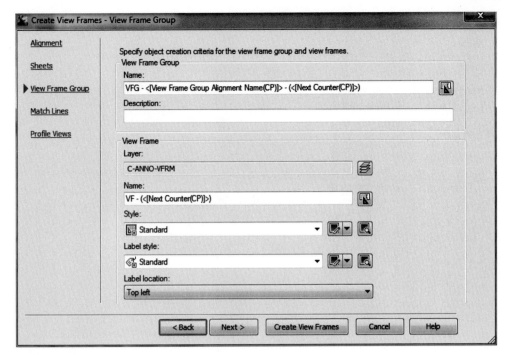

FIGURE 13.33

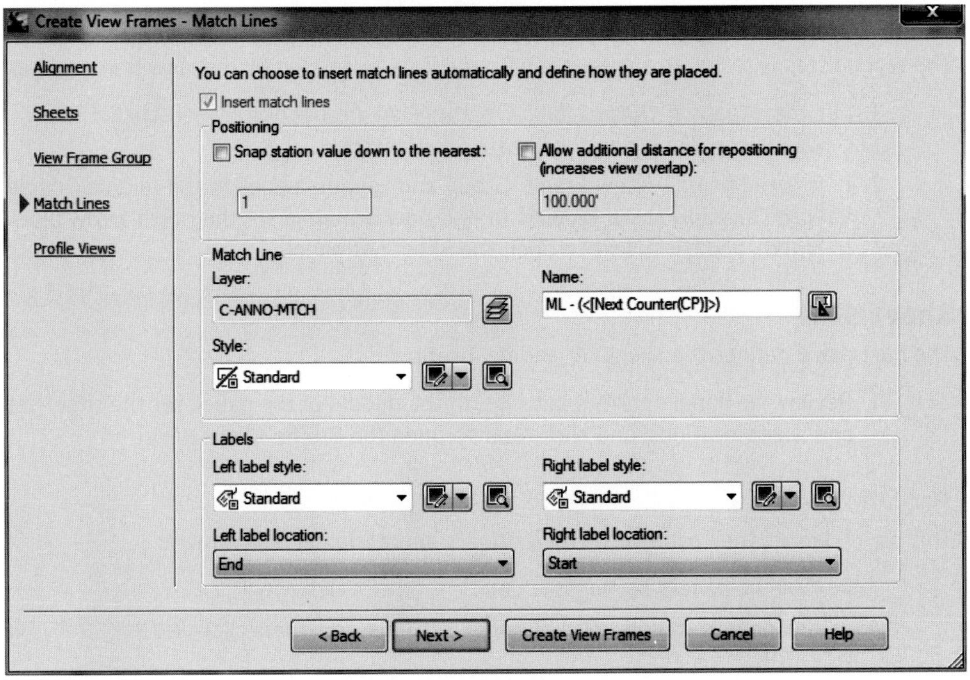

FIGURE 13.34

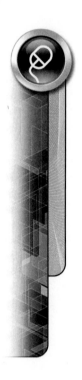

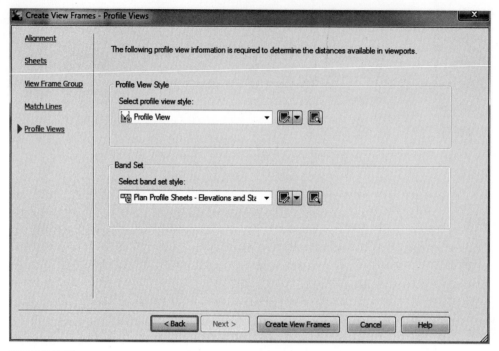

FIGURE 13.35

Create Plan Profile Sheets

The second step is to create a sheet set with the layouts defined by the view frame group.

1. On the Ribbon's **Output** tab. On the Plan Production panel, click CREATE SHEETS icon displaying the Create Sheets wizard.

2. In Create Sheets – View Frame Group and Layouts panel, in the middle, toggle **Layout Creation** to **All layouts in one new drawing**, set the north arrow block to **North**, and click **Next** to open the Sheet Set panel.

Sheet Sets

The next panel defines the sheet set and its location.

1. Review the panel's current settings, in the middle of the panel, set the sheet set and sheet locations, and click **Next** to open the Profile Views panel.

Profile Views

This panel defines the Profile and Profile View's annotation and alignment.

1. Review the panel's settings.

2. In the panel's middle, click **Choose settings** and then click PROFILE VIEW WIZARD....

3. In Profile View Wizard, click **Next** until you view the Profile Display Options panel.

4. Scroll the profiles to the right until you are viewing their label assignments.

5. Double-click the _Rosewood Preliminary – (1) label assignment (_No Labels), in the Pick Profile Label Set dialog box, click the drop-list arrow, select **Complete Label Set**, and click **OK** to return to the Profile Display Options panel.

6. Click **Next** to open the Data Bands panel.

Notice the double band assignment: one for data, the dummy band for spacing.

7. For the first band, set Profile1 to Existing Ground.
8. Click ***Finish*** to return to the Profile Views panel.
9. Click ***Next*** to open the Data References panel.

Data References

The new drawing(s) and their layouts can be linked to the data in the current drawing.

1. There is nothing to change in this panel.
2. Click ***Create Sheets*** to display the automatic save warning. Click **OK** to continue.
3. The command line prompts you for a profile view origin. Enter **0,0** and press ENTER.
4. An Events panorama is displayed; close the panorama.
5. Review the plan and profile sheets in the Sheet Set Manager palette.
6. In Sheet Set Manager, double-click each sheet to review its layout and annotation.
7. After reviewing the sheets, close the drawing and the Sheet Set Manager.
8. At Civil 3D's top left, Quick Access Toolbar, click the ***Save*** icon to save the drawing, and exit Civil 3D.

This ends the exercise on Hydrology and Plan and Profile Sheets.

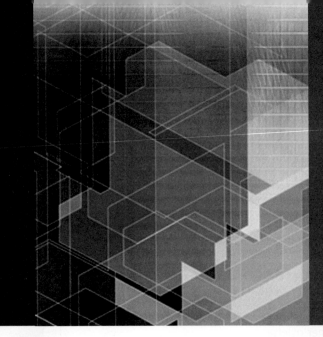

Survey Basics

EXERCISE 14-1

After completing this exercise, you will:

- Be familiar with the Cmdhelp.ref file.
- Be able to read field book files.
- Be able to write a field book file.
- Be able to review and print Cmdhelp.ref.

Locate and Review Cmdhelp.ref

1. From the Windows All Programs, Accessories flyout, select Notepad.
2. In Notepad, from the File menu, select OPEN..., browse to the folder C:\Documents and Settings\All Users\Application Data\Autodesk\C3D 2010\ enu\Survey (XP) (in Windows 7 it is in C:\ProgramData\Autodesk\C3D 2010\ enu\Survey), change the Files of Type to **All Files**, select *Cmdhelp.ref*, and click **Open**.
3. From the File menu, select SAVE AS..., and save a copy to the desktop.
4. Review the command's language Point Creation and Point Location formats.
5. If possible, print the file and use it as a reference.

Review and Print Nichol.txt

1. Open Windows Explorer, browse to the Chapter 14 folder of the CD that accompanies this textbook, locate and select the file *Nichols.doc* or *Nichols.txt*, and click **Open**.
2. Review the survey information and, if possible, print the file for reference.

Convert the Survey to a Field Book — Manually

1. Start a second Notepad file, and in the Notepad file containing the **Nichols.txt** file, review the Benchmark entries.

One benchmark entry does not have an elevation; the remaining entries do.

2. In the empty Notepad file, using the Field Book formats for points with known coordinates, for point 1206, enter its coordinates (the NE format), and enter the coordinates for points 1213 and 1226 (NEZ coordinates with elevation).

   After entering their values, your field book should look like the following:

   ```
   NE 1206 2265.3919 1552.9250 "+"
   NEZ 1213 2601.6437 1497.8024 634.52 "IP"
   NEZ 1226 2623.0725 660.6777 636.43 "+"
   ```

3. In field book Notepad (the entered the NE and NEZ values), from its File menu, select SAVE AS…, browse to the C:\Civil 3D Projects folder, change the file type to **All Files (*.*)**, enter *Nichols.fbk* for the filename, and click Save to save the file.

Next in the survey is setting the station at point 1213 and setting an instrument height.

4. Using STN, set the instrument at 1226 with the instrument height of 5.13.

The backsight (BS entry) is to a known port. All that needs to be done is to enter the backsight point number. Survey assumes it is the 0 (zero) direction.

5. Enter the backsight as BS 1213.

After setting the station and backsight values, you next set the prism height (PRISM).

6. For the prism height, enter 4.65 using the PRISM format.

After entering the values, your field book should look like the following:

```
NE 1206 2265.3919 1552.9250 "+"
NEZ 1213 2601.6437 1497.8024 634.52 "IP"
NEZ 1226 2623.0725 660.6777 636.43 "+"
STN 1226 5.13
BS  1213
PRISM 4.65
```

7. From Notepad's File menu, select SAVE.

Enter Point Observations

Next, you enter point observations. These observations are angles and distances with a vertical difference. These observations use the AD VD command format.

1. Use the AD VD format and for points 200–207, code their observations.

After entering the values, your field book should look like the following:

```
NE 1206 2265.3919 1552.9250 "+"
NEZ 1213 2601.6437 1497.8024 634.52 "IP"
NEZ 1226 2623.0725 660.6777 636.43 "+"
STN 1226 5.13
BS  1213
PRISM 4.65
AD VD 200   92.1224   61.31 -0.89 BC
AD VD 201   87.3738   61.13 -0.82 BW
AD VD 202   84.4515   17.12 -0.43 BW
AD VD 203 101.0837   17.49 -0.51 BC
```

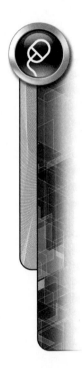

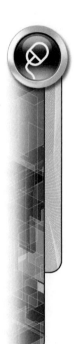

```
AD VD 204 252.2341  13.08 -0.32 BC
AD VD 205 274.3455  12.68 -0.37 BW
AD VD 206 262.1906  33.27 -0.28 BC
AD VD 207 259.3437  33.45 -0.74 INB
```

2. From Notepad's File menu, select SAVE.

3. Close all open Notepad files.

Create a Local Survey Database

1. If you are not in Civil 3D, on the Windows Desktop double-click the **Civil 3D** icon.

2. At Civil 3D's top left, click the Civil 3D drop-list arrow, and from the Application Menu, highlight SAVE AS, from the flyout select AUTOCAD DRAWING, browse to the Civil 3D Projects folder, for the drawing name enter **Survey**, and click **Save** to save the file.

3. If necessary, in Ribbon's Home tab, Palettes panel, click the **Survey Toolspace** icon or click the Toolspace's **Survey** tab.

4. At the Toolspace's top, click the **Survey User Settings** icon.

5. In Survey User Setting, Import Defaults section, use Figure 14.1 as a guide, and set your values to match.

6. Click **OK** to exit the dialog box.

7. In Survey, place your cursor over Survey Databases, press the right mouse button, and from the shortcut menu, select NEW LOCAL SURVEY DATABASE....

8. In New Local Survey Database, for the name, enter **Road Survey**, and click **OK**.

9. In Survey select the survey database Road Survey, right mouse click, and from the shortcut menu, select EDIT SURVEY DATABASE SETTINGS....

10. In Edit Survey Database Settings, if necessary, expand the Units section, set the distance value to US Foot by clicking the drop-list value and selecting US Foot.

11. Click OK to exit the Edit Survey Database dialog box.

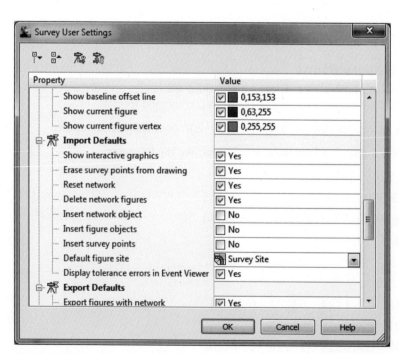

FIGURE 14.1

Create a Network

1. In Road Survey, select Networks, press the right mouse button, and from the shortcut menu, select NEW... to open the New Network dialog box.

2. In New Network, for the network name, enter **Road Survey**, enter today's date as its description, and click **OK**.

3. Use Windows Explorer and from the Chapter 14 folder of the CD that accompanies this textbook, copy the file *10113f2.rw5* to the network folder Road Survey (C:\Civil 3D Projects\Road Survey\Road Survey).

4. Minimize Windows Explorer.

Convert a TDS File to Autodesk–Softdesk FBK

1. On Ribbon's Home tab, click the Create Ground Data panel title to unfold the panel, and from the shortcut menu, select SURVEY DATA COLLECTION LINK to display TDS Survey Link.

2. From the File menu of Survey Link DC 7.5.5, select OPEN....

3. In Open, browse to, select, and open the *10113f2.rw5* file (C:\Civil 3D Projects\Road Survey\Road Survey).

4. In the editor, scroll through and review its data, and when you are done, exit the editor.

5. In Survey Link DC 7.5.5, from the Conversions menu, select CONVERT FILE FORMAT....

6. In the Convert dialog box, change the Input to **Raw Data File**, and set Input Type to **TDS Raw Data (.rw5)**.

7. Below Input File Name:, click **Choose File...**, browse to C:\Civil 3D Projects\Road Survey\Road Survey, select the file *10113f2.rw5*, and click **Open** to return to the Convert dialog box.

8. In the Convert dialog box, change the Output Type to **Autodesk-Softdesk FBK**.

9. Below Output File Name..., click CHOOSE FILE..., browse to the C:\Civil 3D Projects\Road Survey\Road Survey, for the filename enter *10113f2*, and click **Save** to return to the Convert dialog box.

10. In the Convert dialog box, click **Convert**, and when the Successful conversion dialog box appears, click **OK** to exit the message.

11. Click **Close** to exit the Convert dialog box.

12. At the upper right of Survey Link DC 7.5.5, click the **X** to close it.

13. In Survey, select the network Road Survey, right mouse click, and from the shortcut menu, select EDIT FIELD BOOK....

14. Browse to the Road Survey network folder, select *10113f2.fbk*, and click **Open**.

15. In the editor, in the second line, add US to the FOOT entry.

    The second line should now have the following text:

    UNIT USFOOT DMS

16. Review the contents of the field book file.

17. Save the file and close Notepad.

18. At Civil 3D's top left, Quick Access Toolbar, select the **Save** icon to save the drawing.

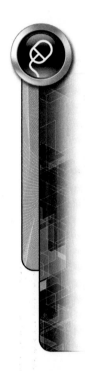

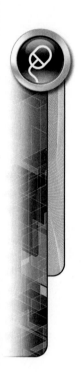

EXERCISE 14-2

After completing this exercise, you will:

- Be familiar with the styles in settings.
- Be familiar with the various settings for Survey.

Exercise Setup

1. If you are not in **Civil 3D**, double-click its desktop icon.
2. When you are at the command prompt, close the open drawing and do not save it.
3. If necessary, click the **Prospector** tab and set Prospector to Master View.
4. In Prospector, expand the Drawing Templates and AutoCAD list, from the list, select **_Autodesk Civil 3D (Imperial) NCS**, right mouse click, and from the shortcut menu, select CREATE NEW DRAWING.
5. Click the **Settings** tab.
6. At the top of Settings, select the drawing name, press the right mouse button, and from the shortcut menu, select EDIT DRAWING SETTINGS....
7. Click the **Object Layers** tab and scroll to the object list's bottom.
8. In Object Layers for Survey Network and Survey Figure, set the Modifier to **Suffix** and its value to **-***.
9. Click **OK** to close the dialog box.
10. At Civil 3D's top left, click Civil 3D's drop-list arrow, from the Application Menu, highlight SAVE AS and from the flyout select AUTOCAD DRAWING, browse to the C:\Civil 3D Projects directory, for the drawing name enter **Nichols**, and click **Save** to save the file.

Survey Database Settings

1. Click the **Survey** tab.
2. At Survey's top, select the **Survey User Settings** icon.
3. If necessary, Expand each section and review its settings.
4. In the Import Defaults section, make sure your settings match those in Figure 14.1.
5. Click **OK** to close the Survey User Settings dialog box.

Create a Local Survey Database

1. In Survey, select the Survey Databases heading, press the right mouse button, and from the shortcut menu, select NEW LOCAL SURVEY DATABASE....
2. In New Local Survey Database, for the Name, enter **Nichols**, and click **OK**.

A Nichols folder structure is built under Survey Databases.

3. From the Survey Databases list, select **Nichols**, press the right mouse button, and from the shortcut menu, select EDIT SURVEY DATABASE SETTINGS....
4. In the Survey Database Settings dialog box, expand the Units section and set the Distance to **US Foot**.
5. At the top of Survey Database Settings, click the **export** icon (the third from left), for the setting's name, enter Local coordinate survey, and click **Save** to save the settings.

When you create the next new local coordinate survey database, simply import these saved settings.

6. Click **OK** to exit Survey Database Settings.

Create a Named Network

1. In Survey, the Nichols survey, select the Networks heading, press the right mouse button, and from the shortcut menu, select NEW....

2. In the Network dialog box, for the Name, enter **Topo**, and click **OK**.

3. Select the Nichols survey database, press the right mouse button, and from the shortcut menu, select CLOSE SURVEY DATABASE.

Survey creates a network folder, Topo, below the Nichols survey folder.

4. At Civil 3D's top left, Quick Access Toolbar, select the **Save** icon to save the drawing.

Copy the Field Book File

1. Using Windows Explorer, from the C:\Civil 3D Projects folder or from the Chapter 14 folder of the CD that accompanies this textbook, copy the file *Nichols.fbk* and place it in the C:\Civil 3D Projects\Nichols\Topo folder.

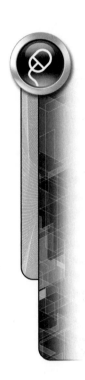

EXERCISE 14-3

After completing this exercise, you will:

- Be able to import a field book.
- Be able to insert network objects.
- Be able to review Survey component values.
- Be able to use the Survey command window.

Exercise Setup

This exercise continues with the field book from the previous exercise and its saved drawing.

1. If necessary, start **Civil 3D** by double-clicking its desktop icon.

2. When you are at the command prompt, close the open drawing and do not save it.

3. Click the **Prospector** tab.

4. If necessary, at Prospector's top, click the view drop-list arrow and from the list of views, select **Master View**.

5. At Civil 3D's top left, click **Civil 3D's** drop-list arrow, in the Application Menu highlight OPEN, in the flyout click AUTOCAD DRAWING, browse to the folder C:\Civil 3D Projects\Nichols, select the drawing named **Nichols**, and click **Open**.

6. Click the **Survey** tab.

Review Field Book Data

1. In Survey, select the Nichols survey, press the right mouse button, and from the shortcut menu, select **Open**.

2. In the Nichols survey, Networks list, select **Topo**, press the right mouse button, from the shortcut menu, select EDIT FIELD BOOK..., browse to the C:\Civil 3D Projects\Nichols\Topo folder, select the *Nichols.fbk*, and click **Open** to edit the file.

3. Review the file's contents, and when you are done, close Notepad.

Import the Field Book

1. In the Nichols survey, select Import Events, press the right mouse button, from the shortcut menu, and select IMPORT SURVEY DATA... to display the Import Survey Data wizard.

2. In Import Survey data, click **Next**.

3. In the Specify Data Source panel, toggle on Field Book File, in the lower right click the **identify source file** icon, browse to the C:\Civil 3D Projects\Nichols\Topo folder, select the *Nichols.fbk*, and click **Open**.

4. Click the **Next**.

5. Select the network Topo; click **Next**.

6. In Import Options, match your settings to Figure 14.2, and when set, click **Finish** to import the file.

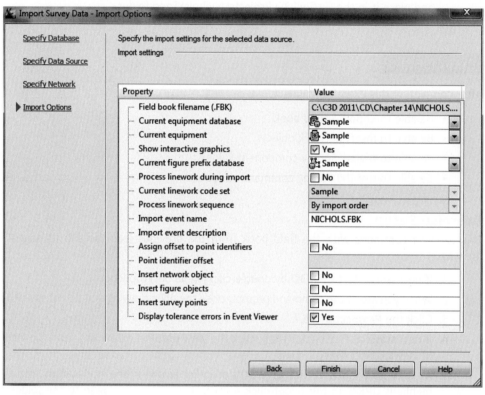

FIGURE 14.2

Review Network Component Values

1. In the Nichols survey, Networks list, expand Topo until you are viewing the Control Points. Select the Control Points heading, press the right mouse button, and from the shortcut menu, select EDIT....

2. In the Control Points Editor vista, review control points values.

3. Close the Panorama.

4. In the Nichols survey, the Topo network, Setups, select the Setup heading, press the right mouse button, and from the shortcut menu, select EDIT....

5. In the panorama, select the only set, right mouse click, and from the shortcut menu select Edit Observations.

A preview setup displays on the screen.

6. In the panorama, select an observation.

The preview shows the setup's observation in the sideshort cluster.

7. Close the panorama.

8. If the Survey goes out-of-date, select the network name Topo, press the right mouse button, and from the shortcut menu, select UPDATE NETWORK.

9. In the Nichols survey, select the Survey Points heading, right click, and in the Points flyout, select EDIT POINTS....

The vista displays all of the survey points. A prism icon next to a point indicates that it is a sideshot. Control points have a triangular icon next to their number.

10. In the vista, review the points and click the **green checkmark** until closing the Panorama.

Insert the Network Components

1. In the Nichols survey, select the network Topo, press the right mouse button, and from the shortcut menu, select INSERT INTO DRAWING.

The network is displayed in the drawing. The red line represents the backsight and the tentacles on the western end represent the observations to point 200–207.

2. Use the ZOOM and PAN commands to better view the observations on the western end.

3. In the Nichols survey, select the network Topo, press the right mouse button, from the shortcut menu, select REMOVE FROM DRAWING..., and click Yes to remove the network.

4. In the Nichols survey, select the Survey Points heading, press the right mouse button, and from the Points flyout menu, select INSERT INTO DRAWING.

Only markers appear in the drawing. The _All Points point group point labels style is set to None.

5. Click the **Prospector** tab and expand the Point Groups branch.

6. In Prospector, from the Points Groups list, select **_All Points**, press the right mouse button, and from the shortcut menu, select PROPERTIES....

7. In Point Group Properties, the Information tab, change the Point Label Style to **Point#-Elevation-Description**, and click **OK**.

8. Change the annotation scale to 1"=20'.

The points now have a marker and point label.

9. Click the **Survey** tab.

10. In the Nichols survey, select the Survey Points heading, press the right mouse button, and from the Points flyout menu, select REMOVE FROM DRAWING....

11. In the **Are you sure** dialog box, click **Yes**.

Use the Survey Command Window

1. In the Nichols survey, select the network **Topo**, press the right mouse button, and from the shortcut menu, select SURVEY COMMAND WINDOW....

2. In the Survey Command Window, from the Point Information menu, select BEARING.

3. In Point Information – Bearing, for the Start Point, enter point number **1213**, for the Ahead Point, enter point number **1226**, and click **OK** to review the Survey Command Window results.

4. In the Survey Command Window Command line, enter **1206 1213** and press ENTER to query the bearing between points.

5. Use a few more commands from the Point Information menu to query the current point data.

6. Close the Survey Command Window.

7. Select the Nichols survey database, press the right mouse button, and from the shortcut menu, select CLOSE SURVEY DATABASE.

8. Close and save the Nichols drawing.

This ends the exercise on importing a field book and querying basic survey data.

EXERCISE 14-4

After completing this exercise, you will:

- Be able to create Survey figure prefixes.
- Be able to set up a local survey database.
- Be able to adjust the Survey settings.
- Be able to insert points.
- Be able to create a point group.
- Be able to create a surface.
- Be able to add figures as breakline data.

Drawing Setup

1. If it is not open, start **Civil 3D** by double-clicking its desktop icon.

2. When you are at the command prompt, close the open drawing and do not save it.

3. At Civil 3D's top left, click Civil 3D's drop-list arrow, and from the Application Menu, click **New**.

4. In the Select Template dialog box, browse to the Chapter 14 folder of the CD that accompanies this textbook, select the file *HC3D-Extended NCS (Imperial) template*, and click **Open**.

5. If necessary, click the **Survey** tab.

Create Figure Prefixes

1. In the Survey panel, expand the Figure Prefix Databases branch until you view the Sample figure prefix database.

2. In Figure Prefixes, select Sample, press the right mouse button, and from the shortcut menu, select MANAGE FIGURE PREFIX DATABASE....

3. In the Figure Prefix Database Manager's top left, click the plus sign (+). A new entry with the name Sample displays. In the Sample entry click into the name cell and for the Name, enter **BC**, toggle on Breakline, assign the layer **C-ROAD-FIG**, and for the style assign **Curb**.

4. Repeat step three and create three additional figure prefixes. Use the values from Table 14.1 to create the figure prefix:

TABLE 14.1

Name	Breakline	Layer	Style
FL	Yes	C-ROAD-FIG	Gutter
FN	Yes	C-ROAD-FIG	Fences
CL	Yes	C-ROAD-FIG	Road Centerline

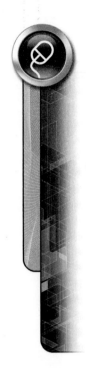

5. After creating the new Prefixes, click **OK** to close the Prefix Manager.

Create the Local Survey Database and Network

1. From the Survey panel, select the Survey Databases heading, press the right mouse button, and from the shortcut menu, select NEW LOCAL SURVEY DATABASE....

2. In New Local Survey Database, for the Name enter **Peoria**, and click **OK**.

A Peoria data structure is displayed under Survey Databases.

3. From the Survey Databases list, select Peoria, press the right mouse button, and from the shortcut menu, select EDIT SURVEY DATABASE SETTINGS....

4. In Survey Database Settings, if necessary, expand the Units section, set the Distance to **US Foot**, and click **OK**.

5. At Civil 3D's top left, click Civil 3D's drop-list arrow, from the Application Menu, highlight SAVE AS, from the flyout select AUTOCAD DRAWING, browse to the C:\Civil 3D Projects\Peoria folder, for the drawing name enter **Peoria**, and click **Save** to save the file.

6. In Survey Databases, the Peoria survey, select its Networks heading, press the right mouse button, and from the shortcut menu, select NEW....

7. In New Network, if necessary, expand Network, for the name enter **Parking Lot Topo**, and click **OK**.

Copy the Field Book to the Parking Lot Topo Folder

1. In Windows Explorer, browse to the Chapter 14 folder of the CD that accompanies this textbook, copy the file *Peoria.fbk*, and place it in the C:\Civil 3D Projects\Peoria\Parking Lot Topo folder.

Review the Field Book File

1. In Survey, the Peoria survey, expand Networks, select Parking Lot Topo, press the right mouse button, and from the shortcut menu, select EDIT FIELD BOOK.... Browse to the C:\Civil 3D Projects\Peoria\Parking Lot Topo folder, select *Peoria.fbk*, and click **Open** to edit the file.

2. In Notepad, scroll to the line containing F1 VA for point 1001.

Note that the previous line starts a figure FLL and that all the points are the same description (FLL). This indicates that the survey crew observed the line from its beginning to end. The same is true for the BCL line. All of the points that describe the FLL and BCL line are sequential observations.

3. In Notepad, scroll to the beginning of the FLR, BCR, and CL figures.

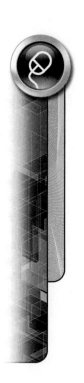

In this section, the field crew crisscrossed three figures and observed (collected information on) additional items (trees). When converting the field data file, the conversion program added the CONT commands when it encountered a point with an active figure name.

    4. Close Notepad and do not save any changes.

Import the Field Book

    1. In Survey, the Peoria survey, select the heading Import Events, press the right mouse button, and from the shortcut menu, select Import survey data....

    2. In the Import Survey Data – Survey Database panel, check the current survey, it should have Peoria highlighted, and click **Next**.

    3. In the Specify Data Source panel, toggle on Field Book File, and at the lower right of the panel, click the ***identify source file*** icon, browse to the C:\Civil 3D Projects\Peoria\Parking Lot Topo folder, select *Peoria.fbk*, and click **Open**.

    4. Click **Next** to continue.

    5. The Specify Network panel allows you to change the current network for the field book. Click **Next** to continue.

    6. In the Import Options panel, make sure your toggles match those in Figure 14.3, and when set, click **Finish** to import the file.

The field book processes, but does not insert, objects in the drawing.

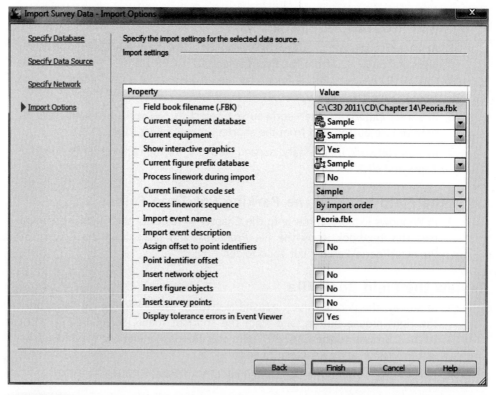

FIGURE 14.3

Review Survey Data

1. In Survey, the Peoria survey, from the Networks list, expand Parking Lot Topo until viewing Control Points. Select the heading Control Points, press the right mouse button, and from the shortcut menu, select EDIT....

2. Review values for each point number and click the ***green checkmark*** to close the Panorama.

3. In the Peoria survey, from Networks - Parking Lot Topo, select the Directions heading, right click, and from the shortcut menu select Edit... to display the Direction Editor.

A vista is displayed with the point 1 to 2 direction expressed as an azimuth.

4. Review the values for the direction from **1 to 2**.

5. When done click the ***green checkmark*** to close the Directions Editor vista.

6. In the Peoria survey, the Network - Parking Lot Topo, select the Setups heading, right click, and from the shortcut menu select Edit... to display the Setups Editor.

A vista appears with all observations made from the setup. In the drawing, Survey inserts the observations as a preview object.

7. Review the values and click the ***green checkmark*** to close the Observations Editor vista.

Review Figures

1. In the Peoria survey, select the heading Figures, right click, and from the shortcut menu select Edit Figures....

2. From the figure list, select **FLL**. A figure preview is displayed in the drawing.

3. Click each figure to preview it in the drawing.

4. While in the Figure Editor, verify that each figure has the breakline option toggled on. If necessary, toggle the breakline option by clicking the figure's Breakline toggle and clicking the ***Apply changes*** icon.

5. Before closing the editor, click on an empty cell to unfocus the panorama.

6. Edit the Figure Editor.

Edit Field Book Figure Commands

There is a problem with two figures. FLL and BCL overlap on the survey's east side. They overlap because the field crew did not take the middle curve observations in the same place for both curves. Also, the crew forgot to add two C3 notes that make curves. You will edit the field book and add the C3 notes in the correct location. After adding the C3s, the field book is correct, but it needs to be reimported to create new survey data, changing the figures.

1. In the Peoria survey, from the Figures list, select **FLL**, press the right mouse button, and from the shortcut menu, select INSERT INTO DRAWING.

2. In the Peoria survey, from the Figures list, select **BCL**, press the right mouse button, and from the shortcut menu, select INSERT INTO DRAWING.

The two figures overlap at their eastern end.

3. In the Peoria survey, from the Figures list, select **FLL**, press the right mouse button, and from the Points flyout, select INSERT INTO DRAWING.

4. Use the ZOOM and PAN commands to better view the figure's eastern end.

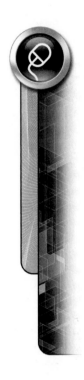

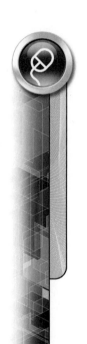

Only markers appear in the drawing. The _All Points point group's point labels style is set in Point Group Properties.

5. Click the **Prospector** tab and expand its Point Groups branch.

6. In Prospector, from the Points Groups list, select _All Points, press the right mouse button, and from the shortcut menu, select PROPERTIES....

7. In Properties, if necessary, click the Information tab, change the Point Label Style to **Point#-Elevation-Description**, and click **OK**.

The figure should have a curve starting at point 1003 and ending at 1005, but there are only line segments connecting the points. We need to review the figure coding at these point numbers to see if the C3 code is missing.

8. Click the **Survey** tab.

9. In the Peoria survey, from the Figures list, select **FLL**, press the right mouse button, and from the Points flyout, select REMOVE FROM DRAWING....

10. In the Are You Sure dialog box, click **Yes**.

11. In the Peoria survey, from the Figures list, select **BCL**, press the right mouse button, and from the Points flyout, select INSERT INTO DRAWING.

12. Use the ZOOM and PAN commands to better view the figure's eastern end.

The figure should have a curve starting at point 1023 and ending at 1025, but there are only line segments connecting the points. We need to review the figure coding at these point numbers to see if the C3 code is missing.

13. In the Peoria survey, from the Figures list, select **BCL**, press the right mouse button, and from the Points flyout, select REMOVE FROM DRAWING....

14. In the Are You Sure dialog box, click **Yes**.

Edit a Field Book

The field book needs to be checked to verify if the C3s need to be added. You display the field book editor from the same shortcut menu you used to import it.

1. In the Peoria survey, from the list of Networks, select Parking Lot Topo, press the right mouse button, and select EDIT FIELD BOOK.... If necessary, browse to the C:\Civil 3D Projects\Peoria\Parking Lot Topo folder, select *Peoria.fbk*, and click **Open**.

2. In Notepad, scroll down to the entries for points 1001–1005.

3. In Notepad, add a line before point 1003, and in the new line, enter C3.

Your field book should look like the following around the entry for point 1002:

```
F1 VA 1002 291.37480 126.320 90.12220 "FLL"
C3
F1 VA 1003 295.56320 125.210 90.26220 "FLL"
F1 VA 1004 300.43480 119.400 90.51180 "FLL"
F1 VA 1005 303.58040 105.880 91.09220 "FLL"
```

The second edit takes place in the line before 1023.

4. In Notepad, scroll to the line containing the observation for point 1023.

5. In Notepad, add a line before point 1023, and in the new line, enter C3.

Your field book should look like the following around the entry for point 1023:

```
F1 VA 1022 317.07040 37.130 91.28120 "BCL"
C3
F1 VA 1023 303.42260 105.520 90.53320 "BCL"
F1 VA 1024 302.09220 113.880 90.46280 "BCL"
F1 VA 1025 296.12240 124.400 90.15540 "BCL"
F1 VA 1026 291.53000 125.680 89.59460 "BCL"
F1 VA 1027 291.10560 110.230 89.53380 "BCL"
```

6. Exit Notepad and save the changes.

Reimport a Field Book

The current survey does not include the field book changes. You need to reimport the field book and remove all current data, replacing it with the edited field book values. To remove the current survey data, the survey must be reset. This is done from a shortcut menu, or from a toggle in the Import Field Book dialog box.

1. In Survey, the Peoria survey, from the list of Networks, select Parking Lot Topo, press the right mouse button, and from the shortcut menu, select RESET....

2. In the Are you sure dialog box, click **Yes**.

3. In the Peoria survey, expand Import Events, select the Peoria.fbk import event, right mouse click, and from the shortcut menu, select Re-import.

4. Your toggles should match Figure 14.3, and when they do, click **OK** to re-import the file.

Survey processes the field book file, but does not insert anything in the drawing.

5. In the Peoria survey, from the Figures list, select **FLL**, press the right mouse button, and from the shortcut menu, select INSERT INTO DRAWING.

6. In the Peoria survey, from the Figures list, select **BCL**, press the right mouse button, and from the shortcut menu, select INSERT INTO DRAWING.

The two figures do not overlap at their eastern ends.

Review Survey Data

Use the Survey Command Window to query the survey data.

1. From the Peoria survey database, select the network Parking Lot Topo, press the right mouse button, and from the shortcut menu, select SURVEY COMMAND WINDOW....

2. In the Survey Command Window, from the Point Information menu, select BEARING.

3. In Point Information – Bearing, for the Start Point, enter point number **1003**, for the Ahead Point, enter point number **1010**, and click **OK** to review the results.

4. In the Survey Command Window's Command line, enter **1006 1033** and press ENTER to query the bearing between the points.

5. Use a few more Point Information commands to query the current point data.

6. Close the Survey Command Window.

Create a Surface from Survey Data

The survey contains points and figures that represent surface data and linear surface features. To use the data, you need to perform two steps. First, you place the points in a point group and assign the figures as surface breakline data.

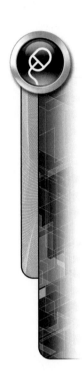

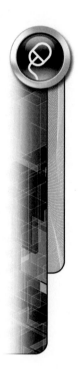

1. In Survey, the Peoria survey, select the Survey Points heading, press the right mouse button, and from the Points flyout, select INSERT INTO DRAWING.

2. If a duplicate point warning appears, change the value to **overwrite**, apply it to all duplicate points, and click **OK**.

3. In the Peoria survey, select the Figures heading, press the right mouse button, and from the Points flyout, select INSERT INTO DRAWING.

4. If a duplicate warning appears, overwrite the figures and click **OK**.

5. Click the **Prospector** tab.

6. In Prospector, select the Point Groups heading, press the right mouse button, and from the shortcut menu, select NEW....

7. In Point Group Properties, the Information tab, for the name, enter **EG**.

8. Click the Include tab, toggle on With Numbers Matching, enter the point number range **1-1075**, and click the Point List tab to view the selected points list.

9. Click the **Exclude** tab, toggle on With Point Numbers Matching, and enter **2**.

10. Click **OK** to create the point group.

11. In Prospector, select the Surfaces heading, press the right mouse button, and from the shortcut menu, select CREATE SURFACE....

12. In Create Surface, for the Name enter **EG**, for the description, enter **Parking Lot Topo – Peoria**, for the Style select **BTP**, and click **OK** until you have exited the dialog boxes.

13. In Prospector, expand Surfaces and EG branches until you are viewing EG's Definition data type list.

14. From EG's Definition data type list, select Point Groups, press the right mouse button, and from the shortcut menu, select ADD....

15. In Point Groups, from the list select **EG**, and click **OK** to assign the point group.

16. If the Event Viewer displays, close it by clicking the green check mark in the upper-right of the panorama.

17. Click the **Survey** tab.

18. In Survey, the Peoria survey, select the Figures heading, press the right mouse button, and from the shortcut menu, select CREATE BREAKLINES....

19. In Create Breaklines, make sure the Surface is **EG** and each breakline is toggled on to be a breakline, and click OK to continue.

20. In Add Breaklines, for the description, enter **Parking Lot Topo Figures**, make sure the Type is **Standard**, set the Mid-ordinate value to **0.01**, and click **OK** to continue.

21. In the drawing, select any triangle leg, press the right mouse button, and from the shortcut menu, select SURFACE PROPERTIES....

22. If necessary, click the Information tab, change the Surface Style to **Contours 1' and 5' (Design)**, and click **OK** to exit.

23. At Civil 3D's top left, Quick Access Toolbar, click the **Save** icon to save the drawing.

24. Close the drawing.

Linework Coding

The Survey panel contains the linework coding values. All of the linework imports will be in a single survey and drawing. After reviewing the import you will delete the import event and update the network.

1. At Civil 3D's top left, Quick Access Toolbar, click the Open icon, browse to the CD that accompanies this textbook, and in the Linework folder for Chapter 14, select the drawing *Linework*, and click **Open**.

2. If necessary, click the **Survey** tab.

3. In the Survey panel, select Survey Databases, press the right mouse button, and from the shortcut menu, select NEW LOCAL SURVEY DATABASE....

4. In the New Survey Database dialog box, for the survey name enter Linework.

5. Open Windows Explorer and from the Chapter 14 folder of the CD that accompanies this textbook, browse to and select the Linework folder, and copy the files to the Linework survey's folder (C:\Civil 3D Projects\Linework).

Linework – Building Offset

Linework coding is for ASCII files or LandXML survey files. The following exercise sections show code snippets and how to import them in to a survey network.

1. In the Linework survey, select the Import Events heading, press the right mouse button, and from the shortcut menu, select IMPORT SURVEY DATA....

2. In the Survey Database panel, make sure Linework is highlighted and click **Next**.

3. In the Specify Data Source panel, set the file type to Point File, at the bottom set the point file format to PNEZD (comma delimited).

4. At the panel's middle right, click the **Browse** icon. In the Select Source File dialog box browse to the C:\Civil 3D Projects\Linework folder, set the Files of type to *.txt, select the file *Bldg Offset.txt*, and click Open.

The panel displays the file path and its format.

5. Click **Next**.

6. In the Specify Network panel, at the bottom, click Create New Network.... In the New Network dialog box, for the network name, enter Linework, and click **OK** to return to the Specify Network panel.

7. Click **Next**.

8. In the Import Options panel, match your toggles to Figure 14.4, and when set click **Finish** to import the file and create the points and their linework.

9. In the Survey, select the Figures heading, right click, and from the shortcut menu select Edit Figures... to display the Figures panorama.

10. In the Figure panorama review the figures. After reviewing the figures, Close the panorama.

11. After reviewing the figure's values, in the Linework survey, select the *Bldg Offset.txt* import event, press the right mouse button, and from the shortcut menu, select **Delete....**

12. In the Linework survey, select the Linework network, right mouse click, and from the shortcut menu, select **Reset....** In the Are you sure dialog box, click **Yes**.

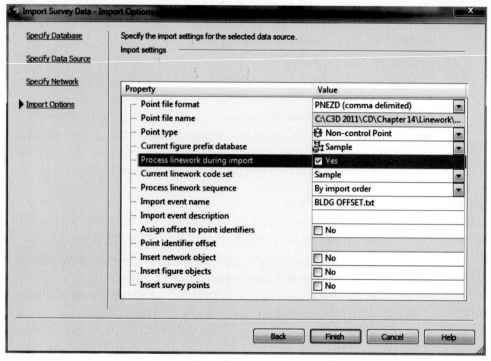

FIGURE 14.4

Linework – Offset Lines

1. In the Linework survey, select the Import Events heading, right mouse click, and from the shortcut menu, select IMPORT SURVEY DATA....

2. In the Survey Database panel, make sure Linework is highlighted and then click **Next**.

3. In the Specify Data Source panel, set the file type to Point File, at the bottom set the point file format to PNEZD (comma delimited).

4. At the panel's middle right, click the **Browse** icon. In the Select Source File dialog box browse to the C:\Civil 3D Projects\Linework folder, select the file *Roadway with Offset.csv*, and click **Open**.

The panel displays the file path and its format.

5. Click **Next** twice.

6. In the Import Options panel, match your settings to Figure 14.4 (except for the file name), and when set click **Finish** to import the file and create the points and their linework.

7. In the Survey, select the Figures heading, right click, and from the shortcut menu select Edit Figures... to display the Figures panorama.

8. In the Figure panorama review the figures. After reviewing the figures, Close the panorama.

9. After reviewing the figure's values, in the Linework survey, select *Roadway with offsets.csv* import event, press the right mouse button, and from the shortcut menu, select **Delete....**

10. In the Linework survey, select the Linework network, right mouse click, and from the shortcut menu, select **Reset....** In the Are you sure dialog box, click **Yes**.

Linework Curves

1. Using the same process from the previous two exercise sections, import the remaining text files: Line Continue.txt, Multi-point Curve.txt, Point on Curve.txt, and TOS.txt.

Linework – Survey

Linework coding applies to unconverted LandXML survey files. These files do not need to be converted to a field book file.

1. Before importing the survey, delete all import events and reset the network (select the network name, right-mouse click, and from the shortcut menu select Reset...).

2. In the Linework survey, select the Import Events heading, right mouse click, and from the shortcut menu, select IMPORT SURVEY DATA....

3. In the Survey Databases panel, make sure Linework is highlighted and then click **Next**.

4. In the Specify Data Source panel, set the file type to LandXML File.

5. At the panel's middle right, click the **Browse** icon. In the Select Source File dialog box browse to the C:\Civil 3D Projects\Linework folder, select the file *Peoria-Linework.xml*, and click **Open**.

6. Click **Next** twice.

7. In the Import Options panel, match your settings to Figure 14.4 (except for the file name), and when set click **Finish** to import the file and create the points and their linework.

8. In the Survey, select the Figures heading, right click, and from the shortcut menu select Edit Figures... to display the Figures panorama.

9. In the Figure panorama review the figures. After reviewing the figures, Close the panorama.

10. Close the survey database.

11. Close the drawing and do not save it.

This ends the exercise on field books, their command language, and their resulting data. Field books are powerful tools for taking handwritten or converted surveys and processing and evaluating their values in Civil 3D Survey.

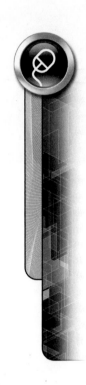

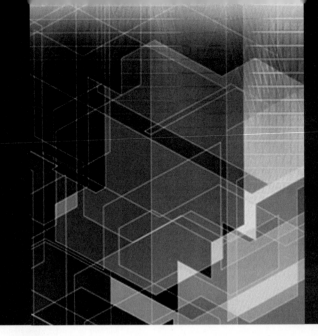

CHAPTER 15

Survey and Traverse Adjustments

EXERCISE 15-1

After completing this exercise, you will:

- Be able to create a new control point.
- Be able to create a direction.
- Be able to define a traverse.
- Be able to enter traverse data.
- Be able to review preliminary traverse adjustments.
- Be able to adjust a traverse.
- Be able to close a survey database.

Drawing Setup

This exercise uses a local survey database.

1. If you are not in **Civil 3D**, on the Windows Desktop double-click the Civil 3D icon.

2. When you are at the command prompt, close the open drawing and do not save it.

3. If necessary, click the **Prospector** tab and set Prospector to **Master View**.

4. If necessary, in Ribbon's Home tab, Palettes panel, click the **Survey Toolspace** icon or click the **Survey** tab.

5. In the Survey panel, at its top select Survey Databases, press the right mouse button, and from the shortcut menu, select NEW LOCAL SURVEY DATABASE....

6. In the New Local Survey Database dialog box, for the survey name enter **Manual Traverse**, and click **OK** to create the survey.

7. From Start, All Programs, Accessories, start Notepad, and from its File menu, select OPEN....

8. In Open, browse to the Chapter 15 folder of the CD that accompanies this textbook, select the *ManTrv.txt* file, and click **Open**.

9. If possible, print the file.

10. Exit Notepad.

The file contains the same information as Table 15.1, which is found in this exercise.

Local Survey Database

1. In Survey, select Manual Traverse, press the right mouse button, and from the shortcut menu, select EDIT SURVEY DATABASE SETTINGS....

2. In Survey Database Settings, expand the Units section and change the Distance units to **US Foot**.

3. In the Measurement Type Defaults, set Angle Type to **Angle**, set Distance Type to **Horizontal**, set Vertical Type to **None**, and set Target Type to **Prism**.

4. Scroll down and, if necessary, expand the Traverse Analysis Defaults. Set the Vertical Adjustment Method to **None**.

5. Scroll down and if necessary, expand the Least Squares Analysis Defaults. Set the Network Adjustment Type to **2-Dimensional**.

6. Click **OK** to set the values and exit the dialog box.

Define the Manual PB Loop Network

1. In Survey, the Manual Traverse survey, select the Networks heading, press the right mouse button, and from the shortcut menu, select NEW....

2. In New Network, for the name enter *Manual PB Loop* and click **OK**.

Define a Control Point and a Direction

To enter a survey, there must be at least one control and a backsight point.

1. In the Manual Traverse survey, expand Networks and then expand the network Manual PB Loop. Select the Control Points heading, press the right mouse button, and from the shortcut menu, select NEW....

2. In New Control Point, assign the following values:

Point Number: 1

Easting: 4464.1621

Northing: 4715.0055

3. After entering the values, click **OK** to exit.

4. In the Manual Traverse survey, the Manual PB Loop network, select Directions, press the right mouse button, and from the shortcut menu, select NEW....

5. In New Direction, for the From Point enter **1**, for the To Point enter **25**, for the Azimuth (Direction) enter **275.0000**, set the measurement type to **Azimuth**, and click **OK** to exit.

6. Select the network Manual PB Loop, press the right mouse button, and from the shortcut menu, select Automatic Update.

Define the PB Loop Traverse

1. In the Manual Traverse survey, the network Manual PB Loop, select the Traverses heading, press the right mouse button, and from the shortcut menu, select NEW....

2. In the New Traverse dialog box, for the name enter *PB Loop*, for the description enter **Preliminary Loop – Manual**, set the Initial Station to 1, set the Initial Backsight to 25, and click **OK**.

Enter Traverse Data

Table 15.1 contains the traverse data. Each setup has two lines of information. The first line is the station point number, the backsight point number, and the angle. The second line is the foresight point number, horizontal angle, horizontal distance, and description.

The Local Survey Settings directly affect manual survey value prompting.

1. In the Manual Traverse survey, the network Manual PB Loop, select the heading Traverses. Right click on the Traverses heading, and from the shortcut menu, select Edit... to display the Traverses Editor.

2. In the Traverses Editor, select the traverse *PB Loop*, press the right mouse button, and from the shortcut menu, select EDIT TRAVERSE....

The Traverse Editor vista displays and shows the first setup and its backsight. The editor's cell-based side is where the turned angles and distance are entered.

The foresight entry for the first station is the second line of Table 15.1.

TABLE 15.1

STN	Pnt#	Angle	Hrz Dist	DescType
1	25	0.0		BS
	6	71.3033	429.7884	TRV-1FS
6	1	0.0		BS
	102	214.0533	438.5154	TRV-2FS
102	6	0.0		BS
	106	235.4233	451.0348	TRV-3FS
106	102	0.0		BS
	107	205.2035	564.9924	TRV-4FS
107	106	0.0		BS
	8	267.4337	369.7415	TRV-5FS
8	107	0.0		BS
	108	170.3710	313.1525	TRV-6FS
108	8	0.0		BS
	2	261.3654	590.6619	TRV-7FS
2	108	0.0		BS
	1	180.1813	405.0081	BM1FS

The following step enters the data from the first line of Table 15.1.

3. In the Traverse Editor for Station 1, and below Station 1, Backsight 25, click in the cell to the right of the prism, and for the foresight point type **6**.

The editor presents a new station based on the foresight of point 6, backsighting point 1. The editor assumes that the first entry is the next forward station's point number.

4. Scroll the editor to the right. For the Angle enter **71.3033**, for the horizontal Distance enter **429.7884**, click the Apply changes icon, and for the Description enter **TRV-1FS**.

5. In the Traverse Editor for Station 6, and below Station 6, Backsight 1, click in the cell to the right of the prism, and for the foresight point enter **102**.

6. Scroll the editor to the right. For the angle enter **214.0533**, for the horizontal distance enter **438.5154**, click the Apply changes icon, and for the description enter **TRV-2FS**.

7. In Table 15.1, using the remaining data, finish entering traverse point numbers, angles, distances, and descriptions.

8. After entering the data, in the Traverse Editor, click the floppy icon to save the edits.

9. Click the green checkmark to close the Traverse Editor vista.

The Traverses Editor vista displays the control point, the backsight point, traverse stations, and the Final Foresight.

10. In the Manual Traverse survey, select the Survey Points heading, press the right mouse button, and from the Points flyout menu, select Edit Points….

11. Review the points for the current traverse.

The only control point is point 1. The remaining points are observed survey points that were occupied not yet adjustment (transit icon).

12. Click the ***green checkmark*** to close the Survey Points Editor vista.

Traverse Adjustments

1. In the Traverses Editor panorama, select the *PB Loop*, press the right mouse button, and from the shortcut menu, select TRAVERSE ANALYSIS….

2. In Traverse Analysis, set the Horizontal adjustment method to Compass, toggle **OFF** Update Survey Database, and click **OK** to produce a preliminary adjustment report.

The Adjustment routine computes and displays the results as three reports in Notepad.

3. From the Notepad group, select the file *PB Loop Raw Closure.trv*.

This is an overall traverse quality report.

4. Close the file *PB Loop Raw Closure.trv*.

5. From the Notepad group, select the file *PB Loop Balanced Angles.trv*.

This report reviews the raw traverse calculations on the right and the potentially correct coordinates on the left with their deltas at the far right.

6. Close the *PB Loop Balanced Angles.trv* file.

7. From the Notepad group, select the *PB Loop Loop.lso* file.

This report lists the traverse coordinates on the left, the adjustment-corrected coordinates on the left, and at the far left, the accumulated amount of change.

8. Close the *PB Loop.lso* file.

Select an Adjustment and Update the Survey

1. In the Traverses Editor, select traverse *PB Loop*, press the right mouse button, and from the shortcut menu, select TRAVERSE ANALYSIS….

2. In Traverse Analysis, for the Horizontal Adjustment Method, select Compass Rule, toggle **ON Update Survey Database**, and click **OK** to adjust the traverse.

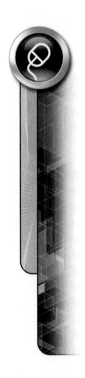

The Adjustment routine computes and displays the results as three reports in Notepad.

3. Review the reports and close them.

4. In the Manual Traverse survey, the Manual PB Loop network, select the Control Points heading, right mouse click, and from the shortcut menu, select Edit...

5. Review the control points list.

The observed and occupy points are now adjusted and considered control points (control triangle with prism pole icon).

6. Close the Panorama.

Close the Survey Database

1. In Survey, select the survey database Manual Traverse, press the right mouse button, and from the shortcut menu, select CLOSE SURVEY DATABASE.

2. Close the drawing.

This ends the exercise on manually entering data in a traverse.

EXERCISE 15-2

After completing this exercise, you will:

- Be able to set local survey settings.
- Be able to define a network.
- Be able to import a field book.
- Be able to define a traverse.
- Be able to identify a traverse error.
- Be able to adjust a traverse definition.

Exercise Setup

This exercise uses a local survey database.

1. If you are not in **Civil 3D**, double-click its desktop icon.

2. When you are at the command prompt, close the open drawing and do not save it.

3. At Civil 3D's top left, click Civil 3D's drop-list arrow, from the Application Menu, select New.

4. In the Select Template dialog box, browse to the Chapter 15 folder of the CD that accompanies this textbook, select the *HC3D-Extended NCS (Imperial)* template, and click **Open**.

5. At Civil 3D's top left, click Civil 3D's drop-list arrow, from the Application Menu, highlight SAVE AS, from the flyout select AUTOCAD DRAWING, browse to the C:\Civil 3D Projects folder, for the drawing name enter **Oloop**, and click **Save** to save the file.

Create a Local Survey Database

This survey is an example of a closed loop traverse.

1. If necessary, in Ribbon's Home tab, Palettes panel, click the **Survey Toolspace** icon or click the Toolspace's **Survey** tab.

2. In the Survey panel, at its top select Survey Databases, press the right mouse button, and from the shortcut menu, select NEW LOCAL SURVEY DATABASE....

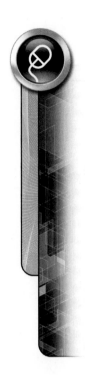

3. In the New Local Survey Database dialog box, for the survey name enter **Oloop**, and click **OK**.

Set Values for the Oloop Survey Database

4. From the Survey Databases list, select Oloop, press the right mouse button, and from the shortcut menu, select EDIT SURVEY DATABASE SETTINGS....

5. In the Survey Database Settings, if necessary, expand the Units section, change the Distance units to **US Foot**, and set Angles to **Degrees DMS (DDD.MMSSS)**.

6. If necessary, expand the Traverse Analysis Defaults section, set the Horizontal Adjustment Method to **Compass Rule**, and set the Vertical Adjustment Method to **None**.

7. If necessary, expand the Least Squares Adjustment Defaults section, set the Network Adjustment Type to **2-Dimensional**, and toggle **ON Perform Blunder Detection**.

8. Click **OK** to exit the dialog box.

Surveyor's Traverse Notes

The surveyor needs to have a survey brought into and adjusted by Civil 3D. The survey contains a topography and a traverse loop. The survey starts at point 10 and ends by occupying point 11 and foresighting to point 10. Figure 15.1 contains the traverse loop with stations and the traverse direction.

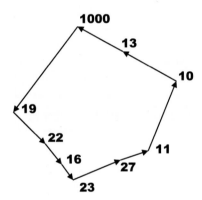

FIGURE 15.1

Create the Network and Import the Field Book

1. In the Oloop survey, select Import Events, press the right mouse button, and from the shortcut menu, select Import Survey Data....

2. In the Import Survey Data dialog box, make sure the Oloop survey is selected and click **Next**.

3. In the Specify Data Source panel, toggle the Data Source Type to **Field Book File**.

4. At the panel's middle right, click the **Browse** icon. In the Field Book Filename dialog box, browse to the Chapter 15 folder of the CD that accompanies this textbook, select the file *OLOOP.fbk*, and click **Open** to return to the Specify Data Source panel.

5. Click **Next**.

6. In the Specify Network panel, at its bottom, click **Create New Network....** In the New Network dialog box, for the Network name enter **Boundary**, and click **OK** to return to the Specify Network panel.

7. Click **Next**.

8. In the Import Options panel make your toggles match those in Figure 15.2, and click **Finish** to import the field book.

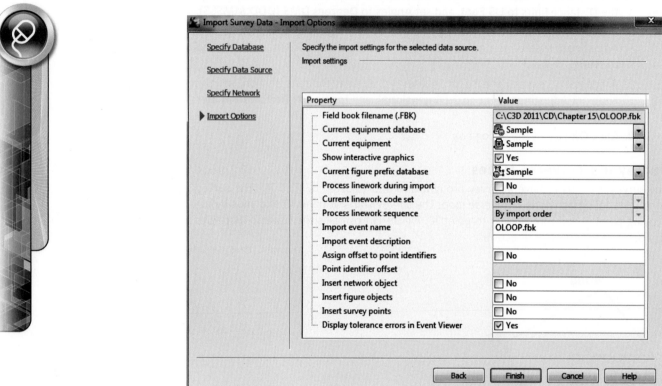

FIGURE 15.2

Review the Survey Data

1. In the Oloop survey, expand the Boundary network, select the Control Points heading, press the right mouse button, and from the shortcut menu, select EDIT....

2. In the Control Points Editor, review the values for the only survey control point.

3. Click the **green checkmark** to exit the Control Points Editor vista.

4. In the Oloop survey, select the Survey Points heading, press the right mouse button, and from the shortcut menu's Points flyout, select EDIT POINTS....

5. In the Survey Points Editor vista, review the point list.

6. Click the Number heading to sort the points by point number.

Each point has an icon that indicates its status: control, observed and occupied, and side shots.

7. In the Survey Points Editor vista, click the question mark and review the icon definitions.

8. Close Help and click the **green checkmark** to close all the vistas.

Review the Setups

If the list of setups matches the survey sketch in Figure 15.1, things look good.

1. In the Oloop survey, the Boundary network, select the heading Setups, right mouse click, and from the shortcut menu, select Edit....

2. Click the Station Point heading to sort the station list.

3. Review the observations for Station 16, backsighting 22.

4. Select the first Station 16, backsight 22, right mouse click, and from the shortcut menu select Edit observations....

The observations for this setup are to points 22 and 24.

5. Click the Setups Editor vista tab.

6. Select the second Station 16, backsight 22, right mouse click, and from the shortcut menu select Edit observations....

7. Click the Number heading to sort the observations and review the observations.

8. Click the Setups Editor vista tab.

9. Select the Station 22, backsight 19 setup, right mouse click, and from the shortcut menu select Edit observations....

10. Review the observations for Station 22.

Point 22 does not observe point 16 as a backsight point. Point 22's backsight point is 19 not 16. Point 16 is a sideshot, not a point in the traverse.

11. Click the **green checkmark** until closing all vistas.

12. In the Oloop survey, select the Boundary network, press the right mouse button, and from the shortcut menu, select INSERT INTO DRAWING.

The survey is displayed in the drawing.

13. In the Oloop survey, select the Boundary network, press the right mouse button, from the shortcut menu, select REMOVE FROM DRAWING..., and in the Are You Sure dialog box, click **Yes**.

Define Traverse per Sketch

1. In the Oloop survey, the Boundary network, select the Traverses heading, press the right mouse button, and from the shortcut menu, select NEW....

2. In New Traverse, for the name enter **Preliminary**, for the Description enter **Jewel Property**, for the Initial Station enter **10**, and click anywhere in the dialog box to have it fill in the remaining entries.

The current traverse definition does not contain point 16.

3. In the New Traverse dialog box, between points **22 and 23**, add point **16**. Your New Traverse dialog box should look like Figure 15.3. When complete, click **OK**.

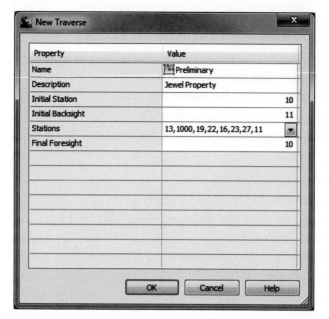

FIGURE 15.3

Review the Traverse

The traverse is broken. One setup's (station) observations break the ahead and back traverse loop observation method. You could find this by looking through the setups in the Boundary network. The easiest place to find broken data is by reviewing it in the Traverse Editor. If a traverse is broken, a setup will have no ahead and/or back loop observations in the Traverse Editor.

1. In the Oloop survey, the Boundary network, select the heading Traverses, right click, and from the shortcut menu select Edit….

2. On the Traverse Editor's right, scroll down the stations list until reaching Station 16.

The next station is 23, but there are no observations between 16 and 23 and because of that the traverse editor indicates incomplete or missing data for this traverse definition.

3. Click the Green check mark to close the Traverse Editor.

Adjust the Traverse

1. In the Traverses Editor, select the traverse, Preliminary. Press the right mouse button, and from the shortcut menu, select TRAVERSE ANALYSIS….

2. In the Traverse Analysis dialog box, set the Horizontal Adjustment Method to **Compass Rule**, set the Vertical Adjustment Method to **None**, toggle **OFF Update Survey Database**, and click **OK**.

Survey issues an error dialog box indicating that there is insufficient angle data to adjust the survey.

3. Click **OK** to exit.

There are no observations at station 16 forward to point 23. This is where the traverse is broken.

The only thing you can do is find out if there are observations from the setup on point 22 to the next ahead point, 23. If there are, point 16 is a sideshot, not a traverse station. The loop would then go from 22 to 23 and bypass point 16 (see Figure 15.4).

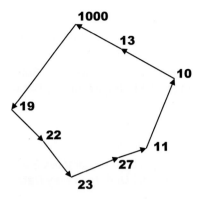

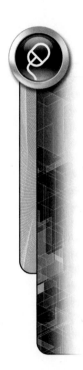

FIGURE 15.4

These alternative observations do not show in the Traverse Editor because the editor's focus is on the current loop definition.

4. Click the **green checkmark** to close the Traverse Editor.

5. In the Oloop survey, Boundary network, select the Setups heading, right click, and from the list select Edit....

6. Click the Station Point heading to sort the list.

7. Scroll down to the Stations for station point 22, backsighting 19, select station point 22, right mouse click, and from the shortcut menu select Edit Observations....

In the Observations Editor, note that, from this setup, the necessary foresight observations to point 23 are present. The station point's observation also includes backsights to point 19. These two observations provide a valid loop definition using this setup.

8. Click the Setup Editor's vista tab.

9. From the list of Station Points, select the first Station Point 23, backsighting point 22. There are no observations of point 16.

10. Click the Setup Editor's vista tab.

11. From the list of Station Points, select the second Station Point 23, backsighting point 22.

12. Click the Station Points heading to sort the list. There are no observations of point 16 in these observations also. But there are observations back to 22 and forward to 27.

With this information, point 16 is a sideshot, not a station along the traverse path and the traverse goes from point 22 to 27 through point 23.

13. Click the Green Checkmark until closing all vistas. If you are prompted to save any changes, just click NO.

Edit the Traverse's Definition

The traverse definition needs to change to correctly define the traverse loop.

1. In the Oloop survey, the Boundary network, select the Traverses heading, right mouse click, and from the shortcut menu select Edit... to display the Traverses Editor.

2. In the Traverses Editor vista, remove station 16 from the list. The edited station list should now be **13, 1000, 19, 22, 23, 27**, and **11** (removing point 16 from the loop), and click the floppy icon to save the edits.

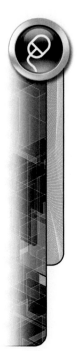

Adjust the Traverse

1. In the Traverses Editor, select the traverse Preliminary, press the right mouse button, and from the shortcut menu, select TRAVERSE ANALYSIS....

2. In Traverse Analysis, set the Horizontal Adjustment Method to **Compass Rule**, set the Vertical Adjustment Method to **None**, toggle **OFF Update Survey Database**, and click **OK**.

3. Review the adjustment reports.

4. Repeat Steps 1 through 3, trying different adjustment methods.

5. In Traverse Analysis, set the Horizontal Adjustment Method to **Compass Rule**, set the Vertical Adjustment Method to **None**, toggle **ON Update Survey Database**, and click **OK**.

6. Review the results and close the Notepad reports after you have reviewed their values.

7. Click the green checkmark to close the Traverses Editor.

Insert Points into a Drawing

1. In the Oloop survey, select the Networks heading, press the right mouse button, and from the Points flyout menu, select INSERT INTO DRAWING.

2. Click the **Prospector** tab.

3. In Prospector, expand the Point Groups branch.

4. From the list of point groups, select _All Points, press the right mouse button, and from the shortcut menu, select PROPERTIES....

5. If necessary, click the Information tab and change the Point Label style to **Point#-Elevation-Description**, and click **OK** to exit.

6. At Civil 3D's top left, Quick Access Toolbar, click the **Save** icon to save the drawing.

7. Click the **Survey** tab.

Close the Survey

1. In Survey, select the Oloop heading, press the right mouse button, and from the shortcut menu, select CLOSE SURVEY DATABASE.

2. Close and save the Oloop drawing.

This completes this close loop and adjustment exercise.

EXERCISE 15-3

After completing this exercise, you will:

- Be able to set local survey settings.
- Be able to define a network.
- Be able to import a field book.
- Be able to define a traverse.
- Be able to identify a traverse error.
- Be able to adjust a traverse definition.
- Be able to adjust a traverse with the Least Squares analysis method.
- Be able to review an adjustment ellipsis.

Exercise Setup
This exercise uses a local survey database.

1. If you are not in **Civil 3D**, double-click its desktop icon.

2. If necessary, click the Survey tab.

3. In the Survey panel, select the Survey Databases heading, press the right mouse button, and from the shortcut menu, select NEW LOCAL SURVEY DATABASE....

4. In the New Local Survey Database dialog box, for the name enter **CC-Loop**, and click **OK**.

5. At Civil 3D's top left, Quick Access Toolbar, click the **Open** icon, browse to the Chapter 15 folder of the CD that accompanies this textbook, select the *Property-CC.dwg* drawing, and click **Open**.

6. At Civil 3D's top left, click **Civil 3D's** drop-list arrow, from the Application Menu, highlight SAVE AS, from the flyout select AUTOCAD DRAWING, browse to the C:\Civil 3D Projects\CC-Loop folder, for the drawing name enter *Property-CC-Work*, and click **Save** to save the file.

Edit the CC-Loop Survey Database Settings

1. From the Survey Databases list, select the CC-Loop heading, press the right mouse button, and from the shortcut menu, select EDIT SURVEY DATABASE SETTINGS....

2. In Survey Database Settings, the Units section, change the Distance units to **US Foot**, and the Angles to **Degrees DMS (DDD.MMSSS)**.

3. If necessary, expand the Traverse Analysis Defaults, set the Horizontal Adjustment Method to **Compass Rule**, and set the Vertical Adjustment Method to **None**.

4. If necessary, expand the Least Squares Adjustment Defaults, set the Network Adjustment type to **2-Dimensional**, and toggle **ON Perform Blunder Detection**.

5. Click **OK** to exit the dialog box.

Surveyor's Notes on the Traverse
The surveyor needs to have a survey brought into Civil 3D to be adjusted. The survey contains a closed connected traverse loop. The traverse is between two known points, points 1250 and 1375. There are redundant observations, as well as observations to an external point, point 1163. Figure 15.6 contains the traverse stations loop and the traverse direction.

The survey in this exercise can be treated as a traditional traverse or as a network Least Squares adjustment, because it contains cross-traverse and external point observations.

Import the Field Book and Create the Network

1. In the CC-Loop survey, select Import Events, press the right mouse button, and from the shortcut menu, select Import Survey Data....

2. In the Import Survey Data dialog box, make sure the CC-Loop survey is selected and click **Next**.

3. In the Specify Data Source panel, toggle the Data Source Type to **Field Book File**.

4. At the panel's middle right, click the **Browse** icon. In the Field Book Filename dialog box, browse to the Chapter 15 folder of the CD that accompanies this

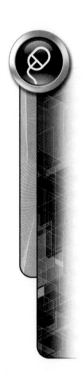

textbook, select the file *CC-Loop.fbk*, and click **Open** to return to the Specify Data Source panel.

5. Click **Next**.

6. In the Specify Network panel, at its bottom, click **Create New Network....** In the New Network dialog box, for the Network name enter **Interior Boundary**, and click **OK** to return to the Specify Network panel.

7. Click **Next**.

8. In the Import Options panel make your toggles match those in Figure 15.5, and then click **Finish** to import the field book.

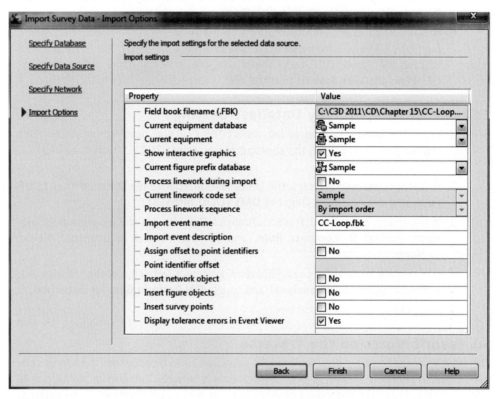

FIGURE 15.5

Review the Survey Data

1. In the CC-Loop survey, expand the Interior Boundary network, and select the heading Control Points, press the right mouse button, and from the shortcut menu, select EDIT....

2. In the Control Points Editor vista, review the control point's values.

3. Click the **green checkmark** to exit the Control Points Editor vista.

4. In the CC-Loop survey, select the Survey Points heading. Press the right mouse button, and from the Point flyout menu, select EDIT POINTS....

5. In the Survey Points Editor vista, review the point list.

Each point has an icon that indicates its status in the survey.

6. Click the **green checkmark** to close the Survey Points Editor vista.

7. In the CC-Loop survey, select the Interior Boundary network. Press the right mouse button, and from the shortcut menu, select INSERT INTO DRAWING.

The network shows the stations and foresights to the exterior reference point 1163.

8. In the CC-Loop survey, select the Interior Boundary network. Press the right mouse button, and from the shortcut menu, select REMOVE FROM DRAWING…. In the Are You Sure dialog box, click Yes to remove the network.

Review the Setups

The traverse follows the setups.

1. In the CC-Loop survey, the Interior Boundary network, select the heading Setups, right mouse click, and from the shortcut menu select, EDIT….

2. From the list, select **Station Point: 100, Backsight: 1250**. Press the right mouse button, and from the shortcut menu, select EDIT OBSERVATIONS….

3. Click the Setups Editor vista tab and select and review the remaining setups by editing their observations.

4. Click the *green checkmark* until you close all the vistas.

Define Traverse per Sketch

1. In the CC-Loop survey, the Interior Boundary network, select the Traverses heading. Press the right mouse button, and from the shortcut menu, select NEW….

2. In New Traverse, for the name enter **Disputed**, for the description enter **Interior Disputed Bound**, for the Initial Station enter **1250**, click anywhere in the dialog box to let survey fill in the remaining stations, and click **OK**.

Preliminary Adjustment Review

1. In the CC-Loop survey, the Interior Boundary network, select the heading Traverses, right mouse click, and from the short cut menu, select Edit….

2. In the Traverses Editor, select Disputed, press the right mouse button, and from the shortcut menu, select TRAVERSE ANALYSIS….

3. In Traverse Analysis, set the Horizontal Adjustment Method to **Compass Rule**, set the Vertical Adjustment Method to **None**, toggle **OFF Update Survey Database**, and click **OK**.

4. Review the preliminary results and close the Notepad reports after reading them.

5. Close the Traverses Editor.

6. In the CC_Loop survey, select the network Interior Boundary, right mouse click, and from the shortcut menu select Reset.

7. In the Are you sure dialog box, click Yes.

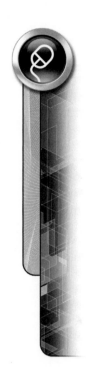

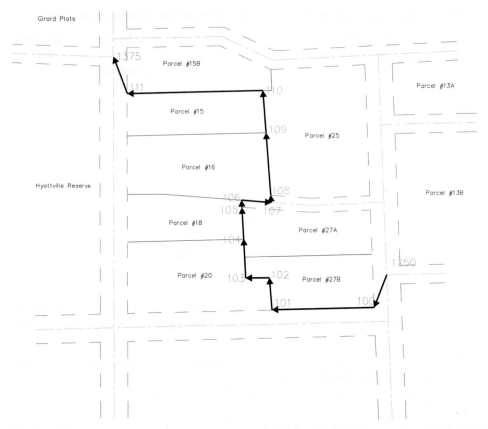

FIGURE 15.6

Least Squares Analysis

Traverse adjustments do not take external observations into consideration. To use these observations, you must use the Least Squares adjustment from the named network short-cut menu.

1. In the CC-Loop survey, the Import Events list, select CC-Loop.fbk, right mouse click, and from the shortcut menu select Re-Import.

2. In the Re-Import Field Book dialog box, click Finish to re-import the field book.

3. In the CC-Loop survey, from the Networks list, select Interior Boundary. Press the right mouse button, and from the Least Squares Analysis flyout, select CREATE INPUT FILE. If the file already exists overwrite it.

4. In the CC-Loop survey, from the Networks list, select Interior Boundary. Press the right mouse button, and from the Least Squares Analysis flyout, select EDIT INPUT FILE....

The input file lists all cross-traverse observations and the observations to point 1163. At the top file's top are the control points (no question marks to the left point's number) and the floating points (those with a question mark).

5. Close the Network.lsi file.

6. In the CC-Loop survey, from the Networks list, select Interior Boundary. Press the right mouse button, and from the Least Squares Analysis flyout, select PROCESS INPUT FILE....

7. In the CC-Loop survey, from the Networks list, select Interior Boundary. Press the right mouse button, and from the Least Squares Analysis flyout, select DISPLAY OUTPUT FILE....

8. Review the output.lsi file. The survey passes at the 95% level. Review all of the report's sections.

9. When done, close the Network.lso file.

10. In the CC-Loop survey, from the Networks list, select Interior Boundary. Press the right mouse button, and from the Least Squares Analysis flyout, select UPDATE SURVEY DATABASE.... The Least Squares Adjusted File Selection dialog box displays.

11. In the dialog box select the *Network.adj* file and click Open.

12. In the CC-Loop survey, from the Networks list, select Interior Boundary. Press the right mouse button, and in the shortcut menu, select INSERT INTO DRAWING.

13. Use the ZOOM and PAN commands to view the error ellipses at stations 100–111. They are very small and in the red dot of the control symbol.

Close the Survey and Save the Drawing

1. In the Survey Databases list, select the CC-Loop heading, press the right mouse button, and from the shortcut menu, select CLOSE SURVEY DATABASE.

2. Click the **Prospector** tab.

3. At Civil 3D's top left, Quick Access Toolbar, click the **Save** icon to save the drawing.

4. Close the drawing.

This completes the closed connected traverses exercise. Survey is the beginning and ending of every project. A survey's data is important and time review at each step of its analysis.

This chapter is the last chapter of this textbook. Each chapter highlights the tools available to the user to complete an engineering project from beginning to end.

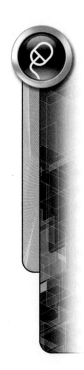